Hands-On Machine Learning with R

Chapman & Hall/CRC
The R Series

Recently Published Titles

Spatial Microsimulation with R
Robin Lovelace, Morgane Dumont

Extending R
John M. Chambers

Using the R Commander: A Point-and-Click Interface for R
John Fox

Computational Actuarial Science with R
Arthur Charpentier

bookdown: Authoring Books and Technical Documents with R Markdown,
Yihui Xie

Testing R Code
Richard Cotton

R Primer, Second Edition
Claus Thorn Ekstrøm

Flexible Regression and Smoothing: Using GAMLSS in R
Mikis D. Stasinopoulos, Robert A. Rigby, Gillian Z. Heller, Vlasios Voudouris, and Fernanda De Bastiani

The Essentials of Data Science: Knowledge Discovery Using R
Graham J. Williams

blogdown: Creating Websites with R Markdown
Yihui Xie, Alison Presmanes Hill, Amber Thomas

Handbook of Educational Measurement and Psychometrics Using R
Christopher D. Desjardins, Okan Bulut

Displaying Time Series, Spatial, and Space-Time Data with R, Second Edition
Oscar Perpinan Lamigueiro

Reproducible Finance with R
Jonathan K. Regenstein, Jr

R Markdown
The Definitive Guide
Yihui Xie, J.J. Allaire, Garrett Grolemund

Practical R for Mass Communication and Journalism
Sharon Machlis

Analyzing Baseball Data with R, Second Edition
Max Marchi, Jim Albert, Benjamin S. Baumer

Spatio-Temporal Statistics with R
Christopher K. Wikle, Andrew Zammit-Mangion, and Noel Cressie

Statistical Computing with R, Second Edition
Maria L. Rizzo

Geocomputation with R
Robin Lovelace, Jakub Nowosad, Jannes Muenchow

Distributions for Modelling Location, Scale, and Shape
Using GAMLSS in R
Robert A. Rigby , Mikis D. Stasinopoulos, Gillian Z. Heller and Fernanda De Bastiani

Advanced Business Analytics in R: Descriptive, Predictive, and Prescriptive
Bradley Boehmke and Brandon Greenwell

For more information about this series, please visit: https://www.crcpress.com/go/the-r-series

Hands-On Machine Learning with R

Brad Boehmke
Brandon Greenwell

CRC Press is an imprint of the
Taylor & Francis Group, an **informa** business
A CHAPMAN & HALL BOOK

CRC Press
Taylor & Francis Group
6000 Broken Sound Parkway NW, Suite 300
Boca Raton, FL 33487-2742

Printed on acid-free paper

International Standard Book Number-13: 978-1-138-49568-5 (Hardback)

Visit the Taylor & Francis Web site at
http://www.taylorandfrancis.com

and the CRC Press Web site at
http://www.crcpress.com

Brad:

To Kate, Alivia, and Jules for making sure I have a life outside of programming and to my mother who, undoubtedly, will try to read the pages that follow.

Brandon:

To my parents for encouragement, to Thaddeus Tarpey for inspiration, and to Julia, Lilly, and Jen for putting up with me while writing this book.

Contents

Preface

Welcome to *Hands-On Machine Learning with R*. This book provides hands-on modules for many of the most common machine learning methods to include:

- Generalized low rank models
- Clustering algorithms
- Autoencoders
- Regularized models
- Random forests
- Gradient boosting machines
- Deep neural networks
- Stacking / super learners
- and more!

You will learn how to build and tune these various models with R packages that have been tested and approved due to their ability to scale well. However, our motivation in almost every case is to describe the techniques in a way that helps develop intuition for its strengths and weaknesses. For the most part, we minimize mathematical complexity when possible but also provide resources to get deeper into the details if desired.

Who should read this

We intend this work to be a practitioner's guide to the machine learning process and a place where one can come to learn about the approach and to gain intuition about the many commonly used, modern, and powerful methods accepted in the machine learning community. If you are familiar with the analytic methodologies, this book may still serve as a reference for how to work with the various R packages for implementation. While an abundance of videos, blog posts, and tutorials exist online, we have long been frustrated by the lack of consistency, completeness, and bias towards singular packages for implementation. This is what inspired this book.

This book is not meant to be an introduction to R or to programming in general; as we assume the reader has familiarity with the R language to include defining functions, managing R objects, controlling the flow of a program, and other basic tasks. If not, we would refer you to R for Data Science[1] (Wickham and Grolemund, 2016) to learn the fundamentals of data science with R such as importing, cleaning, transforming, visualizing, and exploring your data. For those looking to advance their R programming skills and knowledge of the language, we would refer you to Advanced R[2] (Wickham, 2014). Nor is this book designed to be a deep dive into the theory and math underpinning machine learning algorithms. Several books already exist that do great justice in this arena (i.e. Elements of Statistical Learning[3] (Friedman et al., 2001), Computer Age Statistical Inference[4] (Efron and Hastie, 2016), Deep Learning[5] (Goodfellow et al., 2016)).

Instead, this book is meant to help R users learn to use the machine learning stack within R, which includes using various R packages such as **glmnet**, **h2o**, **ranger**, **xgboost**, **lime**, and others to effectively model and gain insight from your data. The book favors a hands-on approach, growing an intuitive understanding of machine learning through concrete examples and just a little bit of theory. While you can read this book without opening R, we highly recommend you experiment with the code examples provided throughout.

Why R

R has emerged over the last couple decades as a first-class tool for scientific computing tasks, and has been a consistent leader in implementing statistical methodologies for analyzing data. The usefulness of R for data science stems from the large, active, and growing ecosystem of third-party packages: **tidyverse** for common data analysis activities; **h2o**, **ranger**, **xgboost**, and others for fast and scalable machine learning; **iml**, **pdp**, **vip**, and others for machine learning interpretability; and many more tools will be mentioned throughout the pages that follow.

[1] http://r4ds.had.co.nz/index.html

[2] http://adv-r.had.co.nz/

[3] https://web.stanford.edu/~hastie/ElemStatLearn/

[4] https://web.stanford.edu/~hastie/CASI/

[5] http://www.deeplearningbook.org/

Conventions used in this book

The following typographical conventions are used in this book:

- ***strong italic***: indicates new terms,
- **bold**: indicates package & file names,
- `inline code`: monospaced highlighted text indicates functions or other commands that could be typed literally by the user,
- code chunk: indicates commands or other text that could be typed literally by the user

```
1 + 2
## [1] 3
```

In addition to the general text used throughout, you will notice the following code chunks with images:

Signifies a tip or suggestion

Signifies a general note

Signifies a warning or caution

Additional resources

There are many great resources available to learn about machine learning. Throughout the chapters we try to include many of the resources that we have found extremely useful for digging deeper into the methodology and applying with code. However, due to print restrictions, the hard copy version of this book limits the concepts and methods discussed. Online supplementary material exists at `https://koalaverse.github.io/homlr/`. The additional material will accumulate over time and include extended chapter material (i.e., random forest package benchmarking) along with brand new content we couldn't fit in (i.e., random hyperparameter search). In addition, you can download the data used throughout the book, find teaching resources (i.e., slides and exercises), and more.

Feedback

Reader comments are greatly appreciated. To report errors or bugs please post an issue at `https://github.com/koalaverse/homlr/issues`.

Acknowledgments

We'd like to thank everyone who contributed feedback, typo corrections, and discussions while the book was being written. GitHub contributors included @agailloty, @asimumba, @benprew, @bfgray3, @bragks, @cunningjames, @liangwu82, @nsharkey, and @tpristavec. We'd also like to thank folks such as Alex Gutman, Greg Anderson, Jay Cunningham, Joe Keller, Mike Pane, Scott Crawford, and several other co-workers who provided great input around much of this machine learning content.

Software information

This book was built with the following packages and R version. All code was executed on 2017 MacBook Pro with a 2.9 GHz Intel Core i7 processor, 16 GB of memory, 2133 MHz speed, and double data rate synchronous dynamic random access memory (DDR3).

TABLE 0.1: Packages mentioned throughout and used to produce this book. *(continued)*

package	version	source

TABLE 0.1: Packages mentioned throughout and used to produce this book.

package	version	source
AmesHousing	0.0.3	CRAN (R 3.6.0)
AppliedPredictiveModeling	1.1-7	CRAN (R 3.6.0)
bookdown	0.11	CRAN (R 3.6.0)
broom	0.5.2	CRAN (R 3.6.0)
caret	6.0-84	CRAN (R 3.6.0)
caretEnsemble	2.0.0	CRAN (R 3.6.0)
cluster	2.1.0	CRAN (R 3.6.1)
cowplot	0.9.4	CRAN (R 3.6.0)
DALEX	0.4	CRAN (R 3.6.0)
data.table	1.12.2	CRAN (R 3.6.0)
doParallel	1.0.14	CRAN (R 3.6.0)
dplyr	0.8.3	CRAN (R 3.6.0)
dslabs	0.5.2	CRAN (R 3.6.0)
e1071	1.7-2	CRAN (R 3.6.0)
earth	5.1.1	CRAN (R 3.6.0)
emo	0.0.0.9000	Github (hadley/emo\@02a5206)
extracat	NA	NA
factoextra	1.0.5	CRAN (R 3.6.0)
foreach	1.4.4	CRAN (R 3.6.0)
forecast	8.7	CRAN (R 3.6.0)
gbm	2.1.5	CRAN (R 3.6.0)
ggbeeswarm	0.6.0	CRAN (R 3.6.0)
ggmap	3.0.0	CRAN (R 3.6.0)
ggplot2	3.2.1	CRAN (R 3.6.0)
ggplotify	0.0.3	CRAN (R 3.6.0)
glmnet	2.0-16	CRAN (R 3.6.0)
gridExtra	2.3	CRAN (R 3.6.0)
h2o	3.22.1.1	CRAN (R 3.6.0)
HDclassif	2.1.0	CRAN (R 3.6.0)
iml	0.9.0	CRAN (R 3.6.0)
ipred	0.9-9	CRAN (R 3.6.0)
kableExtra	1.1.0	CRAN (R 3.6.0)
keras	2.2.4.1.9001	Github (rstudio/keras\@8758aae)
kernlab	0.9-27	CRAN (R 3.6.0)
knitr	1.24	CRAN (R 3.6.0)
lime	0.4.1	CRAN (R 3.6.0)

TABLE 0.1: Packages mentioned throughout and used to produce this book. *(continued)*

package	version	source
markdown	1.1	CRAN (R 3.6.0)
MASS	7.3-51.4	CRAN (R 3.6.1)
Matrix	1.2-17	CRAN (R 3.6.1)
mclust	5.4.3	CRAN (R 3.6.0)
mlbench	2.1-1	CRAN (R 3.6.0)
NbClust	3.0	CRAN (R 3.6.0)
pBrackets	1.0	CRAN (R 3.6.0)
pcadapt	4.1.0	CRAN (R 3.6.0)
pdp	0.7.0	CRAN (R 3.6.0)
plotROC	2.2.1	CRAN (R 3.6.0)
pls	2.7-1	CRAN (R 3.6.0)
pROC	1.14.0	CRAN (R 3.6.0)
purrr	0.3.2	CRAN (R 3.6.0)
ranger	0.11.2	CRAN (R 3.6.0)
readr	1.3.1	CRAN (R 3.6.0)
recipes	0.1.5	CRAN (R 3.6.0)
reshape2	1.4.3	CRAN (R 3.6.0)
ROCR	1.0-7	CRAN (R 3.6.0)
rpart	4.1-15	CRAN (R 3.6.1)
rpart.plot	3.0.7	CRAN (R 3.6.0)
rsample	0.0.4	CRAN (R 3.6.0)
scales	1.0.0	CRAN (R 3.6.0)
sparsepca	0.1.2	CRAN (R 3.6.0)
stringr	1.4.0	CRAN (R 3.6.0)
subsemble	NA	NA
SuperLearner	2.0-25	CRAN (R 3.6.0)
tfestimators	1.9.1	CRAN (R 3.6.0)
tfruns	1.4	CRAN (R 3.6.0)
tidyr	0.8.3	CRAN (R 3.6.0)
vip	0.1.3	CRAN (R 3.6.0)
visdat	0.5.3	CRAN (R 3.6.0)
xgboost	0.82.1	CRAN (R 3.6.0)
yardstick	0.0.3	CRAN (R 3.6.0)

Part I

Fundamentals

1

Introduction to Machine Learning

Machine learning (ML) continues to grow in importance for many organizations across nearly all domains. Some example applications of machine learning in practice include:

- Predicting the likelihood of a patient returning to the hospital (*readmission*) within 30 days of discharge.
- Segmenting customers based on common attributes or purchasing behavior for targeted marketing.
- Predicting coupon redemption rates for a given marketing campaign.
- Predicting customer churn so an organization can perform preventative intervention.
- And many more!

In essence, these tasks all seek to learn from data. To address each scenario, we can use a given set of *features* to train an algorithm and extract insights. These algorithms, or *learners*, can be classified according to the amount and type of supervision needed during training. The two main groups this book focuses on are: ***supervised learners*** which construct predictive models, and ***unsupervised learners*** which build descriptive models. Which type you will need to use depends on the learning task you hope to accomplish.

1.1 Supervised learning

A ***predictive model*** is used for tasks that involve the prediction of a given output (or target) using other variables (or features) in the data set. Or, as stated by Kuhn and Johnson (2013, p. 2), predictive modeling is "...the process of developing a mathematical tool or model that generates an accurate prediction." The learning algorithm in a predictive model attempts to discover and model the relationships among the target variable (the variable being predicted) and the other features (aka predictor variables). Examples of predictive modeling include:

- using customer attributes to predict the probability of the customer churning in the next 6 weeks;
- using home attributes to predict the sales price;
- using employee attributes to predict the likelihood of attrition;
- using patient attributes and symptoms to predict the risk of readmission;
- using production attributes to predict time to market.

Each of these examples has a defined learning task; they each intend to use attributes (X) to predict an outcome measurement (Y).

Throughout this text we'll use various terms interchangeably for

- X: "predictor variable", "independent variable", "attribute", "feature", "predictor"
- Y: "target variable", "dependent variable", "response", "outcome measurement"

The predictive modeling examples above describe what is known as *supervised learning*. The supervision refers to the fact that the target values provide a supervisory role, which indicates to the learner the task it needs to learn. Specifically, given a set of data, the learning algorithm attempts to optimize a function (the algorithmic steps) to find the combination of feature values that results in a predicted value that is as close to the actual target output as possible.

In supervised learning, the training data you feed the algorithm includes the target values. Consequently, the solutions can be used to help *supervise* the training process to find the optimal algorithm parameters.

Most supervised learning problems can be bucketed into one of two categories, *regression* or *classification*, which we discuss next.

1.1.1 Regression problems

When the objective of our supervised learning is to predict a numeric outcome, we refer to this as a ***regression problem*** (not to be confused with linear regression modeling). Regression problems revolve around predicting output that falls on a continuum. In the examples above, predicting home sales prices and time to market reflect a regression problem because the output is numeric and continuous. This means, given the combination of predictor values, the response value could fall anywhere along some continuous spectrum (e.g., the predicted sales price of a particular home could be between $80,000 and $755,000). Figure 1.1 illustrates average home sales prices as a function of two home features: year built and total square footage. Depending on the combination of these two features, the expected home sales price could fall anywhere along a plane.

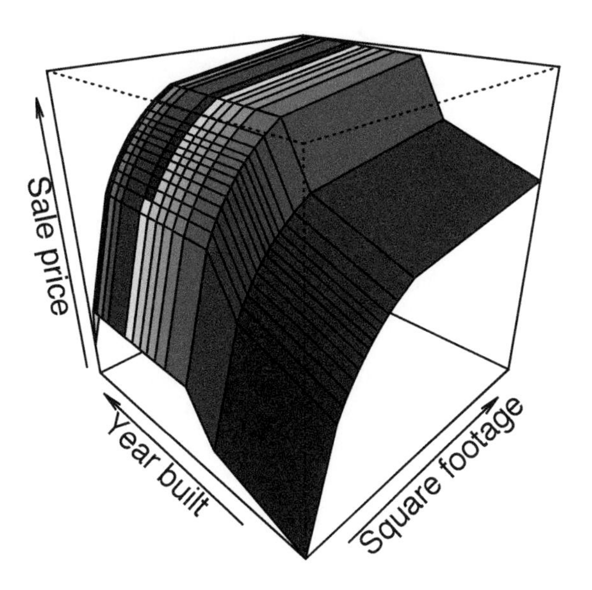

FIGURE 1.1: Average home sales price as a function of year built and total square footage.

1.1.2 Classification problems

When the objective of our supervised learning is to predict a categorical outcome, we refer to this as a ***classification problem***. Classification problems most commonly revolve around predicting a binary or multinomial response measure such as:

- Did a customer redeem a coupon (coded as yes/no or 1/0)?
- Did a customer churn (coded as yes/no or 1/0)?
- Did a customer click on our online ad (coded as yes/no or 1/0)?
- Classifying customer reviews:
  - Binary: positive vs. negative.
  - Multinomial: extremely negative to extremely positive on a 0–5 Likert scale.

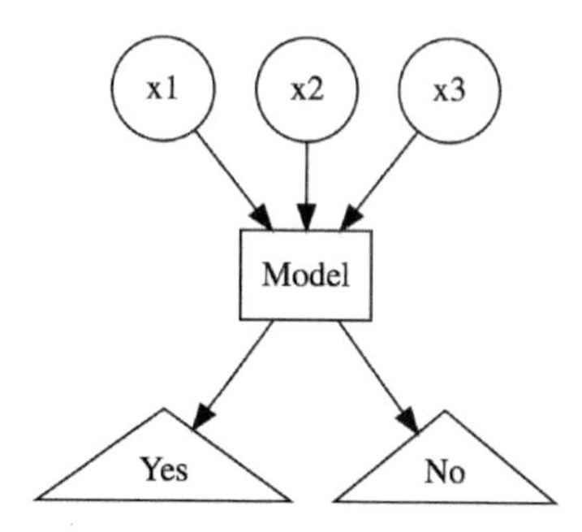

FIGURE 1.2: Classification problem modeling 'Yes'/'No' response based on three features.

However, when we apply machine learning models for classification problems, rather than predict a particular class (i.e., "yes" or "no"), we often want to predict the *probability* of a particular class (i.e., yes: 0.65, no: 0.35). By default, the class with the highest predicted probability becomes the predicted class. Consequently, even though we are performing a classification problem, we are still predicting a numeric output (probability). However, the essence of the problem still makes it a classification problem.

Although there are machine learning algorithms that can be applied to regression problems but not classification and vice versa, most of the supervised learning algorithms we cover in this book can be applied to both. These algorithms have become the most popular machine learning applications in recent years.

1.2 Unsupervised learning

Unsupervised learning, in contrast to supervised learning, includes a set of statistical tools to better understand and describe your data, but performs the analysis without a target variable. In essence, unsupervised learning is concerned with identifying groups in a data set. The groups may be defined by the rows (i.e., *clustering*) or the columns (i.e., *dimension reduction*); however, the motive in each case is quite different.

The goal of ***clustering*** is to segment observations into similar groups based on the observed variables; for example, to divide consumers into different homogeneous groups, a process known as market segmentation. In **dimension reduction**, we are often concerned with reducing the number of variables in a data set. For example, classical linear regression models break down in the presence of highly correlated features. Some dimension reduction techniques can be used to reduce the feature set to a potentially smaller set of uncorrelated variables. Such a reduced feature set is often used as input to downstream supervised learning models (e.g., principal component regression).

Unsupervised learning is often performed as part of an exploratory data analysis (EDA). However, the exercise tends to be more subjective, and there is no simple goal for the analysis, such as prediction of a response. Furthermore, it can be hard to assess the quality of results obtained from unsupervised learning methods. The reason for this is simple. If we fit a predictive model using a supervised learning technique (i.e., linear regression), then it is possible to check our work by seeing how well our model predicts the response Y on observations not used in fitting the model. However, in unsupervised learning, there is no way to check our work because we don't know the true answer—the problem is unsupervised!

Despite its subjectivity, the importance of unsupervised learning should not be overlooked and such techniques are often used in organizations to:

- Divide consumers into different homogeneous groups so that tailored marketing strategies can be developed and deployed for each segment.
- Identify groups of online shoppers with similar browsing and purchase histories, as well as items that are of particular interest to the shoppers within each group. Then an individual shopper can be preferentially shown the items in which he or she is particularly likely to be interested, based on the purchase histories of similar shoppers.
- Identify products that have similar purchasing behavior so that managers can manage them as product groups.

These questions, and many more, can be addressed with unsupervised learning. Moreover, the outputs of unsupervised learning models can be used as inputs to downstream supervised learning models.

1.3 Roadmap

The goal of this book is to provide effective tools for uncovering relevant and useful patterns in your data by using R's ML stack. We begin by providing an overview of the ML modeling process and discussing fundamental concepts that will carry through the rest of the book. These include feature engineering, data splitting, model validation and tuning, and performance measurement. These concepts will be discussed in Chapters 2-3.

Chapters 4-14 focus on common supervised learners ranging from simpler linear regression models to the more complicated gradient boosting machines and deep neural networks. Here we will illustrate the fundamental concepts of each base learning algorithm and how to tune its hyperparameters to maximize predictive performance.

Chapters 15-16 delve into more advanced approaches to maximize effectiveness, efficiency, and interpretation of your ML models. We discuss how to combine multiple models to create a stacked model (aka *super learner*), which allows you to combine the strengths from each base learner and further maximize predictive accuracy. We then illustrate how to make the training and validation process more efficient with automated ML (aka AutoML). Finally, we illustrate many ways to extract insight from your "black box" models with various ML interpretation techniques.

The latter part of the book focuses on unsupervised techniques aimed at reducing the dimensions of your data for more effective data representation (Chapters 17-19) and identifying common groups among your observations with clustering techniques (Chapters 20-22).

1.4 The data sets

The data sets chosen for this book allow us to illustrate the different features of the presented machine learning algorithms. Since the goal of this book is to demonstrate how to implement R's ML stack, we make the assumption that you have already spent significant time cleaning and getting to know your data via EDA. This would allow you to perform many necessary tasks prior to the ML tasks outlined in this book such as:

- Feature selection (i.e., removing unnecessary variables and retaining only those variables you wish to include in your modeling process).
- Recoding variable names and values so that they are meaningful and more interpretable.
- Recoding, removing, or some other approach to handling missing values.

Consequently, the exemplar data sets we use throughout this book have, for the most part, gone through the necessary cleaning processes. In some cases we illustrate concepts with stereotypical data sets (i.e. `mtcars`, `iris`, `geyser`); however, we tend to focus most of our discussion around the following data sets:

- Property sales information as described in De Cock (2011).

  - **problem type**: supervised regression
  - **response variable**: `Sale_Price` (i.e., $195,000, $215,000)
  - **features**: 80
  - **observations**: 2,930
  - **objective**: use property attributes to predict the sale price of a home
  - **access**: provided by the `AmesHousing` package (Kuhn, 2017a)
  - **more details**: See `?AmesHousing::ames_raw`

```
# Access data
ames <- AmesHousing::make_ames()

# Print dimensions
dim(ames)
## [1] 2930   81

# Peek at response variable
head(ames$Sale_Price)
## [1] 215000 105000 172000 244000 189900 195500
```

You can see the entire data cleaning process to transform the raw Ames housing data (`AmesHousing::ames_raw`) to the final clean data (`AmesHousing::make_ames`) that we will use in machine learning algorithms throughout this book by typing `AmesHousing::make_ames` into the R console.

- Employee attrition information originally provided by IBM Watson Analytics Lab[1].
  - **problem type**: supervised binomial classification
  - **response variable**: `Attrition` (i.e., "Yes", "No")
  - **features**: 30
  - **observations**: 1,470
  - **objective**: use employee attributes to predict if they will attrit (leave the company)
  - **access**: provided by the `rsample` package (Kuhn and Wickham, 2019)
  - **more details**: See `?rsample::attrition`

[1]https://www.ibm.com/communities/analytics/watson-analytics-blog/hr-employee-attrition/

```
# Access data
attrition <- rsample::attrition

# Print dimensions
dim(attrition)
## [1] 1470   31

# Peek at response variable
head(attrition$Attrition)
## [1] Yes No  Yes No  No  No
## Levels: No Yes
```

- Image information for handwritten numbers originally presented to AT&T Bell Lab's to help build automatic mail-sorting machines for the USPS. Has been used since early 1990s to compare machine learning performance on pattern recognition (i.e., LeCun et al. (1990); LeCun et al. (1998); Cireşan et al. (2012)).
  - **Problem type**: supervised multinomial classification
  - **response variable**: `V785` (i.e., numbers to predict: 0, 1, ..., 9)
  - **features**: 784
  - **observations**: 60,000 (train) / 10,000 (test)
  - **objective**: use attributes about the "darkness" of each of the 784 pixels in images of handwritten numbers to predict if the number is 0, 1, ..., or 9.
  - **access**: provided by the `dslabs` package (Irizarry, 2018)
  - **more details**: See `?dslabs::read_mnist()` and online MNIST documentation[2]

```
# Access data
mnist <- dslabs::read_mnist()
names(mnist)
## [1] "train" "test"

# Print feature dimensions
dim(mnist$train$images)
## [1] 60000   784

# Peek at response variable
head(mnist$train$labels)
## [1] 5 0 4 1 9 2
```

[2] `http://yann.lecun.com/exdb/mnist/`

- Grocery items and quantities purchased. Each observation represents a single basket of goods that were purchased together.
  - **Problem type**: unsupervised basket analysis
  - **response variable**: NA
  - **features**: 42
  - **observations**: 2,000
  - **objective**: use attributes of each basket to identify common groupings of items purchased together.
  - **access**: available on the companion website for this book

```
# URL to download/read in the data
url <- "https://koalaverse.github.io/homlr/data/my_basket.csv"

# Access data
my_basket <- readr::read_csv(url)

# Print dimensions
dim(my_basket)
## [1] 2000   42

# Peek at response variable
my_basket
## # A tibble: 2,000 x 42
##     '7up' lasagna pepsi   yop red.wine cheese   bbq
##     <dbl>   <dbl> <dbl> <dbl>    <dbl>  <dbl> <dbl>
##  1      0       0     0     0        0      0     0
##  2      0       0     0     0        0      0     0
##  3      0       0     0     0        0      0     0
##  4      0       0     0     2        1      0     0
##  5      0       0     0     0        0      0     0
##  6      0       0     0     0        0      0     0
##  7      1       1     0     0        0      0     1
##  8      0       0     0     0        0      0     0
##  9      0       1     0     0        0      0     0
## 10      0       0     0     0        0      0     0
## # ... with 1,990 more rows, and 35 more variables:
## #   bulmers <dbl>, mayonnaise <dbl>, horlics <dbl>,
## #   chicken.tikka <dbl>, milk <dbl>, mars <dbl>,
## #   coke <dbl>, lottery <dbl>, bread <dbl>,
## #   pizza <dbl>, sunny.delight <dbl>, ham <dbl>,
## #   lettuce <dbl>, kronenbourg <dbl>, leeks <dbl>,
## #   fanta <dbl>, tea <dbl>, whiskey <dbl>, peas <dbl>,
## #   newspaper <dbl>, muesli <dbl>, white.wine <dbl>,
## #   carrots <dbl>, spinach <dbl>, pate <dbl>,
```

```
## #   instant.coffee <dbl>, twix <dbl>, potatoes <dbl>,
## #   fosters <dbl>, soup <dbl>, toad.in.hole <dbl>,
## #   coco.pops <dbl>, kitkat <dbl>, broccoli <dbl>,
## #   cigarettes <dbl>
```

2

Modeling Process

Much like EDA, the ML process is very iterative and heurstic-based. With minimal knowledge of the problem or data at hand, it is difficult to know which ML method will perform best. This is known as the *no free lunch* theorem for ML (Wolpert, 1996). Consequently, it is common for many ML approaches to be applied, evaluated, and modified before a final, optimal model can be determined. Performing this process correctly provides great confidence in our outcomes. If not, the results will be useless and, potentially, damaging [1].

Approaching ML modeling correctly means approaching it strategically by spending our data wisely on learning and validation procedures, properly pre-processing the feature and target variables, minimizing *data leakage* (Section 3.8.2), tuning hyperparameters, and assessing model performance. Many books and courses portray the modeling process as a short sprint. A better analogy would be a marathon where many iterations of these steps are repeated before eventually finding the final optimal model. This process is illustrated in Figure 2.1. Before introducing specific algorithms, this chapter, and the next, introduce concepts that are fundamental to the ML modeling process and that you'll see briskly covered in future modeling chapters.

Although the discussions in this chapter focus on supervised ML modeling, many of the topics also apply to unsupervised methods.

2.1 Prerequisites

This chapter leverages the following packages.

[1] See `https://www.fatml.org/resources/relevant-scholarship` for many discussions regarding implications of poorly applied and interpreted ML.

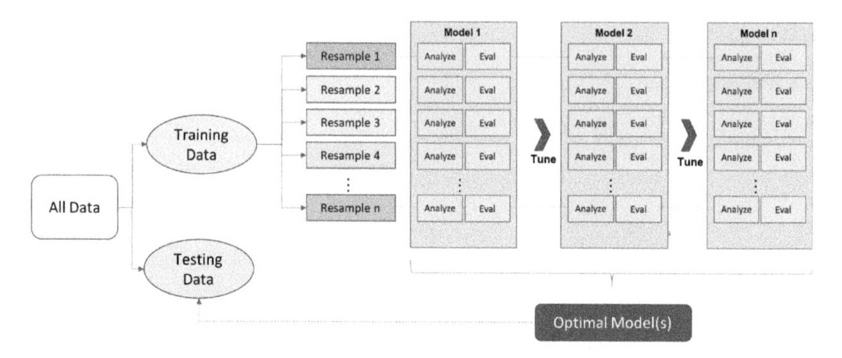

FIGURE 2.1: General predictive machine learning process.

```
# Helper packages
library(dplyr)      # for data manipulation
library(ggplot2)    # for awesome graphics

# Modeling process packages
library(rsample)    # for resampling procedures
library(caret)      # for resampling and model training
library(h2o)        # for resampling and model training

# h2o set-up
h2o.no_progress()  # turn off h2o progress bars
h2o.init()          # launch h2o
```

To illustrate some of the concepts, we'll use the Ames Housing and employee attrition data sets introduced in Chapter 1. Throughout this book, we'll demonstrate approaches with ordinary R data frames. However, since many of the supervised machine learning chapters leverage the **h2o** package, we'll also show how to do some of the tasks with H2O objects. You can convert any R data frame to an H2O object (i.e., import it to the H2O cloud) easily with `as.h2o(<my-data-frame>)`.

If you try to convert the original `rsample::attrition` data set to an H2O object an error will occur. This is because several variables are *ordered factors* and H2O has no way of handling this data type. Consequently, you must convert any ordered factors to unordered; see `?base::ordered` for details.

```
# Ames housing data
ames <- AmesHousing::make_ames()
ames.h2o <- as.h2o(ames)

# Job attrition data
churn <- rsample::attrition %>%
  mutate_if(is.ordered, .funs = factor, ordered = FALSE)
churn.h2o <- as.h2o(churn)
```

2.2 Data splitting

A major goal of the machine learning process is to find an algorithm $f(X)$ that most accurately predicts future values ($\hat{Y}$) based on a set of features (X). In other words, we want an algorithm that not only fits well to our past data, but more importantly, one that predicts a future outcome accurately. This is called the ***generalizability*** of our algorithm. How we "spend" our data will help us understand how well our algorithm generalizes to unseen data.

To provide an accurate understanding of the generalizability of our final optimal model, we can split our data into training and test data sets:

- **Training set**: these data are used to develop feature sets, train our algorithms, tune hyperparameters, compare models, and all of the other activities required to choose a final model (e.g., the model we want to put into production).
- **Test set**: having chosen a final model, these data are used to estimate an unbiased assessment of the model's performance, which we refer to as the *generalization error*.

It is critical that the test set not be used prior to selecting your final model. Assessing results on the test set prior to final model selection biases the model selection process since the testing data will have become part of the model development process.

Given a fixed amount of data, typical recommendations for splitting your data into training-test splits include 60% (training)–40% (testing), 70%–30%, or 80%–20%. Generally speaking, these are appropriate guidelines to follow; however, it is good to keep the following points in mind:

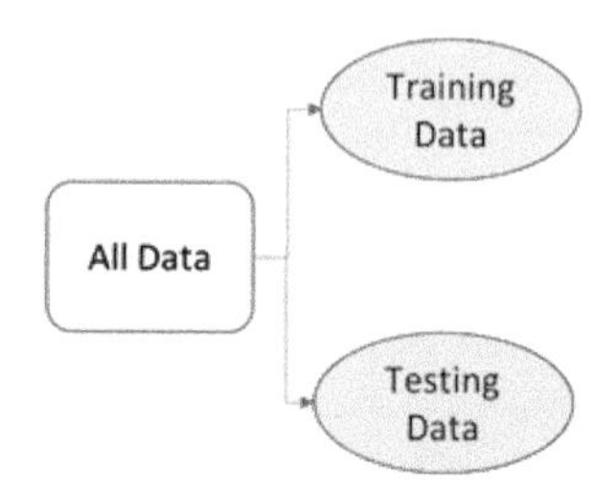

FIGURE 2.2: Splitting data into training and test sets.

- Spending too much in training (e.g., $> 80\%$) won't allow us to get a good assessment of predictive performance. We may find a model that fits the training data very well, but is not generalizable (*overfitting*).
- Sometimes too much spent in testing ($> 40\%$) won't allow us to get a good assessment of model parameters.

Other factors should also influence the allocation proportions. For example, very large training sets (e.g., $n > 100\text{K}$) often result in only marginal gains compared to smaller sample sizes. Consequently, you may use a smaller training sample to increase computation speed (e.g., models built on larger training sets often take longer to score new data sets in production). In contrast, as $p \geq n$ (where p represents the number of features), larger samples sizes are often required to identify consistent signals in the features.

The two most common ways of splitting data include ***simple random sampling*** and ***stratified sampling***.

2.2.1 Simple random sampling

The simplest way to split the data into training and test sets is to take a simple random sample. This does not control for any data attributes, such as the distribution of your response variable (Y). There are multiple ways to split our data in R. Here we show four options to produce a 70–30 split in the Ames housing data:

Sampling is a random process so setting the random number generator with a common seed allows for reproducible results. Throughout this book we'll often use the seed `123` for reproducibility but the number itself has no special meaning.

```r
# Using base R
set.seed(123)  # for reproducibility
index_1 <- sample(1:nrow(ames), round(nrow(ames) * 0.7))
train_1 <- ames[index_1, ]
test_1  <- ames[-index_1, ]

# Using caret package
set.seed(123)  # for reproducibility
index_2 <- createDataPartition(ames$Sale_Price, p = 0.7,
                               list = FALSE)
train_2 <- ames[index_2, ]
test_2  <- ames[-index_2, ]

# Using rsample package
set.seed(123)  # for reproducibility
split_1  <- initial_split(ames, prop = 0.7)
train_3  <- training(split_1)
test_3   <- testing(split_1)

# Using h2o package
split_2 <- h2o.splitFrame(ames.h2o, ratios = 0.7,
                          seed = 123)
train_4 <- split_2[[1]]
test_4  <- split_2[[2]]
```

With sufficient sample size, this sampling approach will typically result in a similar distribution of Y (e.g., `Sale_Price` in the `ames` data) between your training and test sets, as illustrated below.

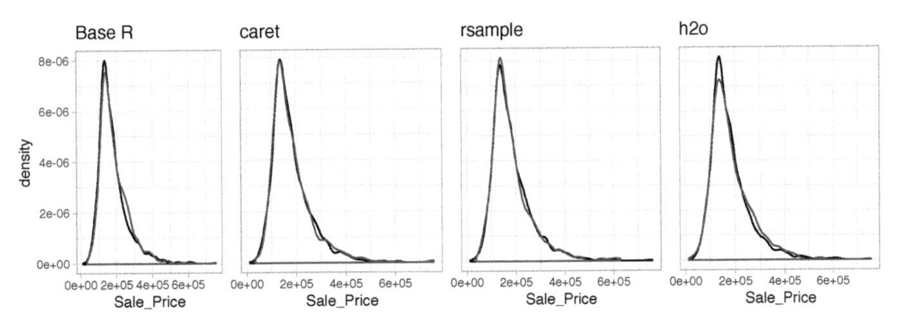

FIGURE 2.3: Training (black) vs. test (red) response distribution.

2.2.2 Stratified sampling

If we want to explicitly control the sampling so that our training and test sets have similar Y distributions, we can use stratified sampling. This is more common with classification problems where the response variable may be severely imbalanced (e.g., 90% of observations with response "Yes" and 10% with response "No"). However, we can also apply stratified sampling to regression problems for data sets that have a small sample size and where the response variable deviates strongly from normality (i.e., positively skewed like `Sale_Price`). With a continuous response variable, stratified sampling will segment Y into quantiles and randomly sample from each. Consequently, this will help ensure a balanced representation of the response distribution in both the training and test sets.

The easiest way to perform stratified sampling on a response variable is to use the **rsample** package, where you specify the response variable to `strata`fy. The following illustrates that in our original employee attrition data we have an imbalanced response (No: 84%, Yes: 16%). By enforcing stratified sampling, both our training and testing sets have approximately equal response distributions.

```
# orginal response distribution
table(churn$Attrition) %>% prop.table()
##
##    No   Yes
## 0.839 0.161

# stratified sampling with the rsample package
set.seed(123)
split_strat  <- initial_split(churn, prop = 0.7,
                              strata = "Attrition")
train_strat  <- training(split_strat)
test_strat   <- testing(split_strat)

# consistent response ratio between train & test
table(train_strat$Attrition) %>% prop.table()
##
##    No   Yes
## 0.839 0.161
table(test_strat$Attrition) %>% prop.table()
##
##    No   Yes
## 0.839 0.161
```

2.2.3 Class imbalances

Imbalanced data can have a significant impact on model predictions and performance (Kuhn and Johnson, 2013). Most often this involves classification problems where one class has a very small proportion of observations (e.g., defaults - 5% versus nondefaults - 95%). Several sampling methods have been developed to help remedy class imbalance and most of them can be categorized as either *up-sampling* or *down-sampling*.

Down-sampling balances the dataset by reducing the size of the abundant class(es) to match the frequencies in the least prevalent class. This method is used when the quantity of data is sufficient. By keeping all samples in the rare class and randomly selecting an equal number of samples in the abundant class, a balanced new dataset can be retrieved for further modeling. Furthermore, the reduced sample size reduces the computation burden imposed by further steps in the ML process.

On the contrary, up-sampling is used when the quantity of data is insufficient. It tries to balance the dataset by increasing the size of rarer samples. Rather than getting rid of abundant samples, new rare samples are generated by using repetition or bootstrapping (described further in Section 2.4.2).

Note that there is no absolute advantage of one sampling method over another. Application of these two methods depends on the use case it applies to and the data set itself. A combination of over- and under-sampling is often successful and a common approach is known as Synthetic Minority Over-Sampling Technique, or SMOTE (Chawla et al., 2002). This alternative sampling approach, as well as others, can be implemented in R (see the `sampling` argument in `?caret::trainControl()`). Furthermore, many ML algorithms implemented in R have class weighting schemes to remedy imbalances internally (e.g., most **h2o** algorithms have a `weights_column` and `balance_classes` argument).

2.3 Creating models in R

The R ecosystem provides a wide variety of ML algorithm implementations. This makes many powerful algorithms available at your fingertips. Moreover, there are almost always more than one package to perform each algorithm (e.g., there are over 20 packages for fitting random forests). There are pros and cons to this wide selection; some implementations may be more computationally efficient while others may be more flexible (i.e., have more hyperparameter tuning options). Future chapters will expose you to many of the packages and algorithms that perform and scale best to the kinds of tabular data and problems encountered by most organizations.

However, this also has resulted in some drawbacks as there are inconsistencies in how algorithms allow you to define the formula of interest and how the results and predictions are supplied.[2] In addition to illustrating the more popular and powerful packages, we'll also show you how to use implementations that provide more consistency.

2.3.1 Many formula interfaces

To fit a model to our data, the model terms must be specified. Historically, there are two main interfaces for doing this. The formula interface uses R's formula rules to specify a symbolic representation of the terms. For example, `Y ~ X` where we say "Y is a function of X". To illustrate, suppose we have some generic modeling function called `model_fn()` which accepts an R formula, as in the following examples:

```
# Sale price as function of neighborhood and year sold
model_fn(Sale_Price ~ Neighborhood + Year_Sold,
         data = ames)

# Variables + interactions
model_fn(Sale_Price ~ Neighborhood + Year_Sold +
           Neighborhood:Year_Sold, data = ames)

# Shorthand for all predictors
model_fn(Sale_Price ~ ., data = ames)

# Inline functions / transformations
model_fn(log10(Sale_Price) ~ ns(Longitude, df = 3) +
           ns(Latitude, df = 3), data = ames)
```

This is very convenient but it has some disadvantages. For example:

- You can't nest in-line functions such as performing principal components analysis on the feature set prior to executing the model (`model_fn(y ~ pca(scale(x1), scale(x2), scale(x3)), data = df)`).
- All the model matrix calculations happen at once and can't be recycled when used in a model function.
- For very wide data sets, the formula method can be extremely inefficient (Kuhn, 2017b).

[2]Many of these drawbacks and inconsistencies were originally organized and presented by Kuhn (2018).

- There are limited roles that variables can take which has led to several re-implementations of formulas.
- Specifying multivariate outcomes is clunky and inelegant.
- Not all modeling functions have a formula method (lack of consistency!).

Some modeling functions have a non-formula (XY) interface. These functions have separate arguments for the predictors and the outcome(s):

```
# Use separate inputs for X and Y
features <- c("Year_Sold", "Longitude", "Latitude")
model_fn(x = ames[, features], y = ames$Sale_Price)
```

This provides more efficient calculations but can be inconvenient if you have transformations, factor variables, interactions, or any other operations to apply to the data prior to modeling.

Overall, it is difficult to determine if a package has one or both of these interfaces. For example, the `lm()` function, which performs linear regression, only has the formula method. Consequently, until you are familiar with a particular implementation you will need to continue referencing the corresponding help documentation.

A third interface, is to use *variable name specification* where we provide all the data combined in one training frame but we specify the features and response with character strings. This is the interface used by the **h2o** package.

```
model_fn(
  x = c("Year_Sold", "Longitude", "Latitude"),
  y = "Sale_Price",
  data = ames.h2o
)
```

One approach to get around these inconsistencies is to use a meta engine, which we discuss next.

2.3.2 Many engines

Although there are many individual ML packages available, there is also an abundance of meta engines that can be used to help provide consistency. For example, the following all produce the same linear regression model output:

```
lm_lm    <- lm(Sale_Price ~ ., data = ames)
lm_glm   <- glm(Sale_Price ~ ., data = ames,
               family = gaussian)
lm_caret <- train(Sale_Price ~ ., data = ames,
                 method = "lm")
```

Here, `lm()` and `glm()` are two different algorithm engines that can be used to fit the linear model and `caret::train()` is a meta engine (aggregator) that allows you to apply almost any direct engine with `method = "<method-name>"`. There are trade-offs to consider when using direct versus meta engines. For example, using direct engines can allow for extreme flexibility but also requires you to familiarize yourself with the unique differences of each implementation. For example, the following highlights the various syntax nuances required to compute and extract predicted class probabilities across different direct engines.[3]

TABLE 2.1: Syntax for computing predicted class probabilities with direct engines.

Algorithm	Package	Code
Linear discriminant analysis	MASS	predict(obj)
Generlized linear model	stats	predict(obj, type = 'response')
Mixture discriminant analysis	mda	predict(obj, type = 'posterior')
Decision tree	rpart	predict(obj, type = 'prob')
Random forest	ranger	predict(obj)$predictions
Gradient boosting macine	gbm	predict(obj, type = 'response', n.trees)

Meta engines provide you with more consistency in how you specify inputs and extract outputs but can be less flexible than direct engines. Future chapters will illustrate both approaches. For meta engines, we'll focus on the **caret** package in the hardcopy of the book while also demonstrating the newer **parsnip** package in the additional online resources.[4]

[3]This table was modified from Kuhn (2019)

[4]The **caret** package has been the preferred meta engine over the years; however, the author is now transitioning to full-time development on **parsnip**, which is designed to be a more robust and tidy meta engine.

2.4 Resampling methods

In Section 2.2 we split our data into training and testing sets. Furthermore, we were very explicit about the fact that we ***do not*** use the test set to assess model performance during the training phase. So how do we assess the generalization performance of the model?

One option is to assess an error metric based on the training data. Unfortunately, this leads to biased results as some models can perform very well on the training data but not generalize well to a new data set (we'll illustrate this in Section 2.5).

A second method is to use a *validation* approach, which involves splitting the training set further to create two parts (as in Section 2.2): a training set and a validation set (or *holdout set*). We can then train our model(s) on the new training set and estimate the performance on the validation set. Unfortunately, validation using a single holdout set can be highly variable and unreliable unless you are working with very large data sets (Molinaro et al., 2005; Hawkins et al., 2003). As the size of your data set reduces, this concern increases.

Although we stick to our definitions of test, validation, and holdout sets, these terms are sometimes used interchangeably in other literature and software. What's important to remember is to always put a portion of the data under lock and key until a final model has been selected (we refer to this as the test data, but others refer to it as the holdout set).

Resampling methods provide an alternative approach by allowing us to repeatedly fit a model of interest to parts of the training data and test its performance on other parts. The two most commonly used resampling methods include *k-fold cross validation* and *bootstrapping*.

2.4.1 k-fold cross validation

k-fold cross-validation (aka k-fold CV) is a resampling method that randomly divides the training data into k groups (aka folds) of approximately equal size. The model is fit on $k-1$ folds and then the remaining fold is used to compute model performance. This procedure is repeated k times; each time, a different fold is treated as the validation set. This process results in k estimates of the generalization error (say $\epsilon_1, \epsilon_2, \dots, \epsilon_k$). Thus, the k-fold CV estimate is

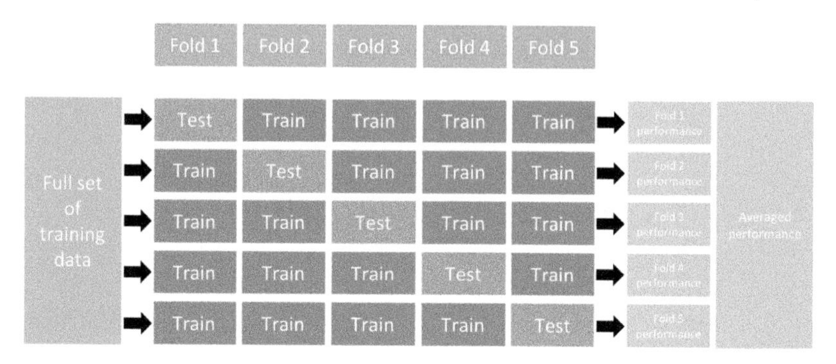

FIGURE 2.4: Illustration of the k-fold cross validation process.

computed by averaging the k test errors, providing us with an approximation of the error we might expect on unseen data.

Consequently, with k-fold CV, every observation in the training data will be held out one time to be included in the test set as illustrated in Figure 2.5. In practice, one typically uses $k = 5$ or $k = 10$. There is no formal rule as to the size of k; however, as k gets larger, the difference between the estimated performance and the true performance to be seen on the test set will decrease. On the other hand, using too large k can introduce computational burdens. Moreover, Molinaro et al. (2005) found that $k = 10$ performed similarly to leave-one-out cross validation (LOOCV) which is the most extreme approach (i.e., setting $k = n$).

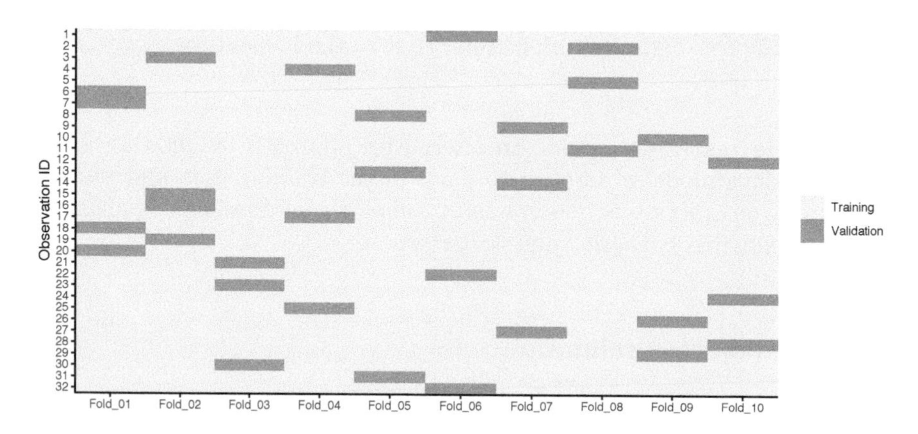

FIGURE 2.5: 10-fold cross validation on 32 observations. Each observation is used once for validation and nine times for training.

Although using $k \geq 10$ helps to minimize the variability in the estimated performance, k-fold CV still tends to have higher variability than bootstrapping (discussed next). Kim (2009) showed that repeating k-fold CV can help to increase the precision of the estimated generalization error. Consequently, for smaller data sets (say $n < 10,000$), 10-fold CV repeated 5 or 10 times will improve the accuracy of your estimated performance and also provide an estimate of its variability.

Throughout this book we'll cover multiple ways to incorporate CV as you can often perform CV directly within certain ML functions:

```
# Example using h2o
h2o.cv <- h2o.glm(
  x = x,
  y = y,
  training_frame = ames.h2o,
  nfolds = 10  # perform 10-fold CV
)
```

Or externally as in the below chunk[5]. When applying it externally to an ML algorithm as below, we'll need a process to apply the ML model to each resample, which we'll also cover.

```
vfold_cv(ames, v = 10)
## #  10-fold cross-validation
## # A tibble: 10 x 2
##    splits             id
##    <named list>       <chr>
##  1 <split [2.6K/293]> Fold01
##  2 <split [2.6K/293]> Fold02
##  3 <split [2.6K/293]> Fold03
##  4 <split [2.6K/293]> Fold04
##  5 <split [2.6K/293]> Fold05
##  6 <split [2.6K/293]> Fold06
##  7 <split [2.6K/293]> Fold07
##  8 <split [2.6K/293]> Fold08
##  9 <split [2.6K/293]> Fold09
## 10 <split [2.6K/293]> Fold10
```

[5] `rsample::vfold_cv()` results in a nested data frame where each element in `splits` is a list containing the training data frame and the observation IDs that will be used for training the model vs. model validation.

2.4.2 Bootstrapping

A bootstrap sample is a random sample of the data taken *with replacement* (Efron and Tibshirani, 1986). This means that, after a data point is selected for inclusion in the subset, it's still available for further selection. A bootstrap sample is the same size as the original data set from which it was constructed. Figure 2.6 provides a schematic of bootstrap sampling where each bootstrap sample contains 12 observations just as in the original data set. Furthermore, bootstrap sampling will contain approximately the same distribution of values (represented by colors) as the original data set.

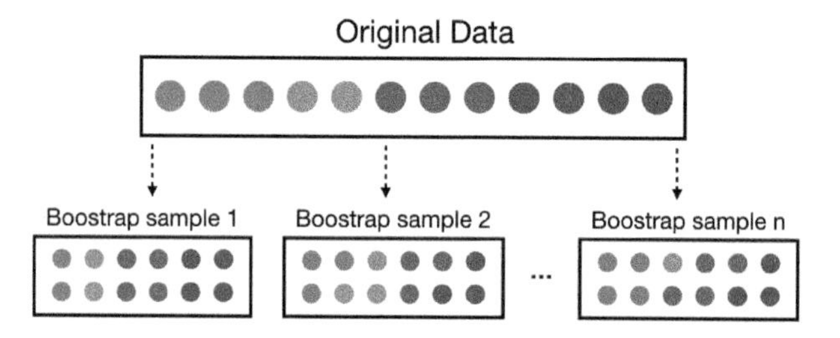

FIGURE 2.6: Illustration of the bootstrapping process.

Since samples are drawn with replacement, each bootstrap sample is likely to contain duplicate values. In fact, on average, $\approx 63.21\%$ of the original sample ends up in any particular bootstrap sample. The original observations not contained in a particular bootstrap sample are considered *out-of-bag* (OOB). When bootstrapping, a model can be built on the selected samples and validated on the OOB samples; this is often done, for example, in random forests (see Chapter 11).

Since observations are replicated in bootstrapping, there tends to be less variability in the error measure compared with k-fold CV (Efron, 1983). However, this can also increase the bias of your error estimate. This can be problematic with smaller data sets; however, for most average-to-large data sets (say $n \geq 1,000$) this concern is often negligible.

Figure 2.7 compares bootstrapping to 10-fold CV on a small data set with $n = 32$ observations. A thorough introduction to the bootstrap and its use in R is provided in Davison et al. (1997).

We can create bootstrap samples easily with `rsample::bootstraps()`, as illustrated in the code chunk below.

```
bootstraps(ames, times = 10)
## # Bootstrap sampling
## # A tibble: 10 x 2
```

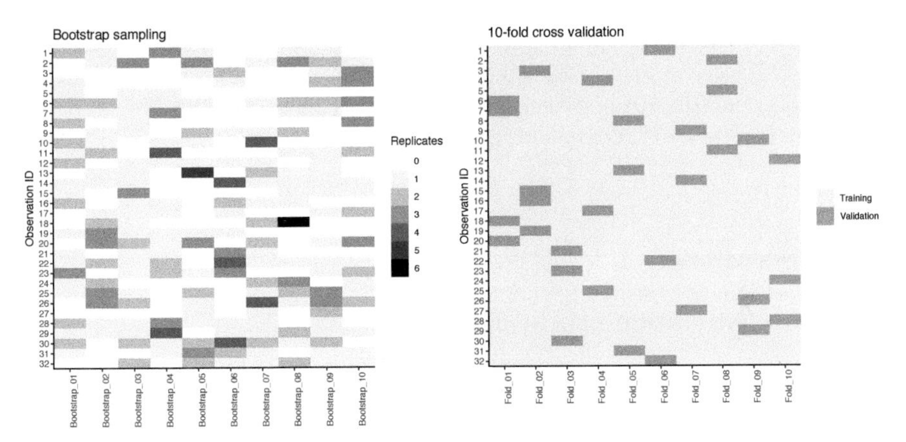

FIGURE 2.7: Bootstrap sampling (left) versus 10-fold cross validation (right) on 32 observations. For bootstrap sampling, the observations that have zero replications (white) are the out-of-bag observations used for validation.

```
##    splits              id
##    <list>              <chr>
##  1 <split [2.9K/1.1K]> Bootstrap01
##  2 <split [2.9K/1.1K]> Bootstrap02
##  3 <split [2.9K/1.1K]> Bootstrap03
##  4 <split [2.9K/1K]>   Bootstrap04
##  5 <split [2.9K/1.1K]> Bootstrap05
##  6 <split [2.9K/1.1K]> Bootstrap06
##  7 <split [2.9K/1.1K]> Bootstrap07
##  8 <split [2.9K/1.1K]> Bootstrap08
##  9 <split [2.9K/1.1K]> Bootstrap09
## 10 <split [2.9K/1K]>   Bootstrap10
```

Bootstrapping is, typically, more of an internal resampling procedure that is naturally built into certain ML algorithms. This will become more apparent in Chapters 10–11 where we discuss bagging and random forests, respectively.

2.4.3 Alternatives

It is important to note that there are other useful resampling procedures. If you're working with time-series specific data then you will want to incorporate rolling origin and other time series resampling procedures. Hyndman and Athanasopoulos (2018) is the dominant, R-focused, time series resource[6].

[6]See their open source book at `https://www.otexts.org/fpp2`

Additionally, Efron (1983) developed the "632 method" and Efron and Tibshirani (1997) discuss the "632+ method"; both approaches seek to minimize biases experienced with bootstrapping on smaller data sets and are available via **caret** (see `?caret::trainControl` for details).

2.5 Bias variance trade-off

Prediction errors can be decomposed into two important subcomponents: error due to "bias" and error due to "variance". There is often a tradeoff between a model's ability to minimize bias and variance. Understanding how different sources of error lead to bias and variance helps us improve the data fitting process resulting in more accurate models.

2.5.1 Bias

Bias is the difference between the expected (or average) prediction of our model and the correct value which we are trying to predict. It measures how far off in general a model's predictions are from the correct value, which provides a sense of how well a model can conform to the underlying structure of the data. Figure 2.8 illustrates an example where the polynomial model does not capture the underlying structure well. Linear models are classical examples of high bias models as they are less flexible and rarely capture non-linear, non-monotonic relationships.

We also need to think of bias-variance in relation to resampling. Models with high bias are rarely affected by the noise introduced by resampling. If a model has high bias, it will have consistency in its resampling performance as illustrated by Figure 2.8.

2.5.2 Variance

On the other hand, error due to *variance* is defined as the variability of a model prediction for a given data point. Many models (e.g., k-nearest neighbor, decision trees, gradient boosting machines) are very adaptable and offer extreme flexibility in the patterns that they can fit to. However, these models offer their own problems as they run the risk of overfitting to the training data. Although you may achieve very good performance on your training data, the model will not automatically generalize well to unseen data.

Since high variance models are more prone to overfitting, using resampling

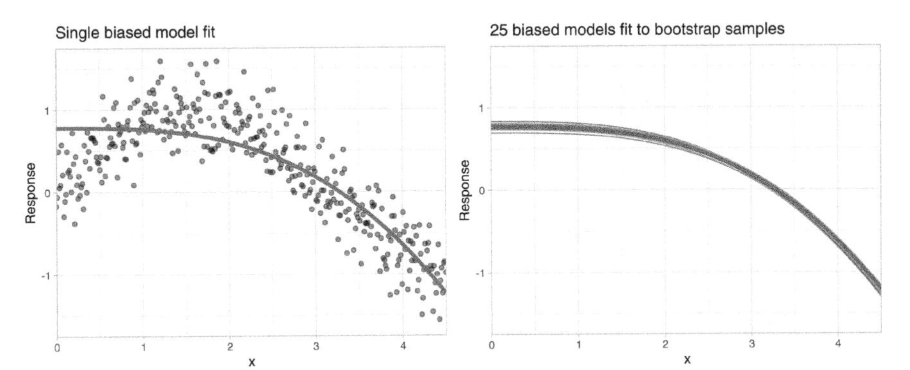

FIGURE 2.8: A biased polynomial model fit to a single data set does not capture the underlying non-linear, non-monotonic data structure (left). Models fit to 25 bootstrapped replicates of the data are underterred by the noise and generates similar, yet still biased, predictions (right).

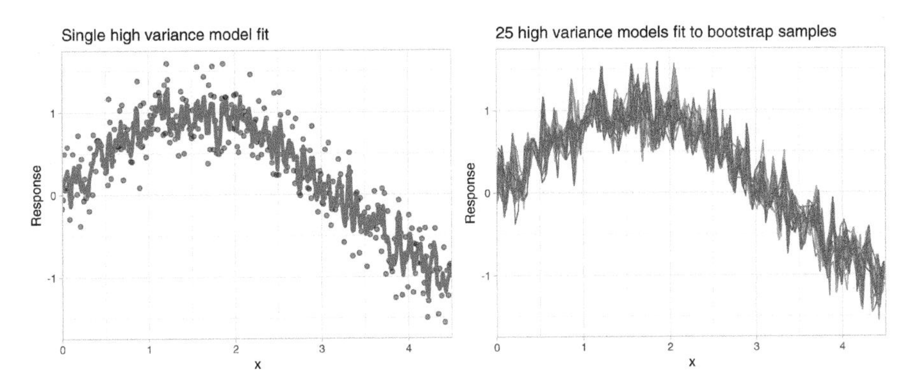

FIGURE 2.9: A high variance k-nearest neighbor model fit to a single data set captures the underlying non-linear, non-monotonic data structure well but also overfits to individual data points (left). Models fit to 25 bootstrapped replicates of the data are deterred by the noise and generate highly variable predictions (right).

procedures are critical to reduce this risk. Moreover, many algorithms that are capable of achieving high generalization performance have lots of *hyperparameters* that control the level of model complexity (i.e., the tradeoff between bias and variance).

2.5.3 Hyperparameter tuning

Hyperparameters (aka *tuning parameters*) are the "knobs to twiddle"[7] to control the complexity of machine learning algorithms and, therefore, the bias-variance trade-off. Not all algorithms have hyperparameters (e.g., ordinary least squares[8]); however, most have at least one or more.

The proper setting of these hyperparameters is often dependent on the data and problem at hand and cannot always be estimated by the training data alone. Consequently, we need a method of identifying the optimal setting. For example, in the high variance example in the previous section, we illustrated a high variance k-nearest neighbor model (we'll discuss k-nearest neighbor in Chapter 8). k-nearest neighbor models have a single hyperparameter (k) that determines the predicted value to be made based on the k nearest observations in the training data to the one being predicted. If k is small (e.g., $k = 3$), the model will make a prediction for a given observation based on the average of the response values for the 3 observations in the training data most similar to the observation being predicted. This often results in highly variable predicted values because we are basing the prediction (in this case, an average) on a very small subset of the training data. As k gets bigger, we base our predictions on an average of a larger subset of the training data, which naturally reduces the variance in our predicted values (remember this for later, averaging often helps to reduce variance!). Figure 2.10 illustrates this point. Smaller k values (e.g., 2, 5, or 10) lead to high variance (but lower bias) and larger values (e.g., 150) lead to high bias (but lower variance). The optimal k value might exist somewhere between 20–50, but how do we know which value of k to use?

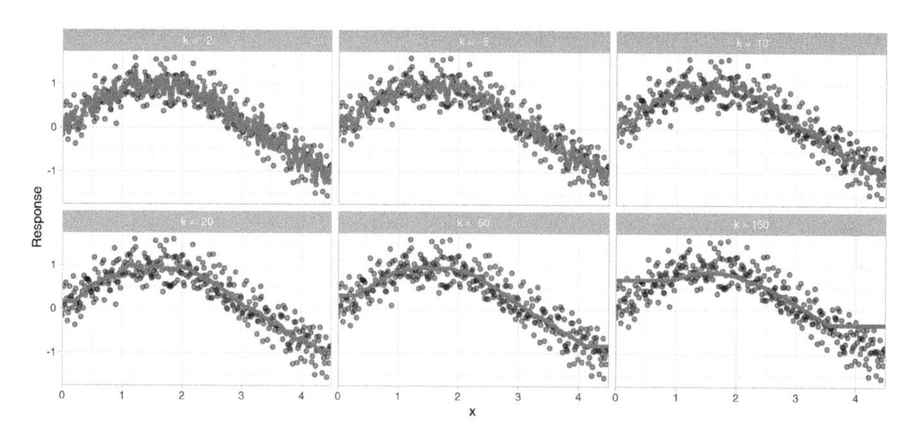

FIGURE 2.10: k-nearest neighbor model with differing values for k.

---

[7]This phrase comes from Brad Efron's comments in Breiman et al. (2001)

[8]At least in the ordinary sense. You could think of polynomial regression as having a single hyperparameter, the degree of the polynomial.

One way to perform hyperparameter tuning is to fiddle with hyperparameters manually until you find a great combination of hyperparameter values that result in high predictive accuracy (as measured using k-fold CV, for instance). However, this can be very tedious work depending on the number of hyperparameters. An alternative approach is to perform a *grid search*. A grid search is an automated approach to searching across many combinations of hyperparameter values.

For our k-nearest neighbor example, a grid search would predefine a candidate set of values for k (e.g., $k = 1, 2, \dots, j$) and perform a resampling method (e.g., k-fold CV) to estimate which k value generalizes the best to unseen data. Figure 2.11 illustrates the results from a grid search to assess $k = 2, 12, 14, \dots, 150$ using repeated 10-fold CV. The error rate displayed represents the average error for each value of k across all the repeated CV folds. On average, $k = 46$ was the optimal hyperparameter value to minimize error (in this case, RMSE which is discussed in Section 2.6)) on unseen data.

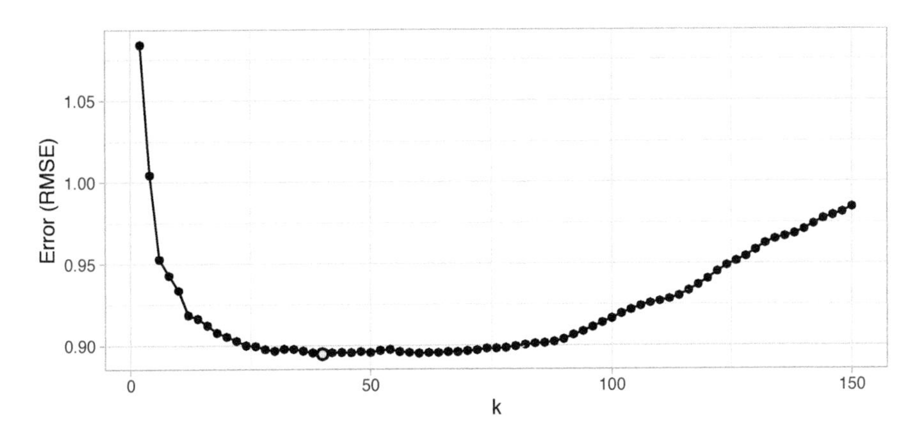

FIGURE 2.11: Results from a grid search for a k-nearest neighbor model assessing values for k ranging from 2-150. We see high error values due to high model variance when k is small and we also see high errors values due to high model bias when k is large. The optimal model is found at k = 46.

Throughout this book you'll be exposed to different approaches to performing grid searches. In the above example, we used a *full cartesian grid search*, which assesses every hyperparameter value manually defined. However, as models get more complex and offer more hyperparameters, this approach can become computationally burdensome and requires you to define the optimal hyperparameter grid settings to explore. Additional approaches we'll illustrate include *random grid searches* (Bergstra and Bengio, 2012) which explores randomly selected hyperparameter values from a range of possible values, *early stopping* which allows you to stop a grid search once reduction in the error stops marginally improving, *adaptive resampling* via futility analysis (Kuhn,

2014) which adaptively resamples candidate hyperparameter values based on approximately optimal performance, and more.

2.6 Model evaluation

Historically, the performance of statistical models was largely based on goodness-of-fit tests and assessment of residuals. Unfortunately, misleading conclusions may follow from predictive models that pass these kinds of assessments (Breiman et al., 2001). Today, it has become widely accepted that a more sound approach to assessing model performance is to assess the predictive accuracy via *loss functions*. Loss functions are metrics that compare the predicted values to the actual value (the output of a loss function is often referred to as the *error* or pseudo *residual*). When performing resampling methods, we assess the predicted values for a validation set compared to the actual target value. For example, in regression, one way to measure error is to take the difference between the actual and predicted value for a given observation (this is the usual definition of a residual in ordinary linear regression). The overall validation error of the model is computed by aggregating the errors across the entire validation data set.

There are many loss functions to choose from when assessing the performance of a predictive model, each providing a unique understanding of the predictive accuracy and differing between regression and classification models. Furthermore, the way a loss function is computed will tend to emphasize certain types of errors over others and can lead to drastic differences in how we interpret the "optimal model". Its important to consider the problem context when identifying the preferred performance metric to use. And when comparing multiple models, we need to compare them across the same metric.

2.6.1 Regression models

- **MSE**: Mean squared error is the average of the squared error ($MSE = \frac{1}{n}\sum_{i=1}^{n}(Y_i - \hat{Y}_i)^2$)[9]. The squared component results in larger errors having larger penalties. This (along with RMSE) is the most common error metric to use. **Objective: minimize**
- **RMSE**: Root mean squared error. This simply takes the square root of the MSE metric ($RMSE = \sqrt{\frac{1}{n}\sum_{i=1}^{n}(y_i - \hat{y}_i)^2}$) so that your error is in

[9]This deviates slightly from the usual definition of MSE in ordinary linear regression, where we divide by $n - p$ (to adjust for bias) as opposed to n.

the same units as your response variable. If your response variable units are dollars, the units of MSE are dollars-squared, but the RMSE will be in dollars. **Objective: minimize**

- **Deviance**: Short for mean residual deviance. In essence, it provides a degree to which a model explains the variation in a set of data when using maximum likelihood estimation. Essentially this computes a saturated model (i.e. fully featured model) to an unsaturated model (i.e. intercept only or average). If the response variable distribution is Gaussian, then it will be approximately equal to MSE. When not, it usually gives a more useful estimate of error. Deviance is often used with classification models. [10] **Objective: minimize**
- **MAE**: Mean absolute error. Similar to MSE but rather than squaring, it just takes the mean absolute difference between the actual and predicted values ($MAE = \frac{1}{n}\sum_{i=1}^{n}(|y_i - \hat{y}_i|)$). This results in less emphasis on larger errors than MSE. **Objective: minimize**
- **RMSLE**: Root mean squared logarithmic error. Similar to RMSE but it performs a `log()` on the actual and predicted values prior to computing the difference ($RMSLE = \sqrt{\frac{1}{n}\sum_{i=1}^{n}(log(y_i + 1) - log(\hat{y}_i + 1))^2}$). When your response variable has a wide range of values, large response values with large errors can dominate the MSE/RMSE metric. RMSLE minimizes this impact so that small response values with large errors can have just as meaningful of an impact as large response values with large errors. **Objective: minimize**
- R^2: This is a popular metric that represents the proportion of the variance in the dependent variable that is predictable from the independent variable(s). Unfortunately, it has several limitations. For example, two models built from two different data sets could have the exact same RMSE but if one has less variability in the response variable then it would have a lower R^2 than the other. You should not place too much emphasis on this metric. **Objective: maximize**

Most models we assess in this book will report most, if not all, of these metrics. We will emphasize MSE and RMSE but it's important to realize that certain situations warrant emphasis on some metrics more than others.

2.6.2 Classification models

- **Misclassification**: This is the overall error. For example, say you are predicting 3 classes (*high*, *medium*, *low*) and each class has 25, 30,

[10]See this StackExchange thread (`http://bit.ly/what-is-deviance`) for a good overview of deviance for different models and in the context of regression versus classification.

35 observations respectively (90 observations total). If you misclassify 3 observations of class *high*, 6 of class *medium*, and 4 of class *low*, then you misclassified 13 out of 90 observations resulting in a 14% misclassification rate. **Objective: minimize**

- **Mean per class error**: This is the average error rate for each class. For the above example, this would be the mean of $\frac{3}{25}, \frac{6}{30}, \frac{4}{35}$, which is 14.5%. If your classes are balanced this will be identical to misclassification. **Objective: minimize**
- **MSE**: Mean squared error. Computes the distance from 1.0 to the probability suggested. So, say we have three classes, A, B, and C, and your model predicts a probability of 0.91 for A, 0.07 for B, and 0.02 for C. If the correct answer was A the $MSE = 0.09^2 = 0.0081$, if it is B $MSE = 0.93^2 = 0.8649$, if it is C $MSE = 0.98^2 = 0.9604$. The squared component results in large differences in probabilities for the true class having larger penalties. **Objective: minimize**
- **Cross-entropy (aka Log Loss or Deviance)**: Similar to MSE but it incorporates a log of the predicted probability multiplied by the true class. Consequently, this metric disproportionately punishes predictions where we predict a small probability for the true class, which is another way of saying having high confidence in the wrong answer is really bad. **Objective: minimize**
- **Gini index**: Mainly used with tree-based methods and commonly referred to as a measure of *purity* where a small value indicates that a node contains predominantly observations from a single class. **Objective: minimize**

When applying classification models, we often use a *confusion matrix* to evaluate certain performance measures. A confusion matrix is simply a matrix that compares actual categorical levels (or events) to the predicted categorical levels. When we predict the right level, we refer to this as a *true positive*. However, if we predict a level or event that did not happen this is called a *false positive* (i.e. we predicted a customer would redeem a coupon and they did not). Alternatively, when we do not predict a level or event and it does happen that this is called a *false negative* (i.e. a customer that we did not predict to redeem a coupon does).

We can extract different levels of performance for binary classifiers. For example, given the classification (or confusion) matrix illustrated in Figure 2.13 we can assess the following:

- **Accuracy**: Overall, how often is the classifier correct? Opposite of misclassification above. Example: $\frac{TP+TN}{total} = \frac{100+50}{165} = 0.91$. **Objective: maximize**

	predicted events	predicted non-events
actual events	correctly forecasted events	missed events
actual non-events	missed non-events	correctly forecasted non-events

→

	predicted events	predicted non-events
actual events	True Positive	False Negative
actual non-events	False Positive	True Negative

FIGURE 2.12: Confusion matrix and relationships to terms such as true-positive and false-negative.

- **Precision**: How accurately does the classifier predict events? This metric is concerned with maximizing the true positives to false positive ratio. In other words, for the number of predictions that we made, how many were correct? Example: $\frac{TP}{TP+FP} = \frac{100}{100+10} = 0.91$. **Objective: maximize**
- **Sensitivity (aka recall)**: How accurately does the classifier classify actual events? This metric is concerned with maximizing the true positives to false negatives ratio. In other words, for the events that occurred, how many did we predict? Example: $\frac{TP}{TP+FN} = \frac{100}{100+5} = 0.95$. **Objective: maximize**
- **Specificity**: How accurately does the classifier classify actual non-events? Example: $\frac{TN}{TN+FP} = \frac{50}{50+10} = 0.83$. **Objective: maximize**

	predicted events	predicted non-events
actual events	100	5
actual non-events	10	50

FIGURE 2.13: Example confusion matrix.

- **AUC**: Area under the curve. A good binary classifier will have high precision and sensitivity. This means the classifier does well when it predicts an event will and will not occur, which minimizes false positives and false negatives. To capture this balance, we often use a ROC curve that plots the false positive rate along the x-axis and the true positive rate along the y-axis. A line that is diagonal from the lower left corner to the upper right corner represents a random guess. The higher the line is in the upper left-hand corner, the better. AUC computes the area under this curve. **Objective: maximize**

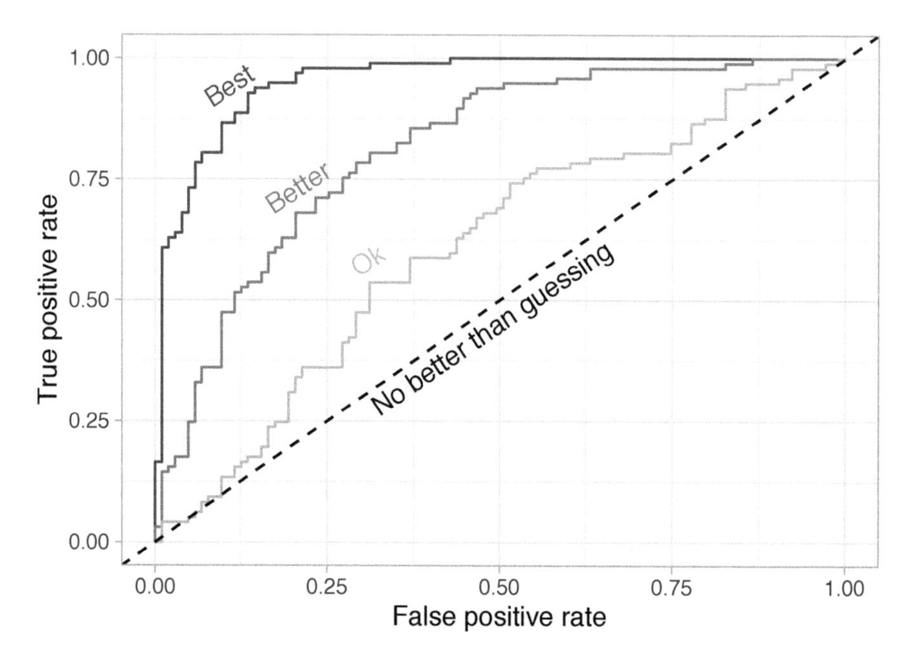

FIGURE 2.14: ROC curve.

2.7 Putting the processes together

To illustrate how this process works together via R code, let's do a simple assessment on the `ames` housing data. First, we perform stratified sampling as illustrated in Section 2.2.2 to break our data into training vs. test data while ensuring we have consistent distributions between the training and test sets.

```
# Stratified sampling with the rsample package
set.seed(123)
split <- initial_split(ames, prop = 0.7,
                       strata = "Sale_Price")
ames_train  <- training(split)
ames_test   <- testing(split)
```

Next, we're going to apply a k-nearest neighbor regressor to our data. To do so, we'll use **caret**, which is a meta-engine to simplify the resampling, grid search, and model application processes. The following defines:

1. **Resampling method**: we use 10-fold CV repeated 5 times.

2. **Grid search**: we specify the hyperparameter values to assess ($k = 2, 4, 6, \dots, 25$).
3. **Model training & Validation**: we train a k-nearest neighbor (`method = "knn"`) model using our pre-specified resampling procedure (`trControl = cv`), grid search (`tuneGrid = hyper_grid`), and preferred loss function (`metric = "RMSE"`).

This grid search takes approximately 3.5 minutes

```
# Specify resampling strategy
cv <- trainControl(
  method = "repeatedcv",
  number = 10,
  repeats = 5
)

# Create grid of hyperparameter values
hyper_grid <- expand.grid(k = seq(2, 25, by = 1))

# Tune a knn model using grid search
knn_fit <- train(
  Sale_Price ~ .,
  data = ames_train,
  method = "knn",
  trControl = cv,
  tuneGrid = hyper_grid,
  metric = "RMSE"
)
```

Looking at our results we see that the best model coincided with $k = 5$, which resulted in an RMSE of 44738. This implies that, on average, our model mispredicts the expected sale price of a home by $44,738. Figure 2.15 illustrates the cross-validated error rate across the spectrum of hyperparameter values that we specified.

```
# Print and plot the CV results
knn_fit
## k-Nearest Neighbors
##
```

```
## 2054 samples
##   80 predictor
##
## No pre-processing
## Resampling: Cross-Validated (10 fold, repeated 5 times)
## Summary of sample sizes: 1849, 1848, 1848, 1849, 1849, 1847, ...
## Resampling results across tuning parameters:
##
##    k  RMSE   Rsquared  MAE
##     2  47138  0.659     30432
##     3  45374  0.681     29403
##     4  45055  0.685     29194
##     5  44738  0.690     28966
##     6  44773  0.691     28926
##     7  44816  0.692     28970
##     8  44911  0.692     29022
##     9  45012  0.693     29047
##    10  45058  0.695     28972
##    11  45057  0.697     28908
##    12  45229  0.696     28952
##    13  45339  0.696     29031
##    14  45492  0.696     29124
##    15  45584  0.696     29188
##    16  45668  0.696     29277
##    17  45822  0.696     29410
##    18  46000  0.694     29543
##    19  46206  0.693     29722
##    20  46417  0.691     29845
##    21  46612  0.690     29955
##    22  46824  0.688     30120
##    23  47009  0.686     30257
##    24  47256  0.684     30413
##    25  47454  0.682     30555
##
## RMSE was used to select the optimal model using
##  the smallest value.
## The final value used for the model was k = 5.
ggplot(knn_fit)
```

The question remains: "Is this the best predictive model we can find?" We may have identified the optimal k-nearest neighbor model for our given data set, but this doesn't mean we've found the best possible overall model. Nor have we considered potential feature and target engineering options. The remainder of

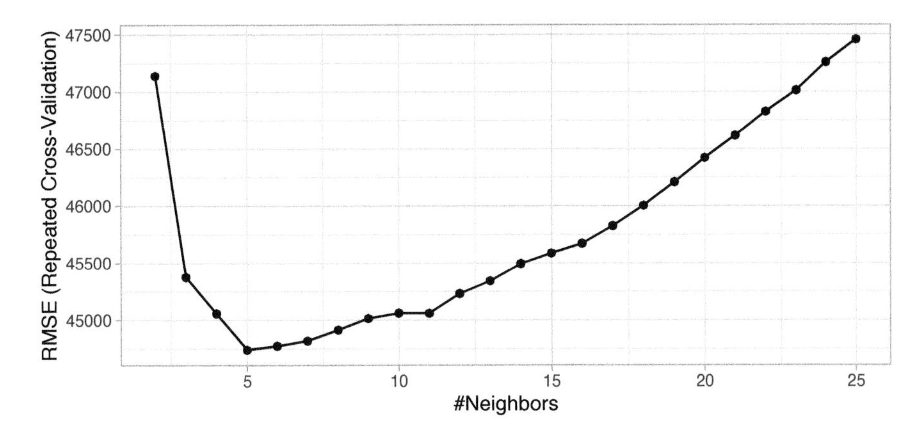

FIGURE 2.15: Results from a grid search for a k-nearest neighbor model on the Ames housing data assessing values for k ranging from 2-25.

this book will walk you through the journey of identifying alternative solutions and, hopefully, a much more optimal model.

3

Feature & Target Engineering

Data preprocessing and engineering techniques generally refer to the addition, deletion, or transformation of data. The time spent on identifying data engineering needs can be significant and requires you to spend substantial time understanding your data...or as Leo Breiman said "live with your data before you plunge into modeling" (Breiman et al., 2001, p. 201). Although this book primarily focuses on applying machine learning algorithms, feature engineering can make or break an algorithm's predictive ability and deserves your continued focus and education.

We will not cover all the potential ways of implementing feature engineering; however, we'll cover several fundamental preprocessing tasks that can potentially significantly improve modeling performance. Moreover, different models have different sensitivities to the type of target and feature values in the model and we will try to highlight some of these concerns. For more in depth coverage of feature engineering, please refer to Kuhn and Johnson (2019) and Zheng and Casari (2018).

3.1 Prerequisites

This chapter leverages the following packages:

```
# Helper packages
library(dplyr)     # for data manipulation
library(ggplot2)   # for awesome graphics
library(visdat)    # for additional visualizations

# Feature engineering packages
library(caret)     # for various ML tasks
library(recipes)   # for feature engineering tasks
```

We'll also continue working with the `ames_train` data set created in Section 2.7:

3.2 Target engineering

Although not always a requirement, transforming the response variable can lead to predictive improvement, especially with parametric models (which require that certain assumptions about the model be met). For instance, ordinary linear regression models assume that the prediction errors (and hence the response) are normally distributed. This is usually fine, except when the prediction target has heavy tails (i.e., *outliers*) or is skewed in one direction or the other. In these cases, the normality assumption likely does not hold. For example, as we saw in the data splitting section (2.2), the response variable for the Ames housing data (`Sale_Price`) is right (or positively) skewed as illustrated in Figure 3.1 (ranging from \$12,789 to \$755,000). A simple linear model, say $\text{Sale\_Price} = \beta_0 + \beta_1 \text{Year\_Built} + \epsilon$, often assumes the error term ϵ (and hence `Sale_Price`) is normally distributed; fortunately, a simple log (or similar) transformation of the response can often help alleviate this concern as Figure 3.1 illustrates.

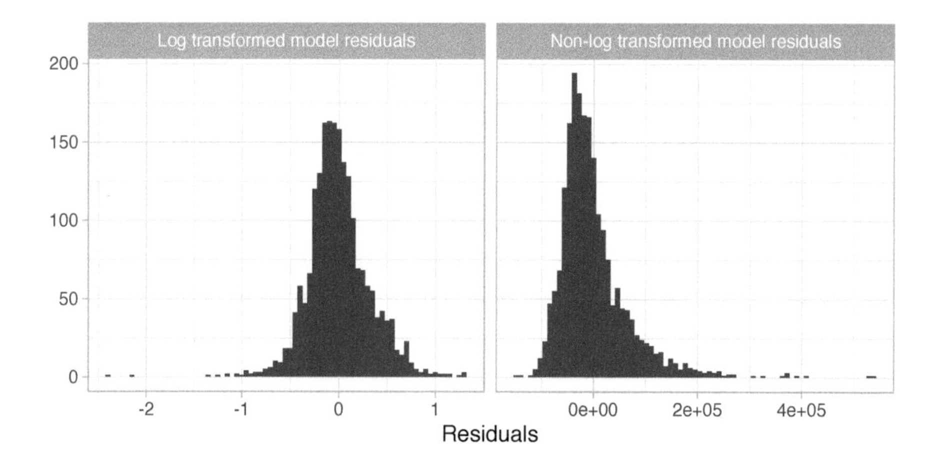

FIGURE 3.1: Transforming the response variable to minimize skewness can resolve concerns with non-normally distributed errors.

Furthermore, using a log (or other) transformation to minimize the response skewness can be used for shaping the business problem as well. For example, in the House Prices: Advanced Regression Techniques Kaggle competition[1], which used the Ames housing data, the competition focused on using a log transformed Sale Price response because "...taking logs means that errors in predicting expensive houses and cheap houses will affect the result equally."

[1] `https://www.kaggle.com/c/house-prices-advanced-regression-techniques`

This would be an alternative to using the root mean squared logarithmic error (RMSLE) loss function as discussed in Section 2.6.

There are two main approaches to help correct for positively skewed target variables:

Option 1: normalize with a log transformation. This will transform most right skewed distributions to be approximately normal. One way to do this is to simply log transform the training and test set in a manual, single step manner similar to:

```
transformed_response <- log(ames_train$Sale_Price)
```

However, we should think of the preprocessing as creating a blueprint to be re-applied strategically. For this, you can use the **recipe** package or something similar (e.g., `caret::preProcess()`). This will not return the actual log transformed values but, rather, a blueprint to be applied later.

```
# log transformation
ames_recipe <- recipe(Sale_Price ~ ., data = ames_train) %>%
  step_log(all_outcomes())

ames_recipe
## Data Recipe
##
## Inputs:
##
##       role #variables
##    outcome          1
##  predictor         80
##
## Operations:
##
## Log transformation on all_outcomes()
```

If your response has negative values or zeros then a log transformation will produce `NaN`s and `-Inf`s, respectively (you cannot take the logarithm of a negative number). If the nonpositive response values are small (say between -0.99 and 0) then you can apply a small offset such as in `log1p()` which adds 1 to the value prior to applying a log transformation (you can do the same within `step_log()` by using the `offset` argument). If your data consists of values ≤ -1, use the Yeo-Johnson transformation mentioned next.

```
log(-0.5)
## [1] NaN
log1p(-0.5)
## [1] -0.693
```

Option 2: use a *Box Cox transformation*. A Box Cox transformation is more flexible than (but also includes as a special case) the log transformation and will find an appropriate transformation from a family of power transforms that will transform the variable as close as possible to a normal distribution (Box and Cox, 1964; Carroll and Ruppert, 1981). At the core of the Box Cox transformation is an exponent, lambda (λ), which varies from -5 to 5. All values of λ are considered and the optimal value for the given data is estimated from the training data; The "optimal value" is the one which results in the best transformation to an approximate normal distribution. The transformation of the response Y has the form:

$$y(\lambda) = \begin{cases} \frac{Y^\lambda - 1}{\lambda}, & \text{if } \lambda \neq 0 \\ \log(Y), & \text{if } \lambda = 0. \end{cases} \tag{3.1}$$

Be sure to compute the `lambda` on the training set and apply that same `lambda` to both the training and test set to minimize *data leakage*. The **recipes** package automates this process for you.

If your response has negative values, the Yeo-Johnson transformation is very similar to the Box-Cox but does not require the input variables to be strictly positive. To apply, use `step_YeoJohnson()`.

Figure 3.2 illustrates that the log transformation and Box Cox transformation both do about equally well in transforming `Sale_Price` to look more normally distributed.

Note that when you model with a transformed response variable, your predictions will also be on the transformed scale. You will likely want to undo (or re-transform) your predicted values back to their normal scale so that decision-makers can more easily interpret the results. This is illustrated in the following code chunk:

```
# Log transform a value
y <- log(10)

# Undo log-transformation
```

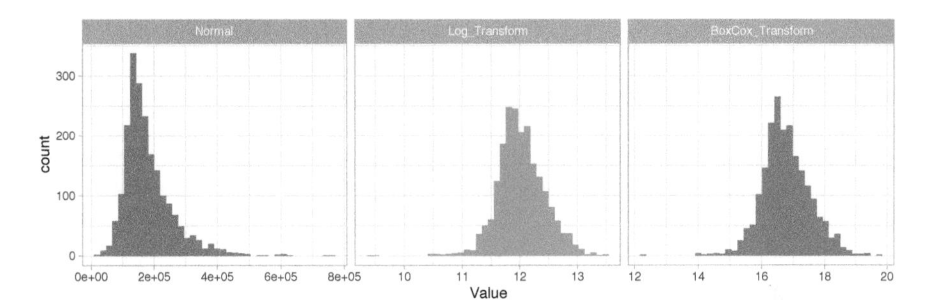

FIGURE 3.2: Response variable transformations.

```
exp(y)
## [1] 10

# Box Cox transform a value
y <- forecast::BoxCox(10, lambda)

# Inverse Box Cox function
inv_box_cox <- function(x, lambda) {
  # for Box-Cox, lambda = 0 --> log transform
  if (lambda == 0) exp(x) else (lambda*x + 1)^(1/lambda)
}

# Undo Box Cox-transformation
inv_box_cox(y, lambda)
## [1] 10
## attr(,"lambda")
## [1] 0.0526
```

3.3 Dealing with missingness

Data quality is an important issue for any project involving analyzing data. Data quality issues deserve an entire book in their own right, and a good reference is The Quartz guide to bad data.[2] One of the most common data quality concerns you will run into is missing values.

Data can be missing for many different reasons; however, these reasons are

[2] https://github.com/Quartz/bad-data-guide

usually lumped into two categories: *informative missingness* (Kuhn and Johnson, 2013) and *missingness at random* (Little and Rubin, 2014). Informative missingness implies a structural cause for the missing value that can provide insight in its own right; whether this be deficiencies in how the data was collected or abnormalities in the observational environment. Missingness at random implies that missing values occur independent of the data collection process[3].

The category that drives missing values will determine how you handle them. For example, we may give values that are driven by informative missingness their own category (e.g., `"None"`) as their unique value may affect predictive performance. Whereas values that are missing at random may deserve deletion[4] or imputation.

Furthermore, different machine learning models handle missingness differently. Most algorithms cannot handle missingness (e.g., generalized linear models and their cousins, neural networks, and support vector machines) and, therefore, require them to be dealt with beforehand. A few models (mainly tree-based), have built-in procedures to deal with missing values. However, since the modeling process involves comparing and contrasting multiple models to identify the optimal one, you will want to handle missing values prior to applying any models so that your algorithms are based on the same data quality assumptions.

3.3.1 Visualizing missing values

It is important to understand the distribution of missing values (i.e., `NA`) in any data set. So far, we have been using a pre-processed version of the Ames housing data set (via the `AmesHousing::make_ames()` function). However, if we use the raw Ames housing data (via `AmesHousing::ames_raw`), there are actually 13,997 missing values—there is at least one missing values in each row of the original data!

```
sum(is.na(AmesHousing::ames_raw))
## [1] 13997
```

It is important to understand the distribution of missing values in a data set in order to determine the best approach for preprocessing. Heat maps are an

[3]Little and Rubin (2014) discuss two different kinds of missingness at random; however, we combine them for simplicity as their nuanced differences are distinguished between the two in practice.

[4]If your data set is large, deleting missing observations that have missing values at random rarely impacts predictive performance. However, as your data sets get smaller, preserving observations is critical and alternative solutions should be explored.

efficient way to visualize the distribution of missing values for small- to medium-sized data sets. The code `is.na(<data-frame-name>)` will return a matrix of the same dimension as the given data frame, but each cell will contain either `TRUE` (if the corresponding value is missing) or `FALSE` (if the corresponding value is not missing). To construct such a plot, we can use R's built-in `heatmap()` or `image()` functions, or **ggplot2**'s `geom_raster()` function, among others; Figure 3.3 illustrates `geom_raster()`. This allows us to easily see where the majority of missing values occur (i.e., in the variables `Alley`, `Fireplace Qual`, `Pool QC`, `Fence`, and `Misc Feature`). Due to their high frequency of missingness, these variables would likely need to be removed prior to statistical analysis, or imputed. We can also spot obvious patterns of missingness. For example, missing values appear to occur within the same observations across all garage variables.

```
AmesHousing::ames_raw %>%
  is.na() %>%
  reshape2::melt() %>%
  ggplot(aes(Var2, Var1, fill=value)) +
    geom_raster() +
    coord_flip() +
    scale_y_continuous(NULL, expand = c(0, 0)) +
    scale_fill_grey(name = "",
                    labels = c("Present",
                               "Missing")) +
    xlab("Observation") +
    theme(axis.text.y  = element_text(size = 4))
```

Digging a little deeper into these variables, we might notice that `Garage_Cars` and `Garage_Area` contain the value `0` whenever the other `Garage_xx` variables have missing values (i.e. a value of `NA`). This might be because they did not have a way to identify houses with no garages when the data were originally collected, and therefore, all houses with no garage were identified by including nothing. Since this missingness is informative, it would be appropriate to impute `NA` with a new category level (e.g., `"None"`) for these garage variables. Circumstances like this tend to only become apparent upon careful descriptive and visual examination of the data!

```
AmesHousing::ames_raw %>%
  filter(is.na(`Garage Type`)) %>%
  select(`Garage Type`, `Garage Cars`, `Garage Area`)
## # A tibble: 157 x 3
##     `Garage Type` `Garage Cars` `Garage Area`
```

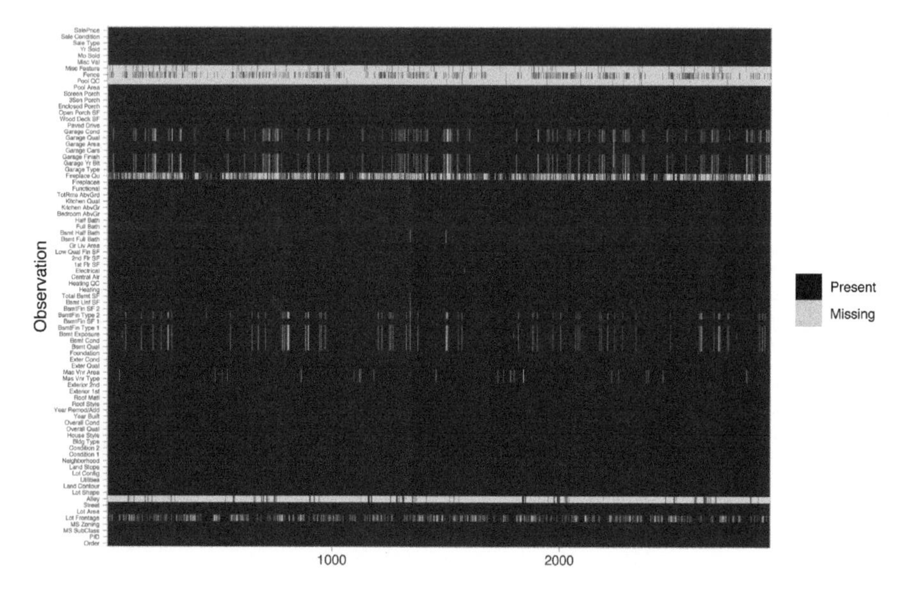

FIGURE 3.3: Heat map of missing values in the raw Ames housing data.

```
##       <chr>                       <int>         <int>
##     1 <NA>                            0             0
##     2 <NA>                            0             0
##     3 <NA>                            0             0
##     4 <NA>                            0             0
##     5 <NA>                            0             0
##     6 <NA>                            0             0
##     7 <NA>                            0             0
##     8 <NA>                            0             0
##     9 <NA>                            0             0
##    10 <NA>                            0             0
## # ... with 147 more rows
```

The `vis_miss()` function in R package `visdat` (Tierney, 2019) also allows for easy visualization of missing data patterns (with sorting and clustering options). We illustrate this functionality below using the raw Ames housing data (Figure 3.4). The columns of the heat map represent the 82 variables of the raw data and the rows represent the observations. Missing values (i.e., `NA`) are indicated via a black cell. The variables and `NA` patterns have been clustered by rows (i.e., `cluster = TRUE`).

```r
vis_miss(AmesHousing::ames_raw, cluster = TRUE)
```

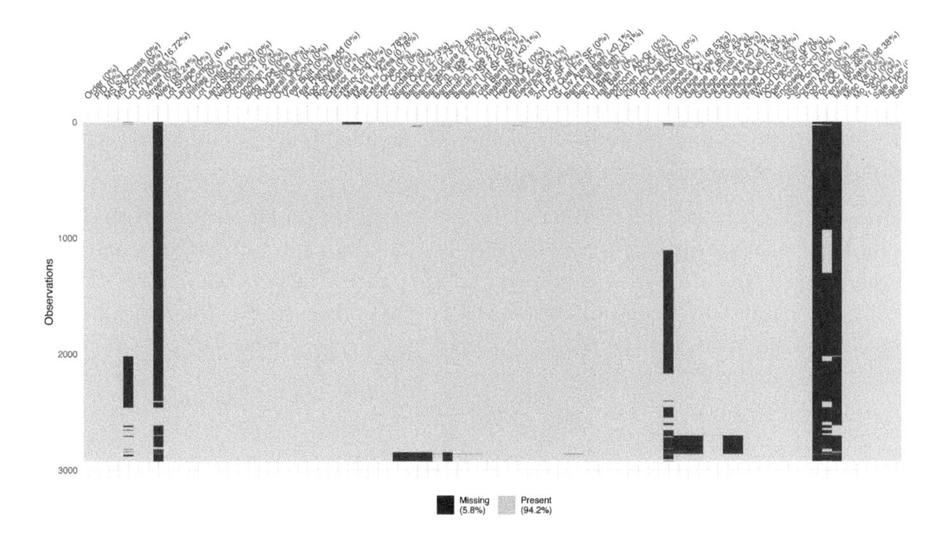

FIGURE 3.4: Visualizing missing data patterns in the raw Ames housing data.

Data can be missing for different reasons. Perhaps the values were never recoded (or lost in translation), or it was recorded in error (a common feature of data entered by hand). Regardless, it is important to identify and attempt to understand how missing values are distributed across a data set as it can provide insight into how to deal with these observations.

3.3.2 Imputation

Imputation is the process of replacing a missing value with a substituted, "best guess" value. Imputation should be one of the first feature engineering steps you take as it will affect any downstream preprocessing[5].

3.3.2.1 Estimated statistic

An elementary approach to imputing missing values for a feature is to compute descriptive statistics such as the mean, median, or mode (for categorical) and use that value to replace `NA`s. Although computationally efficient, this approach

[5]For example, standardizing numeric features will include the imputed numeric values in the calculation and one-hot encoding will include the imputed categorical value.

does not consider any other attributes for a given observation when imputing (e.g., a female patient that is 63 inches tall may have her weight imputed as 175 lbs since that is the average weight across all observations which contains 65% males that average a height of 70 inches).

An alternative is to use grouped statistics to capture expected values for observations that fall into similar groups. However, this becomes infeasible for larger data sets. Modeling imputation can automate this process for you and the two most common methods include K-nearest neighbor and tree-based imputation, which are discussed next.

However, it is important to remember that imputation should be performed **within the resampling process** and as your data set gets larger, repeated model-based imputation can compound the computational demands. Thus, you must weigh the pros and cons of the two approaches. The following would build onto our `ames_recipe` and impute all missing values for the `Gr_Liv_Area` variable with the median value:

```
ames_recipe %>%
  step_medianimpute(Gr_Liv_Area)
## Data Recipe
##
## Inputs:
##
##       role #variables
##    outcome          1
##  predictor         80
##
## Operations:
##
## Box-Cox transformation on all_outcomes()
## Median Imputation for Gr_Liv_Area
```

Use `step_modeimpute()` to impute categorical features with the most common value.

3.3.2.2 *K*-nearest neighbor

K-nearest neighbor (KNN) imputes values by identifying observations with missing values, then identifying other observations that are most similar based on the other available features, and using the values from these nearest neighbor observations to impute missing values.

We discuss KNN for predictive modeling in Chapter 8; the imputation application works in a similar manner. In KNN imputation, the missing value for a given observation is treated as the targeted response and is predicted based on the average (for quantitative values) or the mode (for qualitative values) of the k nearest neighbors.

As discussed in Chapter 8, if all features are quantitative then standard Euclidean distance is commonly used as the distance metric to identify the k neighbors and when there is a mixture of quantitative and qualitative features then Gower's distance (Gower, 1971) can be used. KNN imputation is best used on small to moderate sized data sets as it becomes computationally burdensome with larger data sets (Kuhn and Johnson, 2019).

As we saw in Section 2.7, k is a tunable hyperparameter. Suggested values for imputation are 5–10 (Kuhn and Johnson, 2019). By default, `step_knnimpute()` will use 5 but can be adjusted with the `neighbors` argument.

```
ames_recipe %>%
  step_knnimpute(all_predictors(), neighbors = 6)
## Data Recipe
##
## Inputs:
##
##        role #variables
##     outcome          1
##   predictor         80
##
## Operations:
##
## Box-Cox transformation on all_outcomes()
## 6-nearest neighbor imputation for all_predictors()
```

3.3.2.3 Tree-based

As previously discussed, several implementations of decision trees (Chapter 9) and their derivatives can be constructed in the presence of missing values. Thus, they provide a good alternative for imputation. As discussed in Chapters 9-11, single trees have high variance but aggregating across many trees creates a robust, low variance predictor. Random forest imputation procedures have been studied (Shah et al., 2014; Stekhoven, 2015); however, they require significant computational demands in a resampling environment (Kuhn and Johnson, 2019). Bagged trees (Chapter 10) offer a compromise between predictive accuracy and computational burden.

Similar to KNN imputation, observations with missing values are identified and the feature containing the missing value is treated as the target and predicted using bagged decision trees.

```
ames_recipe %>%
  step_bagimpute(all_predictors())
## Data Recipe
##
## Inputs:
##
##       role #variables
##    outcome          1
##  predictor         80
##
## Operations:
##
## Box-Cox transformation on all_outcomes()
## Bagged tree imputation for all_predictors()
```

Figure 3.5 illustrates the differences between mean, KNN, and tree-based imputation on the raw Ames housing data. It is apparent how descriptive statistic methods (e.g., using the mean and median) are inferior to the KNN and tree-based imputation methods.

3.4 Feature filtering

In many data analyses and modeling projects we end up with hundreds or even thousands of collected features. From a practical perspective, a model with more features often becomes harder to interpret and is costly to compute. Some models are more resistant to non-informative predictors (e.g., the Lasso and tree-based methods) than others as illustrated in Figure 3.6.[6]

Although the performance of some of our models are not significantly affected by non-informative predictors, the time to train these models can be negatively impacted as more features are added. Figure 3.7 shows the increase in time to

[6]See Kuhn and Johnson (2013) Section 19.1 for data set generation.

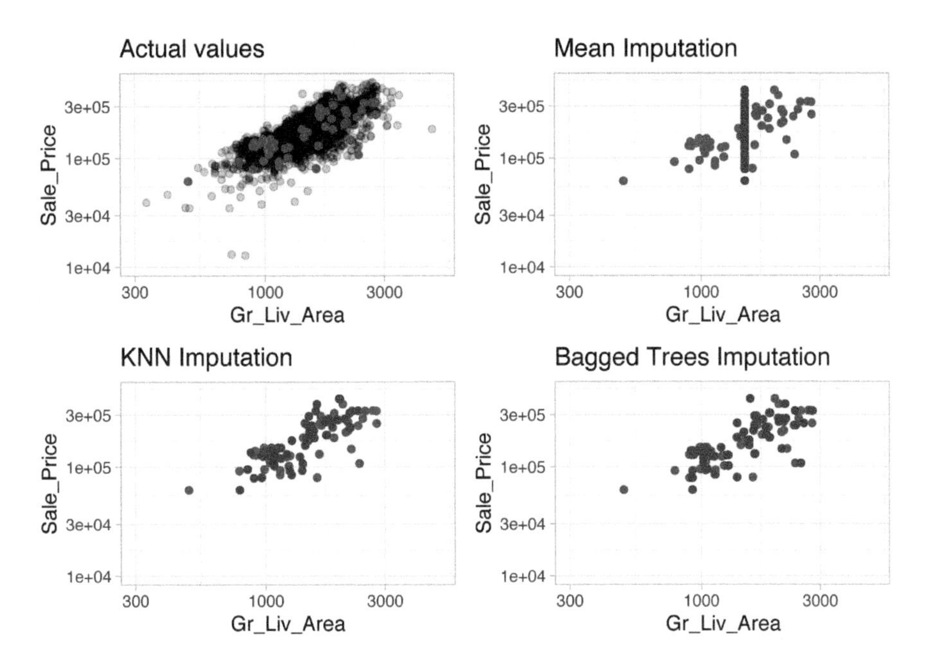

FIGURE 3.5: Comparison of three different imputation methods. The red points represent actual values which were removed and made missing and the blue points represent the imputed values. Estimated statistic imputation methods (i.e. mean, median) merely predict the same value for each observation and can reduce the signal between a feature and the response; whereas KNN and tree-based procedures tend to maintain the feature distribution and relationship.

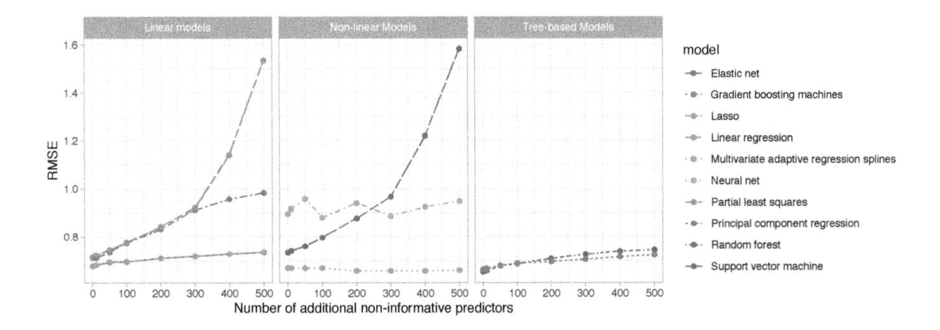

FIGURE 3.6: Test set RMSE profiles when non-informative predictors are added.

perform 10-fold CV on the exemplar data, which consists of 10,000 observations. We see that many algorithms (e.g., elastic nets, random forests, and gradient boosting machines) become extremely time intensive the more predictors we add. Consequently, filtering or reducing features prior to modeling may significantly speed up training time.

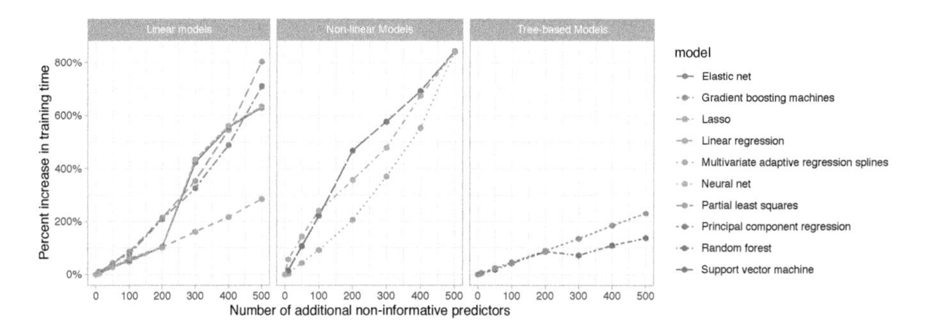

FIGURE 3.7: Impact in model training time as non-informative predictors are added.

Zero and near-zero variance variables are low-hanging fruit to eliminate. Zero variance variables, meaning the feature only contains a single unique value, provides no useful information to a model. Some algorithms are unaffected by zero variance features. However, features that have near-zero variance also offer very little, if any, information to a model. Furthermore, they can cause problems during resampling as there is a high probability that a given sample will only contain a single unique value (the dominant value) for that feature. A rule of thumb for detecting near-zero variance features is:

- The fraction of unique values over the sample size is low (say $\leq 10\%$).
- The ratio of the frequency of the most prevalent value to the frequency of the second most prevalent value is large (say $\geq 20\%$).

If both of these criteria are true then it is often advantageous to remove the variable from the model. For the Ames data, we do not have any zero variance predictors but there are 20 features that meet the near-zero threshold.

```
caret::nearZeroVar(ames_train, saveMetrics= TRUE) %>%
  rownames_to_column() %>%
  filter(nzv)
##            rowname freqRatio percentUnique zeroVar
## 1           Street     255.8        0.0974   FALSE
## 2            Alley      24.7        0.1461   FALSE
## 3     Land_Contour      21.9        0.1947   FALSE
```

```
## 4           Utilities  2052.0          0.1461   FALSE
## 5          Land_Slope    21.7          0.1461   FALSE
## 6         Condition_2   184.7          0.3408   FALSE
## 7           Roof_Matl   112.5          0.2921   FALSE
## 8           Bsmt_Cond    24.0          0.2921   FALSE
## 9      BsmtFin_Type_2    23.1          0.3408   FALSE
## 10            Heating    91.8          0.2921   FALSE
## 11     Low_Qual_Fin_SF  674.3          1.4119   FALSE
## 12       Kitchen_AbvGr   22.8          0.1947   FALSE
## 13         Functional    35.3          0.3895   FALSE
## 14      Enclosed_Porch  109.1          7.2055   FALSE
## 15 Three_season_porch  674.7          1.2658   FALSE
## 16        Screen_Porch  186.9          4.8199   FALSE
## 17           Pool_Area 2046.0          0.4382   FALSE
## 18             Pool_QC  682.0          0.2434   FALSE
## 19        Misc_Feature   29.6          0.2921   FALSE
## 20            Misc_Val  124.0          1.2658   FALSE
##     nzv
## 1  TRUE
## 2  TRUE
## 3  TRUE
## 4  TRUE
## 5  TRUE
## 6  TRUE
## 7  TRUE
## 8  TRUE
## 9  TRUE
## 10 TRUE
## 11 TRUE
## 12 TRUE
## 13 TRUE
## 14 TRUE
## 15 TRUE
## 16 TRUE
## 17 TRUE
## 18 TRUE
## 19 TRUE
## 20 TRUE
```

We can add `step_zv()` and `step_nzv()` to our `ames_recipe` to remove zero or near-zero variance features.

Other feature filtering methods exist; see Saeys et al. (2007) for a thorough review. Furthermore, several wrapper methods exist that evaluate multiple models using procedures that add or remove predictors to find the optimal combination of features that maximizes model performance (see, for example, Kursa et al. (2010), Granitto et al. (2006), Maldonado and Weber (2009)). However, this topic is beyond the scope of this book.

3.5 Numeric feature engineering

Numeric features can create a host of problems for certain models when their distributions are skewed, contain outliers, or have a wide range in magnitudes. Tree-based models are quite immune to these types of problems in the feature space, but many other models (e.g., GLMs, regularized regression, KNN, support vector machines, neural networks) can be greatly hampered by these issues. Normalizing and standardizing heavily skewed features can help minimize these concerns.

3.5.1 Skewness

Similar to the process discussed to normalize target variables, parametric models that have distributional assumptions (e.g., GLMs, and regularized models) can benefit from minimizing the skewness of numeric features. When normalizing many variables, it's best to use the Box-Cox (when feature values are strictly positive) or Yeo-Johnson (when feature values are not strictly positive) procedures as these methods will identify if a transformation is required and what the optimal transformation will be.

Non-parametric models are rarely affected by skewed features; however, normalizing features will not have a negative effect on these models' performance. For example, normalizing features will only shift the optimal split points in tree-based algorithms. Consequently, when in doubt, normalize.

```
# Normalize all numeric columns
recipe(Sale_Price ~ ., data = ames_train) %>%
  step_YeoJohnson(all_numeric())
## Data Recipe
```

```
##
## Inputs:
##
##       role #variables
##    outcome          1
##  predictor         80
##
## Operations:
##
## Yeo-Johnson transformation on all_numeric()
```

3.5.2 Standardization

We must also consider the scale on which the individual features are measured. What are the largest and smallest values across all features and do they span several orders of magnitude? Models that incorporate smooth functions of input features are sensitive to the scale of the inputs. For example, $5X + 2$ is a simple linear function of the input X, and the scale of its output depends directly on the scale of the input. Many algorithms use linear functions within their algorithms, some more obvious (e.g., GLMs and regularized regression) than others (e.g., neural networks, support vector machines, and principal components analysis). Other examples include algorithms that use distance measures such as the Euclidean distance (e.g., k nearest neighbor, k-means clustering, and hierarchical clustering).

For these models and modeling components, it is often a good idea to *standardize* the features. Standardizing features includes *centering* and *scaling* so that numeric variables have zero mean and unit variance, which provides a common comparable unit of measure across all the variables.

Some packages (e.g., **glmnet**, and **caret**) have built-in options to standardize and some do not (e.g., **keras** for neural networks). However, you should standardize your variables within the recipe blueprint so that both training and test data standardization are based on the same mean and variance. This helps to minimize data leakage.

```
ames_recipe %>%
  step_center(all_numeric(), -all_outcomes()) %>%
  step_scale(all_numeric(), -all_outcomes())
## Data Recipe
##
## Inputs:
```

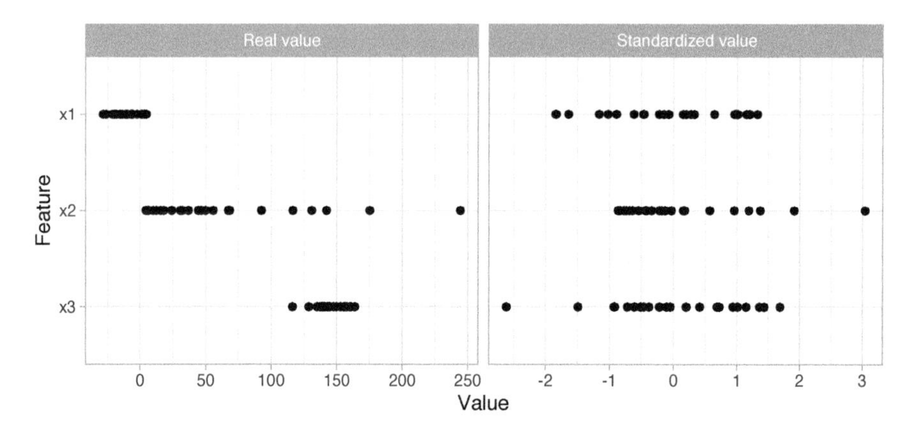

FIGURE 3.8: Standardizing features allows all features to be compared on a common value scale regardless of their real value differences.

```
##
##          role #variables
##       outcome          1
##     predictor         80
##
## Operations:
##
## Box-Cox transformation on all_outcomes()
## Centering for 2 items
## Scaling for 2 items
```

3.6 Categorical feature engineering

Most models require that the predictors take numeric form. There are exceptions; for example, tree-based models naturally handle numeric or categorical features. However, even tree-based models can benefit from preprocessing categorical features. The following sections will discuss a few of the more common approaches to engineer categorical features.

3.6.1 Lumping

Sometimes features will contain levels that have very few observations. For example, there are 28 unique neighborhoods represented in the Ames housing data but several of them only have a few observations.

```
count(ames_train, Neighborhood) %>% arrange(n)
## # A tibble: 28 x 2
##    Neighborhood            n
##    <fct>               <int>
##  1 Green_Hills             1
##  2 Landmark                1
##  3 Blueste                 5
##  4 Greens                  7
##  5 Veenker                16
##  6 Northpark_Villa        17
##  7 Briardale              22
##  8 Bloomington_Heights    23
##  9 Meadow_Village         27
## 10 Clear_Creek            30
## # ... with 18 more rows
```

Even numeric features can have similar distributions. For example, `Screen_Porch` has 92% values recorded as zero (zero square footage meaning no screen porch) and the remaining 8% have unique dispersed values.

```
count(ames_train, Screen_Porch) %>% arrange(n)
## # A tibble: 99 x 2
##    Screen_Porch     n
##           <int> <int>
##  1           40     1
##  2           53     1
##  3           60     1
##  4           63     1
##  5           80     1
##  6           84     1
##  7           88     1
##  8           92     1
##  9           94     1
## 10           95     1
## # ... with 89 more rows
```

Sometimes we can benefit from collapsing, or “lumping” these into a lesser number of categories. In the above examples, we may want to collapse all levels that are observed in less than 10% of the training sample into an "other" category. We can use `step_other()` to do so. However, lumping should be used sparingly as there is often a loss in model performance (Kuhn and Johnson, 2013).

Tree-based models often perform exceptionally well with high cardinality features and are not as impacted by levels with small representation.

```
# Lump levels for two features
lumping <- recipe(Sale_Price ~ ., data = ames_train) %>%
  step_other(Neighborhood, threshold = 0.01,
             other = "other") %>%
  step_other(Screen_Porch, threshold = 0.1,
             other = ">0")

# Apply this blue print --> you will learn about this at
# the end of the chapter
apply_2_training <- prep(lumping, training = ames_train) %>%
  bake(ames_train)

# New distribution of Neighborhood
count(apply_2_training, Neighborhood) %>% arrange(n)
## # A tibble: 23 x 2
##    Neighborhood                                 n
##    <fct>                                    <int>
##  1 Briardale                                   22
##  2 Bloomington_Heights                         23
##  3 Meadow_Village                              27
##  4 Clear_Creek                                 30
##  5 South_and_West_of_Iowa_State_University     33
##  6 Stone_Brook                                 36
##  7 Timberland                                  47
##  8 other                                       47
##  9 Northridge                                  55
## 10 Iowa_DOT_and_Rail_Road                      59
## # ... with 13 more rows

# New distribution of Screen_Porch
count(apply_2_training, Screen_Porch) %>% arrange(n)
## # A tibble: 2 x 2
```

```
##   Screen_Porch     n
##   <fct>        <int>
## 1 >0             185
## 2 0             1869
```

3.6.2 One-hot & dummy encoding

Many models require that all predictor variables be numeric. Consequently, we need to intelligently transform any categorical variables into numeric representations so that these algorithms can compute. Some packages automate this process (e.g., **h2o** and **caret**) while others do not (e.g., **glmnet** and **keras**). There are many ways to recode categorical variables as numeric (e.g., one-hot, ordinal, binary, sum, and Helmert).

The most common is referred to as one-hot encoding, where we transpose our categorical variables so that each level of the feature is represented as a boolean value. For example, one-hot encoding the left data frame in Figure 3.9 results in `X` being converted into three columns, one for each level. This is called less than *full rank* encoding . However, this creates perfect collinearity which causes problems with some predictive modeling algorithms (e.g., ordinary linear regression and neural networks). Alternatively, we can create a full-rank encoding by dropping one of the levels (level `c` has been dropped). This is referred to as *dummy* encoding.

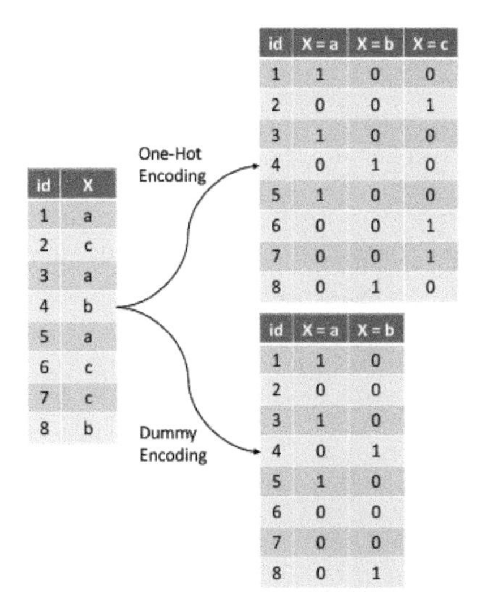

FIGURE 3.9: Eight observations containing a categorical feature X and the difference in how one-hot and dummy encoding transforms this feature.

We can one-hot or dummy encode with the same function (`step_dummy()`). By default, `step_dummy()` will create a full rank encoding but you can change this by setting `one_hot = TRUE`.

```
# Lump levels for two features
recipe(Sale_Price ~ ., data = ames_train) %>%
  step_dummy(all_nominal(), one_hot = TRUE)
## Data Recipe
##
## Inputs:
##
##       role #variables
##    outcome          1
##  predictor         80
##
## Operations:
##
## Dummy variables from all_nominal()
```

Since one-hot encoding adds new features it can significantly increase the dimensionality of our data. If you have a data set with many categorical variables and those categorical variables in turn have many unique levels, the number of features can explode. In these cases you may want to explore label/ordinal encoding or some other alternative.

3.6.3 Label encoding

Label encoding is a pure numeric conversion of the levels of a categorical variable. If a categorical variable is a factor and it has pre-specified levels then the numeric conversion will be in level order. If no levels are specified, the encoding will be based on alphabetical order. For example, the `MS_SubClass` variable has 16 levels, which we can recode numerically with `step_integer()`.

```
# Original categories
count(ames_train, MS_SubClass)
## # A tibble: 16 x 2
##    MS_SubClass                                  n
##    <fct>                                    <int>
##  1 One_Story_1946_and_Newer_All_Styles        749
```

```
##  2 One_Story_1945_and_Older                     97
##  3 One_Story_with_Finished_Attic_All_Ages        4
##  4 One_and_Half_Story_Unfinished_All_Ages       14
##  5 One_and_Half_Story_Finished_All_Ages        192
##  6 Two_Story_1946_and_Newer                    401
##  7 Two_Story_1945_and_Older                     94
##  8 Two_and_Half_Story_All_Ages                  16
##  9 Split_or_Multilevel                          87
## 10 Split_Foyer                                  31
## 11 Duplex_All_Styles_and_Ages                   73
## 12 One_Story_PUD_1946_and_Newer                147
## 13 One_and_Half_Story_PUD_All_Ages               1
## 14 Two_Story_PUD_1946_and_Newer                 94
## 15 PUD_Multilevel_Split_Level_Foyer             12
## 16 Two_Family_conversion_All_Styles_and_Ages    42

# Label encoded
recipe(Sale_Price ~ ., data = ames_train) %>%
  step_integer(MS_SubClass) %>%
  prep(ames_train) %>%
  bake(ames_train) %>%
  count(MS_SubClass)
## # A tibble: 16 x 2
##    MS_SubClass     n
##          <dbl> <int>
##  1           1   749
##  2           2    97
##  3           3     4
##  4           4    14
##  5           5   192
##  6           6   401
##  7           7    94
##  8           8    16
##  9           9    87
## 10          10    31
## 11          11    73
## 12          12   147
## 13          13     1
## 14          14    94
## 15          15    12
## 16          16    42
```

We should be careful with label encoding unordered categorical features because most models will treat them as ordered numeric features. If a categorical feature is naturally ordered then label encoding is a natural choice (most commonly referred to as ordinal encoding). For example, the various quality features in the Ames housing data are ordinal in nature (ranging from `Very_Poor` to `Very_Excellent`).

```
ames_train %>% select(contains("Qual"))
## # A tibble: 2,054 x 6
##    Overall_Qual Exter_Qual Bsmt_Qual Low_Qual_Fin_SF
##    <fct>        <fct>      <fct>               <int>
##  1 Above_Avera~ Typical    Typical                 0
##  2 Average      Typical    Typical                 0
##  3 Above_Avera~ Typical    Typical                 0
##  4 Good         Good       Typical                 0
##  5 Above_Avera~ Typical    Typical                 0
##  6 Very_Good    Good       Good                    0
##  7 Very_Good    Good       Good                    0
##  8 Good         Typical    Typical                 0
##  9 Above_Avera~ Typical    Good                    0
## 10 Above_Avera~ Typical    Good                    0
## # ... with 2,044 more rows, and 2 more variables:
## #   Kitchen_Qual <fct>, Garage_Qual <fct>
```

Ordinal encoding these features provides a natural and intuitive interpretation and can logically be applied to all models.

The various `xxx_Qual` features in the Ames housing are not ordered factors. For ordered factors you could also use `step_ordinalscore()`.

```
# Original categories
count(ames_train, Overall_Qual)
## # A tibble: 10 x 2
##    Overall_Qual       n
##    <fct>          <int>
##  1 Very_Poor          4
##  2 Poor               8
##  3 Fair              23
##  4 Below_Average    169
##  5 Average          582
##  6 Above_Average    497
##  7 Good             425
```

```
## 8 Very_Good 249
## 9 Excellent 75
## 10 Very_Excellent 22

# Label encoded
recipe(Sale_Price ~ ., data = ames_train) %>%
  step_integer(Overall_Qual) %>%
  prep(ames_train) %>%
  bake(ames_train) %>%
  count(Overall_Qual)
## # A tibble: 10 x 2
##    Overall_Qual     n
##           <dbl> <int>
##  1            1     4
##  2            2     8
##  3            3    23
##  4            4   169
##  5            5   582
##  6            6   497
##  7            7   425
##  8            8   249
##  9            9    75
## 10           10    22
```

3.6.4 Alternatives

There are several alternative categorical encodings that are implemented in various R machine learning engines and are worth exploring. For example, target encoding is the process of replacing a categorical value with the mean (regression) or proportion (classification) of the target variable. For example, target encoding the `Neighborhood` feature would change `North_Ames` to 144617.

Target encoding runs the risk of *data leakage* since you are using the response variable to encode a feature. An alternative to this is to change the feature value to represent the proportion a particular level represents for a given feature. In this case, `North_Ames` would be changed to 0.153.

In Chapter 9, we discuss how tree-based models use this approach to order categorical features when choosing a split point.

Several alternative approaches include effect or likelihood encoding (Micci-Barreca, 2001; Zumel and Mount, 2016), empirical Bayes methods (West et al.,

TABLE 3.1: Example of target encoding the Neighborhood feature of the Ames housing data set.

Neighborhood	Avg Sale_Price
North_Ames	147040
College_Creek	202438
Old_Town	121815
Edwards	124297
Somerset	232394
Northridge_Heights	320174
Gilbert	191095
Sawyer	137405
Northwest_Ames	186082
Sawyer_West	183062

TABLE 3.2: Example of categorical proportion encoding the Neighborhood feature of the Ames housing data set.

Neighborhood	Proportion
North_Ames	0.154
College_Creek	0.093
Old_Town	0.081
Edwards	0.064
Somerset	0.062
Northridge_Heights	0.058
Gilbert	0.056
Sawyer	0.052
Northwest_Ames	0.040
Sawyer_West	0.045

2014), word and entity embeddings (Guo and Berkhahn, 2016; Chollet and Allaire, 2018), and more. For more in depth coverage of categorical encodings we highly recommend Kuhn and Johnson (2019).

3.7 Dimension reduction

Dimension reduction is an alternative approach to filter out non-informative features without manually removing them. We discuss dimension reduction topics in depth later in the book (Chapters 17-19) so please refer to those chapters for details.

However, we wanted to highlight that it is very common to include these types of dimension reduction approaches during the feature engineering process. For example, we may wish to reduce the dimension of our features with principal components analysis (Chapter 17) and retain the number of components required to explain, say, 95% of the variance and use these components as features in downstream modeling.

```
recipe(Sale_Price ~ ., data = ames_train) %>%
  step_center(all_numeric()) %>%
  step_scale(all_numeric()) %>%
  step_pca(all_numeric(), threshold = .95)
## Data Recipe
##
## Inputs:
##
##       role #variables
##    outcome          1
##  predictor         80
##
## Operations:
##
## Centering for all_numeric()
## Scaling for all_numeric()
## PCA extraction with all_numeric()
```

3.8 Proper implementation

We stated at the beginning of this chapter that we should think of feature engineering as creating a blueprint rather than manually performing each task individually. This helps us in two ways: (1) thinking sequentially and (2) to apply appropriately within the resampling process.

3.8.1 Sequential steps

Thinking of feature engineering as a blueprint forces us to think of the ordering of our preprocessing steps. Although each particular problem requires you to think of the effects of sequential preprocessing, there are some general suggestions that you should consider:

- If using a log or Box-Cox transformation, don't center the data first or do any operations that might make the data non-positive. Alternatively, use the Yeo-Johnson transformation so you don't have to worry about this.
- One-hot or dummy encoding typically results in sparse data which many algorithms can operate efficiently on. If you standardize sparse data you will create dense data and you loose the computational efficiency. Consequently, it's often preferred to standardize your numeric features and then one-hot/dummy encode.
- If you are lumping infrequently occurring categories together, do so before one-hot/dummy encoding.
- Although you can perform dimension reduction procedures on categorical features, it is common to primarily do so on numeric features when doing so for feature engineering purposes.

While your project's needs may vary, here is a suggested order of potential steps that should work for most problems:

1. Filter out zero or near-zero variance features.
2. Perform imputation if required.
3. Normalize to resolve numeric feature skewness.
4. Standardize (center and scale) numeric features.
5. Perform dimension reduction (e.g., PCA) on numeric features.
6. One-hot or dummy encode categorical features.

3.8.2 Data leakage

Data leakage is when information from outside the training data set is used to create the model. Data leakage often occurs during the data preprocessing period. To minimize this, feature engineering should be done in isolation of each resampling iteration. Recall that resampling allows us to estimate the generalizable prediction error. Therefore, we should apply our feature engineering blueprint to each resample independently as illustrated in Figure 3.10. That way we are not leaking information from one data set to another (each resample is designed to act as isolated training and test data).

For example, when standardizing numeric features, each resampled training data should use its own mean and variance estimates and these specific values should be applied to the same resampled test set. This imitates how real-life prediction occurs where we only know our current data's mean and variance estimates; therefore, on new data that comes in where we need to predict we assume the feature values follow the same distribution of what we've seen in the past.

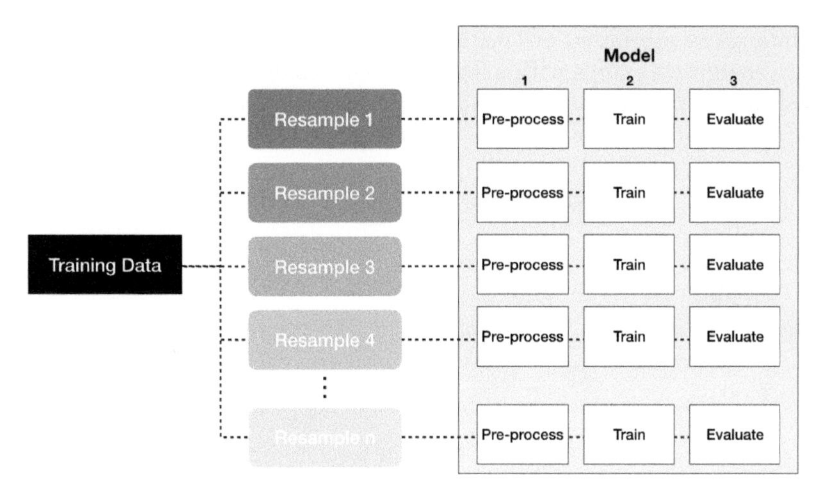

FIGURE 3.10: Performing feature engineering preprocessing within each resample helps to minimize data leakage.

3.8.3 Putting the process together

To illustrate how this process works together via R code, let's do a simple re-assessment on the `ames` data set that we did at the end of the last chapter (Section 2.7) and see if some simple feature engineering improves our prediction error. But first, we'll formally introduce the **recipes** package, which we've been implicitly illustrating throughout.

The **recipes** package allows us to develop our feature engineering blueprint in a sequential nature. The idea behind **recipes** is similar to `caret::preProcess()` where we want to create the preprocessing blueprint but apply it later and within each resample.[7]

There are three main steps in creating and applying feature engineering with **recipes**:

1. `recipe`: where you define your feature engineering steps to create your blueprint.
2. `prepare`: estimate feature engineering parameters based on training data.
3. `bake`: apply the blueprint to new data.

The first step is where you define your blueprint (aka recipe). With this process, you supply the formula of interest (the target variable, features, and

[7] In fact, most of the feature engineering capabilities found in **resample** can also be found in `caret::preProcess()`.

the data these are based on) with `recipe()` and then you sequentially add feature engineering steps with `step_xxx()`. For example, the following defines `Sale_Price` as the target variable and then uses all the remaining columns as features based on `ames_train`. We then:

1. Remove near-zero variance features that are categorical (aka nominal).
2. Ordinal encode our quality-based features (which are inherently ordinal).
3. Center and scale (i.e., standardize) all numeric features.
4. Perform dimension reduction by applying PCA to all numeric features.

```
blueprint <- recipe(Sale_Price ~ ., data = ames_train) %>%
  step_nzv(all_nominal())  %>%
  step_integer(matches("Qual|Cond|QC|Qu")) %>%
  step_center(all_numeric(), -all_outcomes()) %>%
  step_scale(all_numeric(), -all_outcomes()) %>%
  step_pca(all_numeric(), -all_outcomes())

blueprint
## Data Recipe
##
## Inputs:
##
##       role #variables
##    outcome          1
##  predictor         80
##
## Operations:
##
## Sparse, unbalanced variable filter on all_nominal()
## Integer encoding for matches("Qual|Cond|QC|Qu")
## Centering for 2 items
## Scaling for 2 items
## PCA extraction with 2 items
```

Next, we need to train this blueprint on some training data. Remember, there are many feature engineering steps that we do not want to train on the test data (e.g., standardize and PCA) as this would create data leakage. So in this step we estimate these parameters based on the training data of interest.

```
prepare <- prep(blueprint, training = ames_train)
prepare
## Data Recipe
##
## Inputs:
##
##       role #variables
##    outcome          1
##  predictor         80
##
## Training data contained 2054 data points and no missing data.
##
## Operations:
##
## Sparse, unbalanced variable filter removed Street, ... [trained]
## Integer encoding for Condition_1, ... [trained]
## Centering for Lot_Frontage, ... [trained]
## Scaling for Lot_Frontage, ... [trained]
## PCA extraction with Lot_Frontage, ... [trained]
```

Lastly, we can apply our blueprint to new data (e.g., the training data or future test data) with `bake()`.

```
baked_train <- bake(prepare, new_data = ames_train)
baked_test <- bake(prepare, new_data = ames_test)
baked_train
## # A tibble: 2,054 x 27
##    MS_SubClass MS_Zoning Lot_Shape Lot_Config
##    <fct>       <fct>     <fct>     <fct>
##  1 One_Story_~ Resident~ Slightly~ Corner
##  2 One_Story_~ Resident~ Regular   Inside
##  3 One_Story_~ Resident~ Slightly~ Corner
##  4 One_Story_~ Resident~ Regular   Corner
##  5 Two_Story_~ Resident~ Slightly~ Inside
##  6 One_Story_~ Resident~ Slightly~ Inside
##  7 One_Story_~ Resident~ Slightly~ Inside
##  8 Two_Story_~ Resident~ Regular   Inside
##  9 Two_Story_~ Resident~ Slightly~ Corner
## 10 One_Story_~ Resident~ Slightly~ Inside
## # ... with 2,044 more rows, and 23 more variables:
## #   Neighborhood <fct>, Bldg_Type <fct>,
## #   House_Style <fct>, Roof_Style <fct>,
```

```
## #   Exterior_1st <fct>, Exterior_2nd <fct>,
## #   Mas_Vnr_Type <fct>, Foundation <fct>,
## #   Bsmt_Exposure <fct>, BsmtFin_Type_1 <fct>,
## #   Central_Air <fct>, Electrical <fct>,
## #   Garage_Type <fct>, Garage_Finish <fct>,
## #   Paved_Drive <fct>, Fence <fct>, Sale_Type <fct>,
## #   Sale_Price <int>, PC1 <dbl>, PC2 <dbl>, PC3 <dbl>,
## #   PC4 <dbl>, PC5 <dbl>
```

Consequently, the goal is to develop our blueprint, then within each resample iteration we want to apply `prep()` and `bake()` to our resample training and validation data. Luckily, the **caret** package simplifies this process. We only need to specify the blueprint and **caret** will automatically prepare and bake within each resample. We illustrate with the `ames` housing example.

First, we create our feature engineering blueprint to perform the following tasks:

1. Filter out near-zero variance features for categorical features.
2. Ordinally encode all quality features, which are on a 1–10 Likert scale.
3. Standardize (center and scale) all numeric features.
4. One-hot encode our remaining categorical features.

```
blueprint <- recipe(Sale_Price ~ ., data = ames_train) %>%
  step_nzv(all_nominal()) %>%
  step_integer(matches("Qual|Cond|QC|Qu")) %>%
  step_center(all_numeric(), -all_outcomes()) %>%
  step_scale(all_numeric(), -all_outcomes()) %>%
  step_dummy(all_nominal(), -all_outcomes(),
             one_hot = TRUE)
```

Next, we apply the same resampling method and hyperparameter search grid as we did in Section 2.7. The only difference is when we train our resample models with `train()`, we supply our blueprint as the first argument and then **caret** takes care of the rest.

```
# Specify resampling plan
cv <- trainControl(
  method = "repeatedcv",
```

```
  number = 10,
  repeats = 5
)

# Construct grid of hyperparameter values
hyper_grid <- expand.grid(k = seq(2, 25, by = 1))

# Tune a knn model using grid search
knn_fit2 <- train(
  blueprint,
  data = ames_train,
  method = "knn",
  trControl = cv,
  tuneGrid = hyper_grid,
  metric = "RMSE"
)
```

Looking at our results we see that the best model was associated with $k = 11$, which resulted in a cross-validated RMSE of 32,938. Figure 3.11 illustrates the cross-validated error rate across the spectrum of hyperparameter values that we specified.

```
# print model results
knn_fit2
## k-Nearest Neighbors
##
## 2054 samples
##   80 predictor
##
## Recipe steps: nzv, integer, center, scale, dummy
## Resampling: Cross-Validated (10 fold, repeated 5 times)
## Summary of sample sizes: 1849, 1847, 1849, 1847, 1850, 1849, ...
## Resampling results across tuning parameters:
##
##   k  RMSE   Rsquared  MAE
##    2  36656  0.796     22496
##    3  35576  0.809     21758
##    4  34562  0.820     21281
##    5  34074  0.827     20966
##    6  33660  0.832     20793
##    7  33332  0.838     20659
##    8  33209  0.841     20575
```

```
##     9  33049  0.845      20543
##    10  32959  0.847      20530
##    11  32938  0.848      20556
##    12  32961  0.849      20557
##    13  32991  0.849      20631
##    14  33069  0.849      20683
##    15  33091  0.850      20719
##    16  33095  0.850      20734
##    17  33096  0.851      20740
##    18  33154  0.851      20784
##    19  33241  0.851      20850
##    20  33306  0.851      20899
##    21  33423  0.850      20989
##    22  33475  0.850      21053
##    23  33546  0.850      21106
##    24  33580  0.850      21164
##    25  33617  0.851      21226
##
## RMSE was used to select the optimal model using
##  the smallest value.
## The final value used for the model was k = 11.

# plot cross validation results
ggplot(knn_fit2)
```

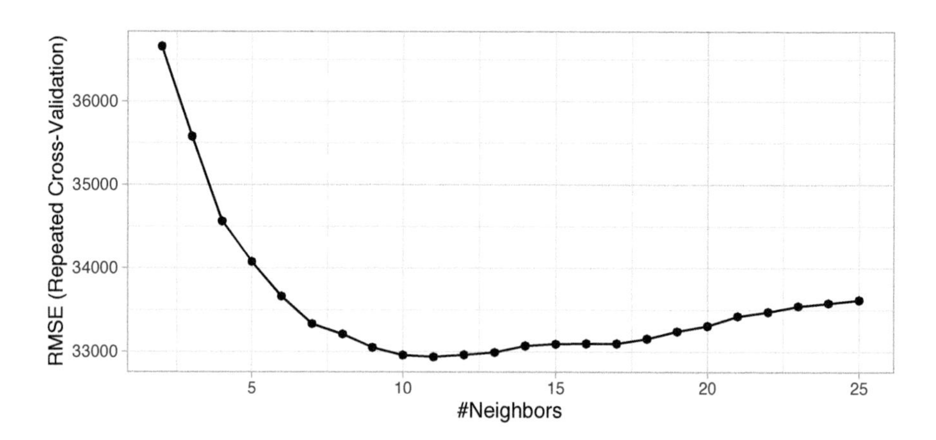

FIGURE 3.11: Results from the same grid search performed in Section 2.7 but with feature engineering performed within each resample.

By applying a handful of the preprocessing techniques discussed throughout this chapter, we were able to reduce our prediction error by over $10,000. The chapters that follow will look to see if we can continue reducing our error by applying different algorithms and feature engineering blueprints.

Part II

Supervised Learning

4

Linear Regression

Linear regression, a staple of classical statistical modeling, is one of the simplest algorithms for doing supervised learning. Though it may seem somewhat dull compared to some of the more modern statistical learning approaches described in later chapters, linear regression is still a useful and widely applied statistical learning method. Moreover, it serves as a good starting point for more advanced approaches; as we will see in later chapters, many of the more sophisticated statistical learning approaches can be seen as generalizations to or extensions of ordinary linear regression. Consequently, it is important to have a good understanding of linear regression before studying more complex learning methods. This chapter introduces linear regression with an emphasis on prediction, rather than inference. An excellent and comprehensive overview of linear regression is provided in Kutner et al. (2005). See Faraway (2016b) for a discussion of linear regression in R (the book's website also provides Python scripts).

4.1 Prerequisites

This chapter leverages the following packages:

```
# Helper packages
library(dplyr)    # for data manipulation
library(ggplot2)  # for awesome graphics

# Modeling packages
library(caret)    # for cross-validation, etc.

# Model interpretability packages
library(vip)      # variable importance
```

We'll also continue working with the `ames_train` data set created in Section 2.7.

4.2 Simple linear regression

Pearson's correlation coefficient is often used to quantify the strength of the linear association between two continuous variables. In this section, we seek to fully characterize that linear relationship. *Simple linear regression* (SLR) assumes that the statistical relationship between two continuous variables (say X and Y) is (at least approximately) linear:

$$Y_i = \beta_0 + \beta_1 X_i + \epsilon_i, \quad \text{for } i = 1, 2, \dots, n, \tag{4.1}$$

where Y_i represents the i-th response value, X_i represents the i-th feature value, β_0 and β_1 are fixed, but unknown constants (commonly referred to as coefficients or parameters) that represent the intercept and slope of the regression line, respectively, and ϵ_i represents noise or random error. In this chapter, we'll assume that the errors are normally distributed with mean zero and constant variance σ^2, denoted $\stackrel{iid}{\sim} (0, \sigma^2)$. Since the random errors are centered around zero (i.e., $E(\epsilon) = 0$), linear regression is really a problem of estimating a *conditional mean*:

$$E(Y_i | X_i) = \beta_0 + \beta_1 X_i. \tag{4.2}$$

For brevity, we often drop the conditional piece and write $E(Y|X) = E(Y)$. Consequently, the interpretation of the coefficients is in terms of the average, or mean response. For example, the intercept β_0 represents the average response value when $X = 0$ (it is often not meaningful or of interest and is sometimes referred to as a *bias term*). The slope β_1 represents the increase in the average response per one-unit increase in X (i.e., it is a *rate of change*).

4.2.1 Estimation

Ideally, we want estimates of β_0 and β_1 that give us the "best fitting" line. But what is meant by "best fitting"? The most common approach is to use the method of *least squares* (LS) estimation; this form of linear regression is often referred to as ordinary least squares (OLS) regression. There are multiple ways to measure "best fitting", but the LS criterion finds the "best fitting" line by minimizing the *residual sum of squares* (RSS):

$$RSS(\beta_0, \beta_1) = \sum_{i=1}^{n} \left[Y_i - (\beta_0 + \beta_1 X_i)\right]^2 = \sum_{i=1}^{n} (Y_i - \beta_0 - \beta_1 X_i)^2 . \tag{4.3}$$

The LS estimates of β_0 and β_1 are denoted as $\widehat{\beta}_0$ and $\widehat{\beta}_1$, respectively. Once obtained, we can generate predicted values, say at $X = X_{new}$, using the estimated regression equation:

$$\widehat{Y}_{new} = \widehat{\beta}_0 + \widehat{\beta}_1 X_{new}, \tag{4.4}$$

where $\widehat{Y}_{new} = E(Y_{new} \widehat{|X = X_{new}})$ is the estimated mean response at $X = X_{new}$.

With the Ames housing data, suppose we wanted to model a linear relationship between the total above ground living space of a home (`Gr_Liv_Area`) and sale price (`Sale_Price`). To perform an OLS regression model in R we can use the `lm()` function:

```
model1 <- lm(Sale_Price ~ Gr_Liv_Area, data = ames_train)
```

The fitted model (`model1`) is displayed in the left plot in Figure 4.1 where the points represent the values of `Sale_Price` in the training data. In the right plot of Figure 4.1, the vertical lines represent the individual errors, called *residuals*, associated with each observation. The OLS criterion in Equation (4.3) identifies the "best fitting" line that minimizes the sum of squares of these residuals.

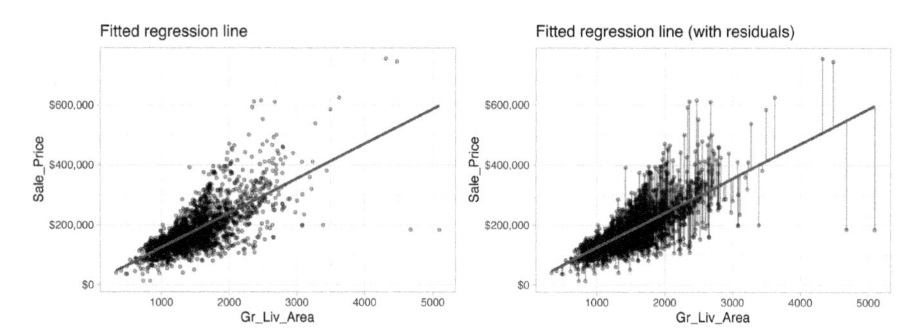

FIGURE 4.1: The least squares fit from regressing sale price on living space for the the Ames housing data. Left: Fitted regression line. Right: Fitted regression line with vertical grey bars representing the residuals.

The `coef()` function extracts the estimated coefficients from the model. We can also use `summary()` to get a more detailed report of the model results.

```
summary(model1)
##
## Call:
```

```
## lm(formula = Sale_Price ~ Gr_Liv_Area, data = ames_train)
##
## Residuals:
##     Min      1Q  Median      3Q     Max
## -413052  -30218   -1612   23383  330421
##
## Coefficients:
##             Estimate Std. Error t value Pr(>|t|)
## (Intercept)  7989.35    3892.40    2.05     0.04 *
## Gr_Liv_Area   115.59       2.46   46.90   <2e-16 ***
## ---
## Signif. codes:
## 0 '***' 0.001 '**' 0.01 '*' 0.05 '.' 0.1 ' ' 1
##
## Residual standard error: 55800 on 2052 degrees of freedom
## Multiple R-squared:  0.517,  Adjusted R-squared:  0.517
## F-statistic: 2.2e+03 on 1 and 2052 DF,  p-value: <2e-16
```

The estimated coefficients from our model are $\widehat{\beta}_0 = 7989.35$ and $\widehat{\beta}_1 = 115.59$. To interpret, we estimate that the mean selling price increases by 115.59 for each additional one square foot of above ground living space. This simple description of the relationship between the sale price and square footage using a single number (i.e., the slope) is what makes linear regression such an intuitive and popular modeling tool.

One drawback of the LS procedure in linear regression is that it only provides estimates of the coefficients; it does not provide an estimate of the error variance σ^2! LS also makes no assumptions about the random errors. These assumptions are important for inference and in estimating the error variance which we're assuming is a constant value σ^2. One way to estimate σ^2 (which is required for characterizing the variability of our fitted model), is to use the method of *maximum likelihood* (ML) estimation (see Kutner et al. (2005) Section 1.7 for details). The ML procedure requires that we assume a particular distribution for the random errors. Most often, we assume the errors to be normally distributed. In practice, under the usual assumptions stated above, an unbiased estimate of the error variance is given as the sum of the squared residuals divided by $n - p$ (where p is the number of regression coefficients or parameters in the model):

$$\widehat{\sigma}^2 = \frac{1}{n-p}\sum_{i=1}^{n} r_i^2, \tag{4.5}$$

where $r_i = \left(Y_i - \widehat{Y}_i\right)$ is referred to as the ith residual (i.e., the difference between the ith observed and predicted response value). The quantity $\widehat{\sigma}^2$ is also referred to as the *mean square error* (MSE) and its square root is denoted RMSE (see Section 2.6 for discussion on these metrics). In R, the RMSE of a linear model can be extracted using the `sigma()` function:

Typically, these error metrics are computed on a separate validation set or using cross-validation as discussed in Section 2.4; however, they can also be computed on the same training data the model was trained on as illustrated here.

```
sigma(model1)    # RMSE
## [1] 55753
sigma(model1)^2  # MSE
## [1] 3.11e+09
```

Note that the RMSE is also reported as the `Residual standard error` in the output from `summary()`.

4.2.2 Inference

How accurate are the LS of β_0 and β_1? Point estimates by themselves are not very useful. It is often desirable to associate some measure of an estimates variability. The variability of an estimate is often measured by its *standard error* (SE)—the square root of its variance. If we assume that the errors in the linear regression model are $\stackrel{iid}{\sim} \left(0, \sigma^2\right)$, then simple expressions for the SEs of the estimated coefficients exist and are displayed in the column labeled `Std. Error` in the output from `summary()`. From this, we can also derive simple t-tests to understand if the individual coefficients are statistically significant from zero. The t-statistics for such a test are nothing more than the estimated coefficients divided by their corresponding estimated standard errors (i.e., in the output from `summary()`, `t value` = `Estimate` / `Std. Error`). The reported t-statistics measure the number of standard deviations each coefficient is away from 0. Thus, large t-statistics (greater than two in absolute value, say) roughly indicate statistical significance at the $\alpha = 0.05$ level. The p-values for these tests are also reported by `summary()` in the column labeled `Pr(>|t|)`.

Under the same assumptions, we can also derive confidence intervals for the coefficients. The formula for the traditional $100\left(1 - \alpha\right)\%$ confidence interval for β_j is

$$\widehat{\beta}_j \pm t_{1-\alpha/2, n-p} \widehat{SE}\left(\widehat{\beta}_j\right). \tag{4.6}$$

In R, we can construct such (one-at-a-time) confidence intervals for each coefficient using `confint()`. For example, a 95% confidence intervals for the coefficients in our SLR example can be computed using

```
confint(model1, level = 0.95)
##               2.5 % 97.5 %
## (Intercept)    356  15623
## Gr_Liv_Area    111    120
```

To interpret, we estimate with 95% confidence that the mean selling price increases between 110.75 and 120.42 for each additional one square foot of above ground living space. We can also conclude that the slope β_1 is significantly different from zero (or any other pre-specified value not included in the interval) at the $\alpha = 0.05$ level. This is also supported by the output from `summary()`.

Most statistical software, including R, will include estimated standard errors, t-statistics, etc. as part of its regression output. However, it is important to remember that such quantities depend on three major assumptions of the linear regresion model:

1. Independent observations
2. The random errors have mean zero, and constant variance
3. The random errors are normally distributed

If any or all of these assumptions are violated, then remdial measures need to be taken. For instance, *weighted least squares* (and other procedures) can be used when the constant variance assumption is violated. Transformations (of both the response and features) can also help to correct departures from these assumptions. The residuals are extremely useful in helping to identify how parametric models depart from such assumptions.

4.3 Multiple linear regression

In practice, we often have more than one predictor. For example, with the Ames housing data, we may wish to understand if above ground square footage (`Gr_Liv_Area`) and the year the house was built (`Year_Built`) are (linearly) related to sale price (`Sale_Price`). We can extend the SLR model so that it can directly accommodate multiple predictors; this is referred to as the *multiple linear regression* (MLR) model. With two predictors, the MLR model becomes:

$$Y = \beta_0 + \beta_1 X_1 + \beta_2 X_2 + \epsilon, \tag{4.7}$$

where X_1 and X_2 are features of interest. In our Ames housing example, X_1 represents `Gr_Liv_Area` and X_2 represents `Year_Built`.

In R, multiple linear regression models can be fit by separating all the features of interest with a `+`:

```
(model2 <- lm(Sale_Price ~ Gr_Liv_Area + Year_Built,
              data = ames_train))
##
## Call:
## lm(formula = Sale_Price ~ Gr_Liv_Area + Year_Built...
##
## Coefficients:
## (Intercept)  Gr_Liv_Area   Year_Built
##    -2.07e+06     9.92e+01     1.07e+03
```

Alternatively, we can use `update()` to update the model formula used in `model1`. The new formula can use a `.` as shorthand for keep everything on either the left or right hand side of the formula, and a `+` or `-` can be used to add or remove terms from the original model, respectively. In the case of adding `Year_Built` to `model1`, we could've used:

```
(model2 <- update(model1, . ~ . + Year_Built))
##
## Call:
## lm(formula = Sale_Price ~ Gr_Liv_Area + Year_Built...
##
## Coefficients:
## (Intercept)  Gr_Liv_Area   Year_Built
##    -2.07e+06     9.92e+01     1.07e+03
```

The LS estimates of the regression coefficients are $\widehat{\beta}_1 = 99.169$ and $\widehat{\beta}_2 = 1067.108$ (the estimated intercept is -2.071×10^6. In other words, every one square foot increase to above ground square footage is associated with an additional \$99.17 in **mean selling price** when holding the year the house was built constant. Likewise, for every year newer a home is there is approximately an increase of $1,067.11 in selling price when holding the above ground square footage constant.

A contour plot of the fitted regression surface is displayed in the left side of Figure 4.2 below. Note how the fitted regression surface is flat (i.e., it does not twist or bend). This is true for all linear models that include only *main effects* (i.e., terms involving only a single predictor). One way to model curvature is to include *interaction effects*. An interaction occurs when the effect of one predictor on the response depends on the values of other predictors. In linear regression, interactions can be captured via products of features (i.e., $X_1 \times X_2$). A model with two main effects can also include a two-way interaction. For example, to include an interaction between $X_1 =$ `Gr_Liv_Area` and $X_2 =$ `Year_Built`, we introduce an additional product term:

$$Y = \beta_0 + \beta_1 X_1 + \beta_2 X_2 + \beta_3 X_1 X_2 + \epsilon. \tag{4.8}$$

Note that in R, we use the `:` operator to include an interaction (technically, we could use `*` as well, but `x1 * x2` is shorthand for `x1 + x2 + x1:x2` so is slightly redundant):

```
lm(Sale_Price ~ Gr_Liv_Area + Year_Built + Gr_Liv_Area:Year_Built,
   data = ames_train)
##
## Call:
## lm(formula = Sale_Price ~ Gr_Liv_Area + Year_Built...
##      data = ames_train)
##
## Coefficients:
##             (Intercept)             Gr_Liv_Area
##               30353.414               -1243.040
##              Year_Built  Gr_Liv_Area:Year_Built
##                   0.149                   0.681
```

A contour plot of the fitted regression surface with interaction is displayed in the right side of Figure 4.2. Note the curvature in the contour lines.

Interaction effects are quite prevalent in predictive modeling. Since linear models are an example of parametric modeling, it is up to the analyst to decide if and when to include interaction effects. In later chapters, we'll discuss algorithms that can automatically detect and incorporate interaction effects (albeit in different ways). It is also important to understand a concept called the *hierarchy principle*—which demands that all lower-order terms corresponding to an interaction be retained in the model—when considering interaction effects in linear regression models.

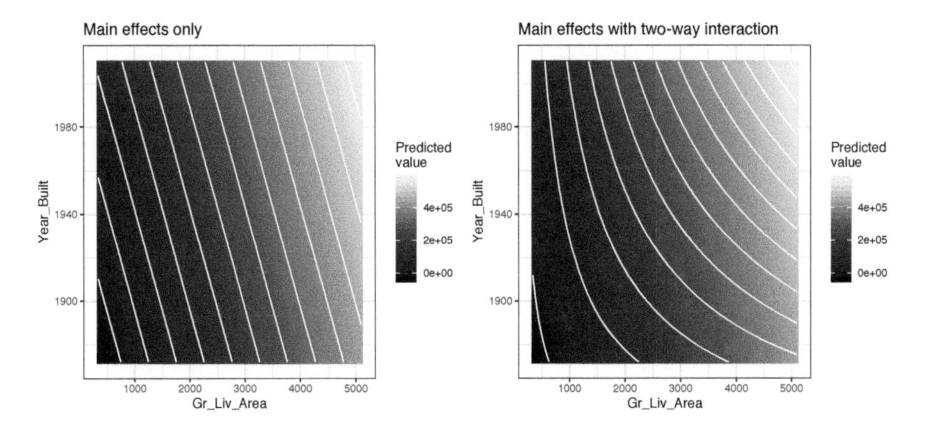

FIGURE 4.2: In a three-dimensional setting, with two predictors and one response, the least squares regression line becomes a plane. The 'best-fit' plane minimizes the sum of squared errors between the actual sales price (individual dots) and the predicted sales price (plane).

In general, we can include as many predictors as we want, as long as we have more rows than parameters! The general multiple linear regression model with p distinct predictors is

$$Y = \beta_0 + \beta_1 X_1 + \beta_2 X_2 + \cdots + \beta_p X_p + \epsilon, \tag{4.9}$$

where X_i for $i = 1, 2, \dots, p$ are the predictors of interest. Note some of these may represent interactions (e.g., $X_3 = X_1 \times X_2$) between or transformations[1] (e.g., $X_4 = \sqrt{X_1}$) of the original features. Unfortunately, visualizing beyond three dimensions is not practical as our best-fit plane becomes a hyperplane. However, the motivation remains the same where the best-fit hyperplane is identified by minimizing the RSS. The code below creates a third model where we use all features in our data set as main effects (i.e., no interaction terms) to predict `Sale_Price`.

[1] Transformations of the features serve a number of purposes (e.g., modeling nonlinear relationships or alleviating departures from common regression assumptions). See Kutner et al. (2005) for details.

```
# include all possible main effects
model3 <- lm(Sale_Price ~ ., data = ames_train)

# print estimated coefficients in a tidy data frame
broom::tidy(model3)
## # A tibble: 292 x 5
##    term                estimate std.error statistic p.value
##    <chr>                  <dbl>     <dbl>     <dbl>   <dbl>
##  1 (Intercept)          -1.20e7 10949313.    -1.09    0.274
##  2 MS_SubClassOne_~      3.37e3     3655.     0.921   0.357
##  3 MS_SubClassOne_~      1.21e4    11926.     1.02    0.309
##  4 MS_SubClassOne_~      1.16e4    12833.     0.902   0.367
##  5 MS_SubClassOne_~      6.67e3     6552.     1.02    0.309
##  6 MS_SubClassTwo_~     -1.81e3     6018.    -0.301   0.763
##  7 MS_SubClassTwo_~      1.02e4     6612.     1.54    0.124
##  8 MS_SubClassTwo_~     -1.62e4    10468.    -1.54    0.123
##  9 MS_SubClassSpli~     -1.03e4    11585.    -0.888   0.375
## 10 MS_SubClassSpli~     -3.13e3     7577.    -0.413   0.680
## # ... with 282 more rows
```

4.4 Assessing model accuracy

We've fit three main effects models to the Ames housing data: a single predictor, two predictors, and all possible predictors. But the question remains, which model is "best"? To answer this question we have to define what we mean by "best". In our case, we'll use the RMSE metric and cross-validation (Section 2.4) to determine the "best" model. We can use the `caret::train()` function to train a linear model (i.e., `method = "lm"`) using cross-validation (or a variety of other validation methods). In practice, a number of factors should be considered in determining a "best" model (e.g., time constraints, model production cost, predictive accuracy, etc.). The benefit of **caret** is that it provides built-in cross-validation capabilities, whereas the `lm()` function does not[2]. The following code chunk uses `caret::train()` to refit `model1` using 10-fold cross-validation:

[2]Although general cross-validation is not available in `lm()` alone, a simple metric called the *PRESS* statistic, for **PRE**dictive **S**um of **S**quare, (equivalent to a *leave-one-out* cross-validated RMSE) can be computed by summing the PRESS residuals which are available using `rstandard(<lm-model-name>, type = "predictive")`. See `?rstandard` for details.

```
# Train model using 10-fold cross-validation
set.seed(123)  # for reproducibility
(cv_model1 <- train(
  form = Sale_Price ~ Gr_Liv_Area,
  data = ames_train,
  method = "lm",
  trControl = trainControl(method = "cv", number = 10)
))
## Linear Regression
##
## 2054 samples
##    1 predictor
##
## No pre-processing
## Resampling: Cross-Validated (10 fold)
## Summary of sample sizes: 1848, 1847, 1848, 1849, 1848, 1850, ...
## Resampling results:
##
##   RMSE   Rsquared  MAE
##   55670  0.521     38380
##
## Tuning parameter 'intercept' was held constant at
##  a value of TRUE
```

The resulting cross-validated RMSE is $55,670.37 (this is the average RMSE across the 10 CV folds). How should we interpret this? When applied to unseen data, the predictions this model makes are, on average, about $55,670.37 off from the actual sale price.

We can perform cross-validation on the other two models in a similar fashion, which we do in the code chunk below.

```
# model 2 CV
set.seed(123)
cv_model2 <- train(
  Sale_Price ~ Gr_Liv_Area + Year_Built,
  data = ames_train,
  method = "lm",
  trControl = trainControl(method = "cv", number = 10)
)

# model 3 CV
set.seed(123)
```

```
cv_model3 <- train(
  Sale_Price ~ .,
  data = ames_train,
  method = "lm",
  trControl = trainControl(method = "cv", number = 10)
)

# Extract out of sample performance measures
summary(resamples(list(
  model1 = cv_model1,
  model2 = cv_model2,
  model3 = cv_model3
)))
##
## Call:
## summary.resamples(object = resamples(list(model1
##  = cv_model1, model2 = cv_model2, model3 = cv_model3)))
##
## Models: model1, model2, model3
## Number of resamples: 10
##
## MAE
##         Min. 1st Qu. Median  Mean 3rd Qu.  Max. NA's
## model1 36295   36806  37005 38380   40034 42096    0
## model2 28076   30690  31325 31479   32620 34536    0
## model3 14257   15855  16131 17080   16689 25677    0
##
## RMSE
##         Min. 1st Qu. Median  Mean 3rd Qu.  Max. NA's
## model1 50003   52413  54193 55670   60344 62415    0
## model2 40456   42957  45597 46133   49114 53745    0
## model3 20945   25674  33769 37304   42967 80339    0
##
## Rsquared
##        Min. 1st Qu. Median  Mean 3rd Qu.  Max. NA's
## model1 0.38   0.483  0.528 0.521   0.588 0.640    0
## model2 0.53   0.667  0.686 0.672   0.708 0.737    0
## model3 0.42   0.746  0.829 0.797   0.910 0.922    0
```

Extracting the results for each model, we see that by adding more information via more predictors, we are able to improve the out-of-sample cross validation performance metrics. Specifically, our cross-validated RMSE reduces from \$46,132.74 (the model with two predictors) down to \$37,304.33 (for our full model). In this case, the model with all possible main effects performs the "best" (compared with the other two).

4.5 Model concerns

As previously stated, linear regression has been a popular modeling tool due to the ease of interpreting the coefficients. However, linear regression makes several strong assumptions that are often violated as we include more predictors in our model. Violation of these assumptions can lead to flawed interpretation of the coefficients and prediction results.

1. Linear relationship: Linear regression assumes a linear relationship between the predictor and the response variable. However, as discussed in Chapter 3, non-linear relationships can be made linear (or near-linear) by applying transformations to the response and/or predictors. For example, Figure 4.3 illustrates the relationship between sale price and the year a home was built. The left plot illustrates the non-linear relationship that exists. However, we can achieve a near-linear relationship by log transforming sale price, although some non-linearity still exists for older homes.

```
p1 <- ggplot(ames_train, aes(Year_Built, Sale_Price)) +
  geom_point(size = 1, alpha = .4) +
  geom_smooth(se = FALSE) +
  scale_y_continuous("Sale price", labels = scales::dollar) +
  xlab("Year built") +
  ggtitle(paste("Non-transformed variables with a\n",
                "non-linear relationship."))

p2 <- ggplot(ames_train, aes(Year_Built, Sale_Price)) +
  geom_point(size = 1, alpha = .4) +
  geom_smooth(method = "lm", se = FALSE) +
  scale_y_log10("Sale price", labels = scales::dollar,
                breaks = seq(0, 400000, by = 100000)) +
  xlab("Year built") +
  ggtitle(paste("Transforming variables can provide a\n",
                "near-linear relationship."))

gridExtra::grid.arrange(p1, p2, nrow = 1)
```

2. Constant variance among residuals: Linear regression assumes the

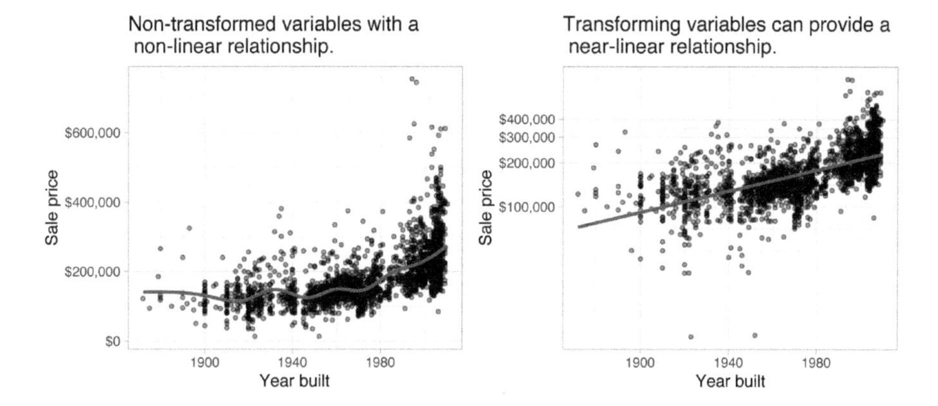

FIGURE 4.3: Linear regression assumes a linear relationship between the predictor(s) and the response variable; however, non-linear relationships can often be altered to be near-linear by applying a transformation to the variable(s).

variance among error terms $(\epsilon_1, \epsilon_2, \dots, \epsilon_p)$ are constant (this assumption is referred to as homoscedasticity). If the error variance is not constant, the p-values and confidence intervals for the coefficients will be invalid. Similar to the linear relationship assumption, non-constant variance can often be resolved with variable transformations or by including additional predictors. For example, Figure 4.4 shows the residuals vs. predicted values for `model1` and `model3`. `model1` displays a classic violation of constant variance as indicated by the cone-shaped pattern. However, `model3` appears to have near-constant variance.

The `broom::augment` function is an easy way to add model results to each observation (i.e. predicted values, residuals).

```
df1 <- broom::augment(cv_model1$finalModel, data = ames_train)

p1 <- ggplot(df1, aes(.fitted, .resid)) +
  geom_point(size = 1, alpha = .4) +
  xlab("Predicted values") +
  ylab("Residuals") +
  ggtitle("Model 1", subtitle = "Sale_Price ~ Gr_Liv_Area")

df2 <- broom::augment(cv_model3$finalModel, data = ames_train)
```

```
p2 <- ggplot(df2, aes(.fitted, .resid)) +
  geom_point(size = 1, alpha = .4)  +
  xlab("Predicted values") +
  ylab("Residuals") +
  ggtitle("Model 3", subtitle = "Sale_Price ~ .")

gridExtra::grid.arrange(p1, p2, nrow = 1)
```

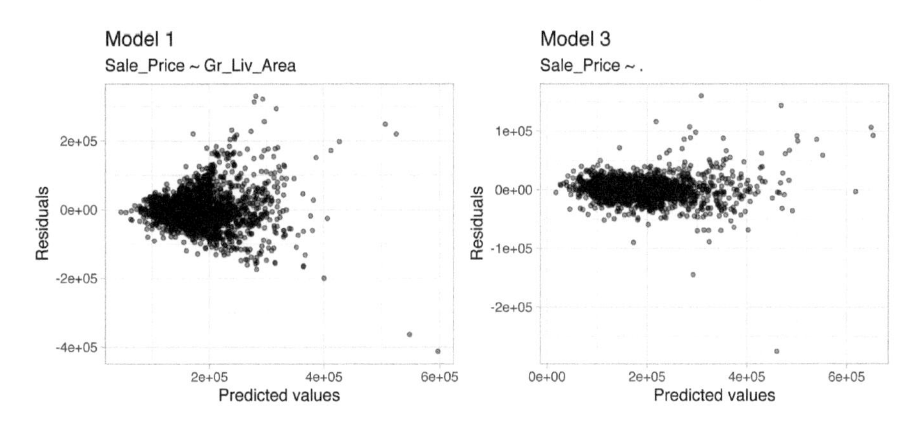

FIGURE 4.4: Linear regression assumes constant variance among the residuals. 'model1' (left) shows definitive signs of heteroskedasticity whereas 'model3' (right) appears to have constant variance.

3. No autocorrelation: Linear regression assumes the errors are independent and uncorrelated. If in fact, there is correlation among the errors, then the estimated standard errors of the coefficients will be biased leading to prediction intervals being narrower than they should be. For example, the left plot in Figure 4.5 displays the residuals (y-axis) vs. the observation ID (x-axis) for `model1`. A clear pattern exists suggesting that information about ϵ_1 provides information about ϵ_2.

This pattern is a result of the data being ordered by neighborhood, which we have not accounted for in this model. Consequently, the residuals for homes in the same neighborhood are correlated (homes within a neighborhood are typically the same size and can often contain similar features). Since the `Neighborhood` predictor is included in `model3` (right plot), the correlation in the errors is reduced.

```
df1 <- mutate(df1, id = row_number())
```

```
df2 <- mutate(df2, id = row_number())

p1 <- ggplot(df1, aes(id, .resid)) +
  geom_point(size = 1, alpha = .4) +
  xlab("Row ID") +
  ylab("Residuals") +
  ggtitle("Model 1", subtitle = "Correlated residuals.")

p2 <- ggplot(df2, aes(id, .resid)) +
  geom_point(size = 1, alpha = .4) +
  xlab("Row ID") +
  ylab("Residuals") +
  ggtitle("Model 3", subtitle = "Uncorrelated residuals.")

gridExtra::grid.arrange(p1, p2, nrow = 1)
```

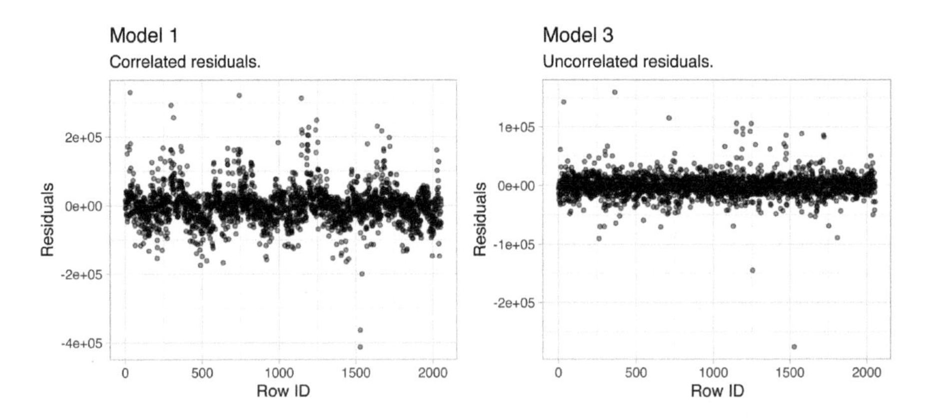

FIGURE 4.5: Linear regression assumes uncorrelated errors. The residuals in 'model1' (left) have a distinct pattern suggesting that information about ϵ_1 provides information about ϵ_2. Whereas 'model3' has no signs of autocorrelation.

4. More observations than predictors: Although not an issue with the Ames housing data, when the number of features exceeds the number of observations ($p > n$), the OLS estimates are not obtainable. To resolve this issue an analyst can remove variables one-at-a-time until $p < n$. Although pre-processing tools can be used to guide this manual approach (Kuhn and Johnson, 2013, 43-47), it can be cumbersome and prone to errors. In Chapter 6 we'll introduce regularized regression which provides an alternative to OLS that can be used when $p > n$.

5. No or little multicollinearity: *Collinearity* refers to the situation in which two or more predictor variables are closely related to one another. The presence of collinearity can pose problems in the OLS, since it can be difficult to separate out the individual effects of collinear variables on the response. In fact, collinearity can cause predictor variables to appear as statistically insignificant when in fact they are significant. This obviously leads to an inaccurate interpretation of coefficients and makes it difficult to identify influential predictors.

In `ames`, for example, `Garage_Area` and `Garage_Cars` are two variables that have a correlation of 0.89 and both variables are strongly related to our response variable (`Sale_Price`). Looking at our full model where both of these variables are included, we see that `Garage_Cars` is found to be statistically significant but `Garage_Area` is not:

```
# fit with two strongly correlated variables
summary(cv_model3) %>%
  broom::tidy() %>%
  filter(term %in% c("Garage_Area", "Garage_Cars"))
## # A tibble: 2 x 5
##   term        estimate std.error statistic p.value
##   <chr>          <dbl>     <dbl>     <dbl>   <dbl>
## 1 Garage_Cars  4962.     1803.        2.75 0.00599
## 2 Garage_Area     9.47      5.97      1.58 0.113
```

However, if we refit the full model without `Garage_Cars`, the coefficient estimate for `Garage_Area` increases two fold and becomes statistically significant.

```
# model without Garage_Area
set.seed(123)
mod_wo_Garage_Cars <- train(
  Sale_Price ~ .,
  data = select(ames_train, -Garage_Cars),
  method = "lm",
  trControl = trainControl(method = "cv", number = 10)
)

summary(mod_wo_Garage_Cars) %>%
  broom::tidy() %>%
  filter(term == "Garage_Area")
## # A tibble: 1 x 5
##   term        estimate std.error statistic      p.value
##   <chr>          <dbl>     <dbl>     <dbl>        <dbl>
## 1 Garage_Area     21.6      4.02      5.38 0.0000000846
```

This reflects the instability in the linear regression model caused by between-predictor relationships; this instability also gets propagated directly to the model predictions. Considering 16 of our 34 numeric predictors have a medium to strong correlation (Chapter 17), the biased coefficients of these predictors are likely restricting the predictive accuracy of our model. How can we control for this problem? One option is to manually remove the offending predictors (one-at-a-time) until all pairwise correlations are below some pre-determined threshold. However, when the number of predictors is large such as in our case, this becomes tedious. Moreover, multicollinearity can arise when one feature is linearly related to two or more features (which is more difficult to detect[3]). In these cases, manual removal of specific predictors may not be possible. Consequently, the following sections offers two simple extensions of linear regression where dimension reduction is applied prior to performing linear regression. Chapter 6 offers a modified regression approach that helps to deal with the problem. And future chapters provide alternative methods that are less affected by multicollinearity.

4.6 Principal component regression

As mentioned in Section 3.7 and fully discussed in Chapter 17, principal components analysis can be used to represent correlated variables with a smaller number of uncorrelated features (called principle components) and the resulting components can be used as predictors in a linear regression model. This two-step process is known as *principal component regression* (PCR) (Massy, 1965) and is illustrated in Figure 4.6.

Performing PCR with **caret** is an easy extension from our previous model. We simply specify `method = "pcr"` within `train()` to perform PCA on all our numeric predictors prior to fitting the model. Often, we can greatly improve performance by only using a small subset of all principal components as predictors. Consequently, you can think of the number of principal components as a tuning parameter (see Section 2.5.3). The following performs cross-validated PCR with $1, 2, \dots, 20$ principal components, and Figure 4.7 illustrates the cross-validated RMSE. You can see a significant drop in prediction error from our previous linear models using just five principal components followed by a gradual decrease thereafter. Using 17 principal components corresponds to the lowest RMSE (see `cv_model_pcr` for a comparison of the cross-validated results).

[3]In such cases we can use a statistic called the *variance inflation factor* which tries to capture how strongly each feature is linearly related to all the others predictors in a model.

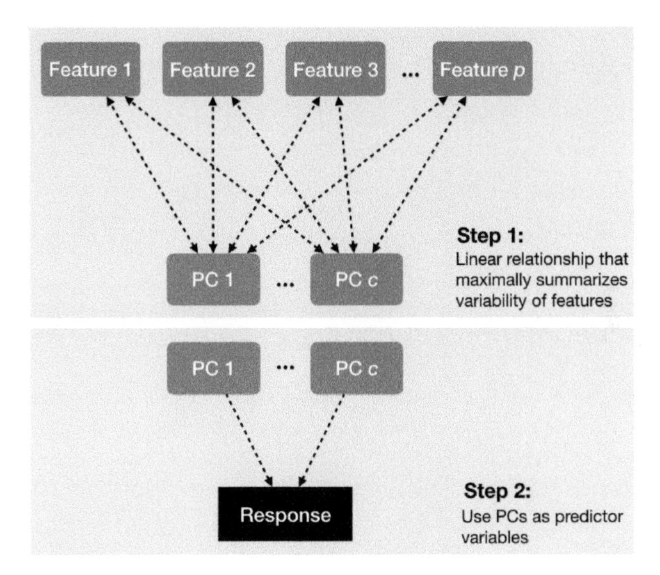

FIGURE 4.6: A depiction of the steps involved in performing principal component regression.

Note in the below example we use `preProcess` to remove near-zero variance features and center/scale the numeric features. We then use `method = "pcr"`. This is equivalent to creating a blueprint as illustrated in Section 3.8.3 to remove near-zero variance features, center/scale the numeric features, perform PCA on the numeric features, then feeding that blueprint into `train()` with `method = "lm"`.

```
# perform 10-fold cross validation on a PCR model tuning the
# number of principal components to use as predictors from 1-20
set.seed(123)
cv_model_pcr <- train(
  Sale_Price ~ .,
  data = ames_train,
  method = "pcr",
  trControl = trainControl(method = "cv", number = 10),
  preProcess = c("zv", "center", "scale"),
  tuneLength = 20
  )

# model with lowest RMSE
```

```
cv_model_pcr$bestTune
##    ncomp
## 19    19

# plot cross-validated RMSE
ggplot(cv_model_pcr)
```

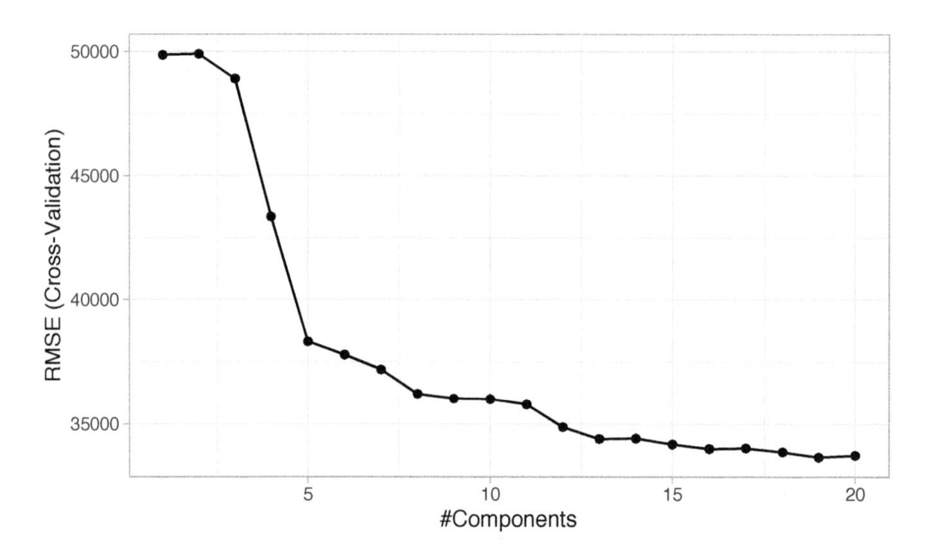

FIGURE 4.7: The 10-fold cross valdation RMSE obtained using PCR with 1-20 principal components.

By controlling for multicollinearity with PCR, we can experience significant improvement in our predictive accuracy compared to the previously obtained linear models (reducing the cross-validated RMSE from about \$37,000 to below \$35,000); however, we still do not improve upon the k-nearest neighbor model illustrated in Section 3.8.3. It's important to note that since PCR is a two step process, the PCA step does not consider any aspects of the response when it selects the components. Consequently, the new predictors produced by the PCA step are not designed to maximize the relationship with the response. Instead, it simply seeks to reduce the variability present throughout the predictor space. If that variability happens to be related to the response variability, then PCR has a good chance to identify a predictive relationship, as in our case. If, however, the variability in the predictor space is not related to the variability of the response, then PCR can have difficulty identifying a predictive relationship when one might actually exists (i.e., we may actually experience a decrease in our predictive accuracy). An alternative approach to reduce the impact of multicollinearity is partial least squares.

4.7 Partial least squares

Partial least squares (PLS) can be viewed as a supervised dimension reduction procedure (Kuhn and Johnson, 2013). Similar to PCR, this technique also constructs a set of linear combinations of the inputs for regression, but unlike PCR it uses the response variable to aid the construction of the principal components as illustrated in Figure 4.8[4]. Thus, we can think of PLS as a supervised dimension reduction procedure that finds new features that not only captures most of the information in the original features, but also are related to the response.

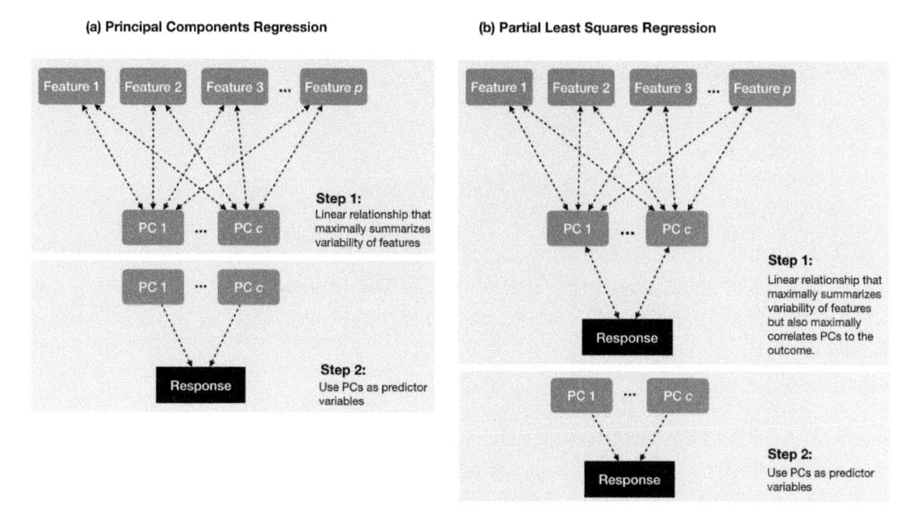

FIGURE 4.8: A diagram depicting the differences between PCR (left) and PLS (right). PCR finds principal components (PCs) that maximally summarize the features independent of the response variable and then uses those PCs as predictor variables. PLS finds components that simultaneously summarize variation of the predictors while being optimally correlated with the outcome and then uses those PCs as predictors.

We illustrate PLS with some exemplar data[5]. Figure 4.9 illustrates that the first two PCs when using PCR have very little relationship to the response variable; however, the first two PCs when using PLS have a much stronger association to the response.

Referring to Equation (17.1) in Chapter 17, PLS will compute the first prin-

[4]Figure 4.8 was inspired by, and modified from, Chapter 6 in Kuhn and Johnson (2013).

[5]This is actually using the solubility data that is provided by the **AppliedPredictiveModeling** package (Kuhn and Johnson, 2018).

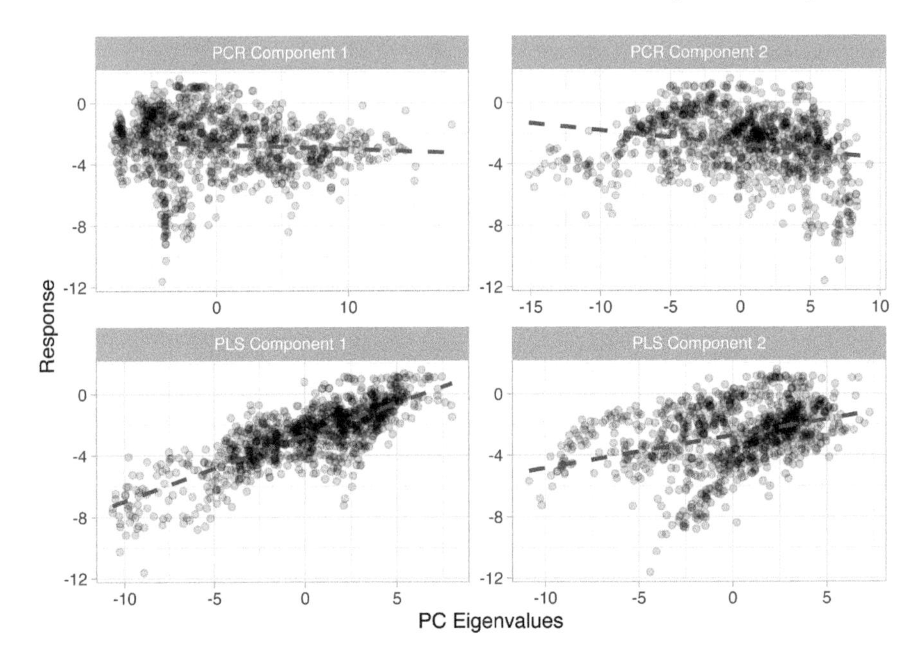

FIGURE 4.9: Illustration showing that the first two PCs when using PCR have very little relationship to the response variable (top row); however, the first two PCs when using PLS have a much stronger association to the response (bottom row).

cipal (z_1) by setting each ϕ_{j1} to the coefficient from a SLR model of y onto that respective x_j. One can show that this coefficient is proportional to the correlation between y and x_j. Hence, in computing $z_1 = \sum_{j=1}^{p} \phi_{j1} x_j$, PLS places the highest weight on the variables that are most strongly related to the response.

To compute the second PC (z_2), we first regress each variable on z_1. The residuals from this regression capture the remaining signal that has not been explained by the first PC. We substitute these residual values for the predictor values in Equation (17.2) in Chapter 17. This process continues until all m components have been computed and then we use OLS to regress the response on $z_1, \dots, z_m$.

See Friedman et al. (2001) and Geladi and Kowalski (1986) for a thorough discussion of PLS.

Similar to PCR, we can easily fit a PLS model by changing the `method` argument in `train()`. As with PCR, the number of principal components to use is a tuning parameter that is determined by the model that maximizes predictive accuracy (minimizes RMSE in this case). The following performs cross-validated PLS with $1, 2, \ldots, 20$ PCs, and Figure 4.10 shows the cross-validated RMSEs. You can see a greater drop in prediction error compared to PCR. Using PLS with $m = 3$ principal components corresponded with the lowest cross-validated RMSE of $29,970.

```
# perform 10-fold cross validation on a PLS model tuning the
# number of principal components to use as predictors from 1-20
set.seed(123)
cv_model_pls <- train(
  Sale_Price ~ .,
  data = ames_train,
  method = "pls",
  trControl = trainControl(method = "cv", number = 10),
  preProcess = c("zv", "center", "scale"),
  tuneLength = 20
)

# model with lowest RMSE
cv_model_pls$bestTune
##   ncomp
## 3     3

# plot cross-validated RMSE
ggplot(cv_model_pls)
```

4.8 Feature interpretation

Once we've found the model that minimizes the predictive accuracy, our next goal is to interpret the model structure. Linear regression models provide a very intuitive model structure as they assume a *monotonic linear relationship* between the predictor variables and the response. The *linear* relationship part of that statement just means, for a given predictor variable, it assumes for every one unit change in a given predictor variable there is a constant change in the response. As discussed earlier in the chapter, this constant rate of change is provided by the coefficient for a predictor. The *monotonic* relationship means that a given predictor variable will always have a positive or negative relationship. But how do we determine the most influential variables?

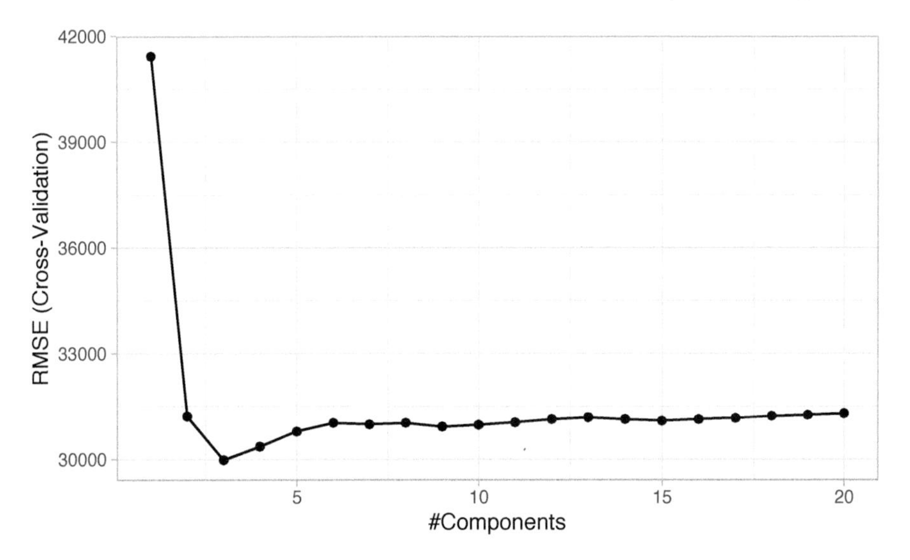

FIGURE 4.10: The 10-fold cross valdation RMSE obtained using PLS with 1-20 principal components.

Variable importance seeks to identify those variables that are most influential in our model. For linear regression models, this is most often measured by the absolute value of the t-statistic for each model parameter used; though simple, the results can be hard to interpret when the model includes interaction effects and complex transformations (in Chapter 16 we'll discuss *model-agnostic* approaches that don't have this issue). For a PLS model, variable importance can be computed using the weighted sums of the absolute regression coefficients. The weights are a function of the reduction of the RSS across the number of PLS components and are computed separately for each outcome. Therefore, the contribution of the coefficients are weighted proportionally to the reduction in the RSS.

We can use `vip::vip()` to extract and plot the most important variables. The importance measure is normalized from 100 (most important) to 0 (least important). Figure 4.11 illustrates that the top 4 most important variables are `Gr_liv_Area`, `First_Flr_SF`, `Total_Bsmt_SF`, and `Garage_Cars` respectively.

```
vip(cv_model_pls, num_features = 20, method = "model")
```

As stated earlier, linear regression models assume a monotonic linear relationship. To illustrate this, we can construct partial dependence plots (PDPs). PDPs plot the change in the average predicted value ($\hat{y}$) as specified feature(s) vary over their marginal distribution. As you will see in later chapters, PDPs

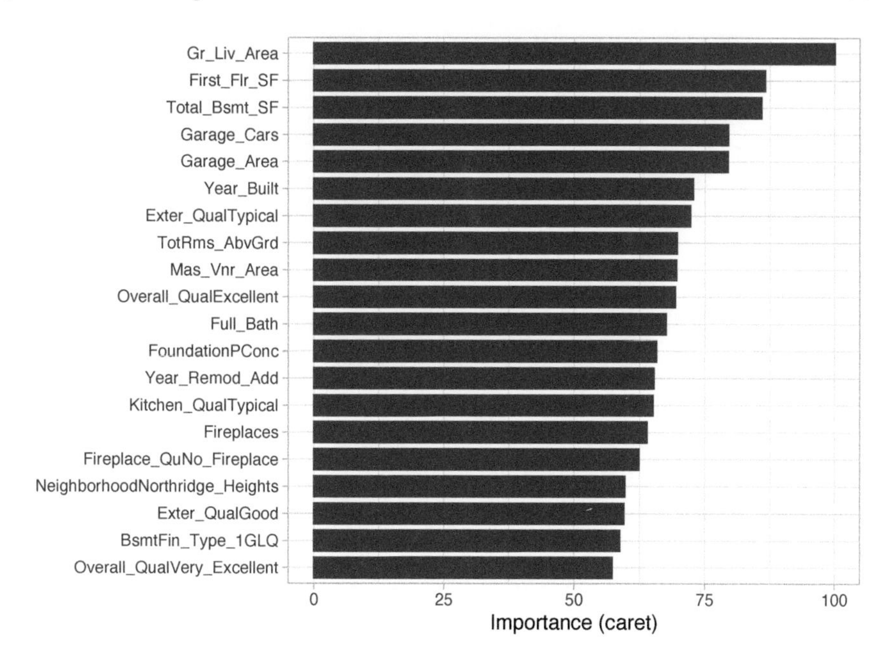

FIGURE 4.11: Top 20 most important variables for the PLS model.

become more useful when non-linear relationships are present (we discuss PDPs and other ML interpretation techniques in Chapter 16). However, PDPs of linear models help illustrate how a fixed change in x_i relates to a fixed linear change in $\widehat{y}_i$ while taking into account the average effect of all the other features in the model (for linear models, the slope of the PDP is equal to the corresponding features LS coefficient).

The **pdp** package (Greenwell, 2018) provides convenient functions for computing and plotting PDPs. For example, the following code chunk would plot the PDP for the `Gr_Liv_Area` predictor.

```
pdp::partial(cv_model_pls, "Gr_Liv_Area", grid.resolution = 20, plot
= TRUE)
```

All four of the most important predictors have a positive relationship with sale price; however, we see that the slope ($\widehat{beta}_i$) is steepest for the most important predictor and gradually decreases for less important variables.

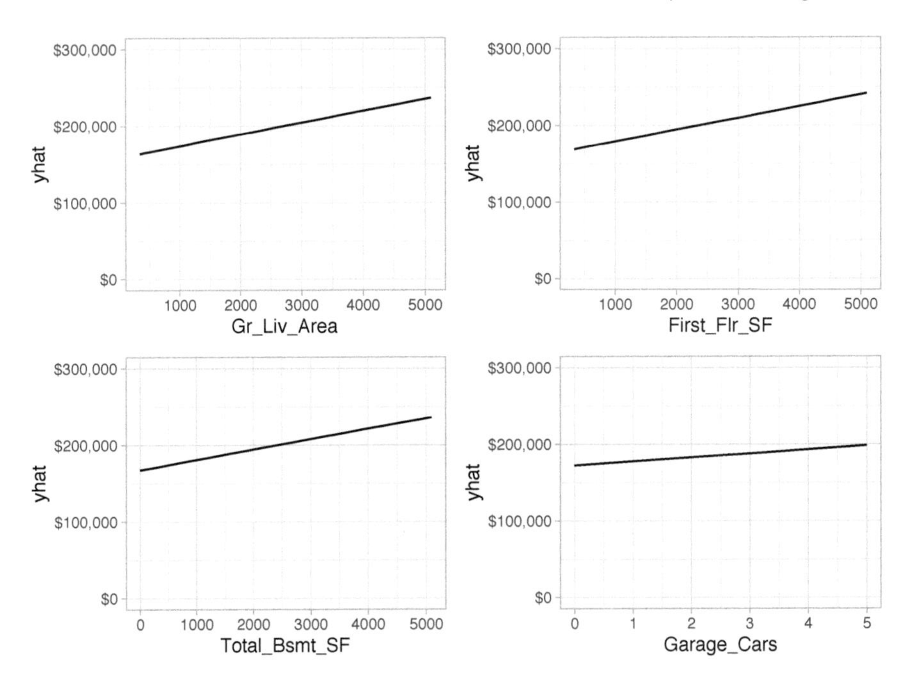

FIGURE 4.12: Partial dependence plots for the first four most important variables.

4.9 Final thoughts

Linear regression is usually the first supervised learning algorithm you will learn. The approach provides a solid fundamental understanding of the supervised learning task; however, as we've discussed there are several concerns that result from the assumptions required. Although extensions of linear regression that integrate dimension reduction steps into the algorithm can help address some of the problems with linear regression, more advanced supervised algorithms typically provide greater flexibility and improved accuracy. Nonetheless, understanding linear regression provides a foundation that will serve you well in learning these more advanced methods.

5

Logistic Regression

Linear regression is used to approximate the (linear) relationship between a continuous response variable and a set of predictor variables. However, when the response variable is binary (i.e., Yes/No), linear regression is not appropriate. Fortunately, analysts can turn to an analogous method, *logistic regression*, which is similar to linear regression in many ways. This chapter explores the use of logistic regression for binary response variables. Logistic regression can be expanded for multinomial problems (see Faraway (2016a) for discussion of multinomial logistic regression in R); however, that goes beyond our intent here.

5.1 Prerequisites

For this section we'll use the following packages:

```
# Helper packages
library(dplyr)     # for data wrangling
library(ggplot2)   # for awesome plotting
library(rsample)   # for data splitting

# Modeling packages
library(caret)     # for logistic regression modeling

# Model interpretability packages
library(vip)       # variable importance
```

To illustrate logistic regression concepts we'll use the employee attrition data, where our intent is to predict the `Attrition` response variable (coded as `"Yes"`/`"No"`). As in the previous chapter, we'll set aside 30% of our data as a test set to assess our generalizability error.

```
df <- attrition %>% mutate_if(is.ordered, factor, ordered = FALSE)

# Create training (70%) and test (30%) sets for the
# rsample::attrition data.
set.seed(123)  # for reproducibility
churn_split <- initial_split(df, prop = .7, strata = "Attrition")
churn_train <- training(churn_split)
churn_test  <- testing(churn_split)
```

5.2 Why logistic regression

To provide a clear motivation for logistic regression, assume we have credit card default data for customers and we want to understand if the current credit card balance of a customer is an indicator of whether or not they'll default on their credit card. To classify a customer as a high- vs. low-risk defaulter based on their balance we could use linear regression; however, the left plot in Figure 5.1 illustrates how linear regression would predict the probability of defaulting. Unfortunately, for balances close to zero we predict a negative probability of defaulting; if we were to predict for very large balances, we would get values bigger than 1. These predictions are not sensible, since of course the true probability of defaulting, regardless of credit card balance, must fall between 0 and 1. These inconsistencies only increase as our data become more imbalanced and the number of outliers increase. Contrast this with the logistic regression line (right plot) that is nonlinear (sigmoidal-shaped).

To avoid the inadequacies of the linear model fit on a binary response, we must model the probability of our response using a function that gives outputs between 0 and 1 for all values of X. Many functions meet this description. In logistic regression, we use the logistic function, which is defined in Equation (5.1) and produces the S-shaped curve in the right plot above.

$$p\left(X\right) = \frac{e^{\beta_0 + \beta_1 X}}{1 + e^{\beta_0 + \beta_1 X}} \tag{5.1}$$

The β_i parameters represent the coefficients as in linear regression and $p\left(X\right)$ may be interpreted as the probability that the positive class (default in the above example) is present. The minimum for $p\left(x\right)$ is obtained at $\lim_{a \rightarrow -\infty} \left[\frac{e^a}{1+e^a} \right] = 0$, and the maximum for $p\left(x\right)$ is obtained at $\lim_{a \rightarrow \infty} \left[\frac{e^a}{1+e^a} \right] = 1$ which restricts the output probabilities to 0–1. Rearranging Equation (5.1) yields the *logit transformation* (which is where logistic regression gets its name):

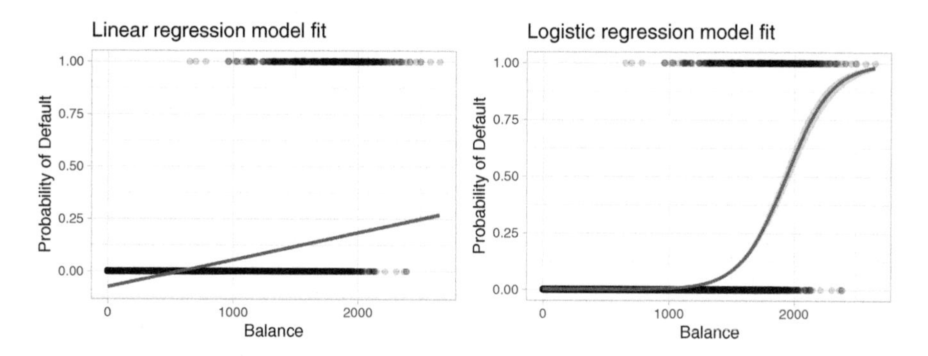

FIGURE 5.1: Comparing the predicted probabilities of linear regression (left) to logistic regression (right). Predicted probabilities using linear regression results in flawed logic whereas predicted values from logistic regression will always lie between 0 and 1.

$$g(X) = \ln\left[\frac{p(X)}{1-p(X)}\right] = \beta_0 + \beta_1 X \tag{5.2}$$

Applying a logit transformation to $p(X)$ results in a linear equation similar to the mean response in a simple linear regression model. Using the logit transformation also results in an intuitive interpretation for the magnitude of β_1: the odds (e.g., of defaulting) increase multiplicatively by $\exp(\beta_1)$ for every one-unit increase in X. A similar interpretation exists if X is categorical; see Agresti (2003), Chapter 5, for details.

5.3 Simple logistic regression

We will fit two logistic regression models in order to predict the probability of an employee attriting. The first predicts the probability of attrition based on their monthly income (`MonthlyIncome`) and the second is based on whether or not the employee works overtime (`OverTime`). The `glm()` function fits generalized linear models, a class of models that includes both logistic regression and simple linear regression as special cases. The syntax of the `glm()` function is similar to that of `lm()`, except that we must pass the argument `family = "binomial"` in order to tell R to run a logistic regression rather than some other type of generalized linear model (the default is `family = "gaussian"`, which is equivalent to ordinary linear regression assuming normally distributed errors).

```
model1 <- glm(Attrition ~ MonthlyIncome, family = "binomial",
              data = churn_train)
model2 <- glm(Attrition ~ OverTime, family = "binomial",
              data = churn_train)
```

In the background `glm()`, uses ML estimation to estimate the unknown model parameters. The basic intuition behind using ML estimation to fit a logistic regression model is as follows: we seek estimates for β_0 and β_1 such that the predicted probability $\widehat{p}\left(X_i\right)$ of attrition for each employee corresponds as closely as possible to the employee's observed attrition status. In other words, we try to find $\widehat{\beta}_0$ and $\widehat{\beta}_1$ such that plugging these estimates into the model for $p\left(X\right)$ (Equation (5.1)) yields a number close to one for all employees who attrited, and a number close to zero for all employees who did not. This intuition can be formalized using a mathematical equation called a *likelihood function*:

$$\ell\left(\beta_0, \beta_1\right) = \prod_{i:y_i=1} p\left(X_i\right) \prod_{i':y_i'=0} \left[1 - p\left(x_i'\right)\right] \tag{5.3}$$

The estimates $\widehat{\beta}_0$ and $\widehat{\beta}_1$ are chosen to *maximize* this likelihood function. What results is the predicted probability of attrition. Figure 5.2 illustrates the predicted probabilities for the two models.

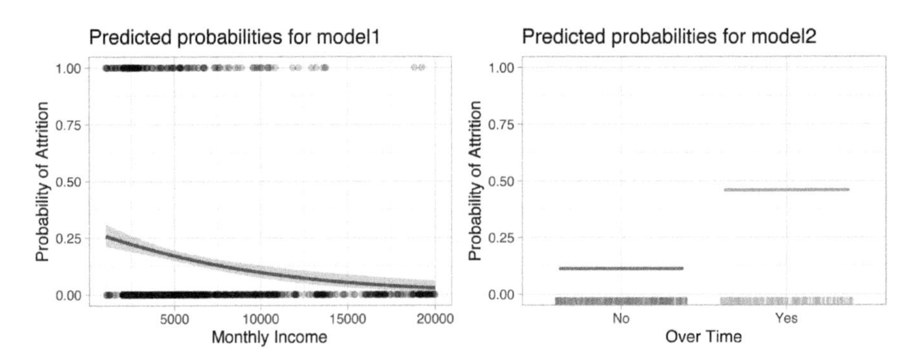

FIGURE 5.2: Predicted probablilities of employee attrition based on monthly income (left) and overtime (right). As monthly income increases, 'model1' predicts a decreased probability of attrition and if employees work overtime 'model2' predicts an increased probability.

The table below shows the coefficient estimates and related information that result from fitting a logistic regression model in order to predict the probability of *Attrition* = *Yes* for our two models. Bear in mind that the coefficient estimates from logistic regression characterize the relationship between the predictor and response variable on a *log-odds* (i.e., logit) scale.

For `model1`, the estimated coefficient for `MonthlyIncome` is $\hat{\beta}_1$ = -0.000130, which is negative, indicating that an increase in `MonthlyIncome` is associated with a decrease in the probability of attrition. Similarly, for `model2`, employees who work `OverTime` are associated with an increased probability of attrition compared to those that do not work `OverTime`.

```
tidy(model1)
## # A tibble: 2 x 5
##   term           estimate std.error statistic     p.value
##   <chr>             <dbl>     <dbl>     <dbl>       <dbl>
## 1 (Intercept)   -0.924    0.155         -5.96     2.59e-9
## 2 MonthlyInco~ -0.000130 0.0000264      -4.93     8.36e-7
tidy(model2)
## # A tibble: 2 x 5
##   term         estimate std.error statistic  p.value
##   <chr>           <dbl>     <dbl>     <dbl>    <dbl>
## 1 (Intercept)     -2.18     0.122     -17.9 6.76e-72
## 2 OverTimeYes      1.41     0.176      8.00 1.20e-15
```

As discussed earlier, it is easier to interpret the coefficients using an exp() transformation:

```
exp(coef(model1))
##   (Intercept) MonthlyIncome
##         0.397         1.000
exp(coef(model2))
## (Intercept) OverTimeYes
##       0.113       4.081
```

Thus, the odds of an employee attriting in `model1` increase multiplicatively by 1 for every one dollar increase in `MonthlyIncome`, whereas the odds of attriting in `model2` increase multiplicatively by 4.081 for employees that work `OverTime` compared to those that do not.

Many aspects of the logistic regression output are similar to those discussed for linear regression. For example, we can use the estimated standard errors to get confidence intervals as we did for linear regression in Chapter 4:

```
confint(model1)  # for odds, you can use 'exp(confint(model1))'
##                   2.5 %    97.5 %
## (Intercept)   -1.226775 -6.18e-01
```

```
## MonthlyIncome -0.000185 -8.11e-05
confint(model2)
##              2.5 % 97.5 %
## (Intercept) -2.43   -1.95
## OverTimeYes  1.06    1.75
```

5.4 Multiple logistic regression

We can also extend our model as seen in Equation 1 so that we can predict a binary response using multiple predictors:

$$p(X) = \frac{e^{\beta_0 + \beta_1 X + \cdots + \beta_p X_p}}{1 + e^{\beta_0 + \beta_1 X + \cdots + \beta_p X_p}} \tag{5.4}$$

Let's go ahead and fit a model that predicts the probability of `Attrition` based on the `MonthlyIncome` and `OverTime`. Our results show that both features are statistically significant (at the 0.05 level) and Figure 5.3 illustrates common trends between `MonthlyIncome` and `Attrition`; however, working `OverTime` tends to nearly double the probability of attrition.

```
model3 <- glm(
  Attrition ~ MonthlyIncome + OverTime,
  family = "binomial",
  data = churn_train
  )

tidy(model3)
## # A tibble: 3 x 5
##   term           estimate std.error statistic  p.value
##   <chr>             <dbl>     <dbl>     <dbl>    <dbl>
## 1 (Intercept)   -1.43     0.176         -8.11 5.25e-16
## 2 MonthlyIncome -0.000139 0.0000270     -5.15 2.62e- 7
## 3 OverTimeYes    1.47     0.180          8.16 3.43e-16
```

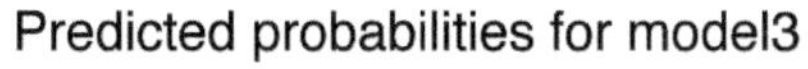

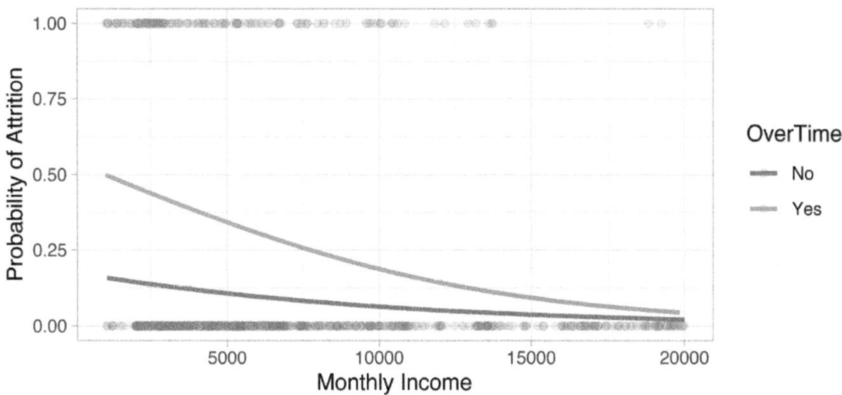

FIGURE 5.3: Predicted probability of attrition based on monthly income and whether or not employees work overtime.

5.5 Assessing model accuracy

With a basic understanding of logistic regression under our belt, similar to linear regression our concern now shifts to how well do our models predict. As in the last chapter, we'll use `caret::train()` and fit three 10-fold cross validated logistic regression models. Extracting the accuracy measures (in this case, classification accuracy), we see that both `cv_model1` and `cv_model2` had an average accuracy of 83.88%. However, `cv_model3` which used all predictor variables in our data achieved an average accuracy rate of 87.58%.

```
set.seed(123)
cv_model1 <- train(
  Attrition ~ MonthlyIncome,
  data = churn_train,
  method = "glm",
  family = "binomial",
  trControl = trainControl(method = "cv", number = 10)
)

set.seed(123)
cv_model2 <- train(
  Attrition ~ MonthlyIncome + OverTime,
  data = churn_train,
```

```
  method = "glm",
  family = "binomial",
  trControl = trainControl(method = "cv", number = 10)
)

set.seed(123)
cv_model3 <- train(
  Attrition ~ .,
  data = churn_train,
  method = "glm",
  family = "binomial",
  trControl = trainControl(method = "cv", number = 10)
)

# extract out of sample performance measures
summary(
  resamples(
    list(
      model1 = cv_model1,
      model2 = cv_model2,
      model3 = cv_model3
    )
  )
)$statistics$Accuracy
##         Min. 1st Qu. Median  Mean 3rd Qu.  Max. NA's
## model1 0.835   0.835  0.837 0.839   0.843 0.845    0
## model2 0.835   0.835  0.837 0.839   0.843 0.845    0
## model3 0.837   0.850  0.879 0.876   0.891 0.931    0
```

We can get a better understanding of our model's performance by assessing the confusion matrix (see Section 2.6). We can use `caret::confusionMatrix()` to compute a confusion matrix. We need to supply our model's predicted class and the actuals from our training data. The confusion matrix provides a wealth of information. Particularly, we can see that although we do well predicting cases of non-attrition (note the high specificity), our model does particularly poor predicting actual cases of attrition (note the low sensitivity).

By default the `predict()` function predicts the response class for a **caret** model; however, you can change the `type` argument to predict the probabilities (see `?caret::predict.train`).

```
# predict class
pred_class <- predict(cv_model3, churn_train)

# create confusion matrix
confusionMatrix(
  data = relevel(pred_class, ref = "Yes"),
  reference = relevel(churn_train$Attrition, ref = "Yes")
)
## Confusion Matrix and Statistics
##
##           Reference
## Prediction Yes  No
##        Yes  93  25
##        No   73 839
##
##                Accuracy : 0.905
##                  95% CI : (0.885, 0.922)
##     No Information Rate : 0.839
##     P-Value [Acc > NIR] : 5.36e-10
##
##                   Kappa : 0.602
##
##  Mcnemar's Test P-Value : 2.06e-06
##
##             Sensitivity : 0.5602
##             Specificity : 0.9711
##          Pos Pred Value : 0.7881
##          Neg Pred Value : 0.9200
##              Prevalence : 0.1612
##          Detection Rate : 0.0903
##    Detection Prevalence : 0.1146
##       Balanced Accuracy : 0.7657
##
##        'Positive' Class : Yes
##
```

One thing to point out, in the confusion matrix above you will note the metric `No Information Rate: 0.839`. This represents the ratio of non-attrition vs. attrition in our training data (`table(churn_train$Attrition) %>% prop.table()`). Consequently, if we simply predicted `"No"` for every employee we would still get an accuracy rate of 83.9%. Therefore, our goal is to maximize our accuracy rate over and above this no information baseline while also trying to balance sensitivity and specificity. To that end, we plot the ROC curve (section 2.6) which is displayed in Figure 5.4. If we compare our simple model (`cv_model1`) to our full model (cv_model3), we see the lift achieved with the more accurate model.

```
library(ROCR)

# Compute predicted probabilities
m1_prob <- predict(cv_model1, churn_train, type = "prob")$Yes
m3_prob <- predict(cv_model3, churn_train, type = "prob")$Yes

# Compute AUC metrics for cv_model1 and cv_model3
perf1 <- prediction(m1_prob, churn_train$Attrition) %>%
  performance(measure = "tpr", x.measure = "fpr")
perf2 <- prediction(m3_prob, churn_train$Attrition) %>%
  performance(measure = "tpr", x.measure = "fpr")

# Plot ROC curves for cv_model1 and cv_model3
plot(perf1, col = "black", lty = 2)
plot(perf2, add = TRUE, col = "blue")
legend(0.8, 0.2, legend = c("cv_model1", "cv_model3"),
       col = c("black", "blue"), lty = 2:1, cex = 0.6)
```

Similar to linear regression, we can perform a PLS logistic regression to assess if reducing the dimension of our numeric predictors helps to improve accuracy. There are 16 numeric features in our data set so the following code performs a 10-fold cross-validated PLS model while tuning the number of principal components to use from 1–16. The optimal model uses 14 principal components, which is not reducing the dimension by much. However, the mean accuracy of 0.876 is no better than the average CV accuracy of cv_model3 (0.876).

```
# Perform 10-fold CV on a PLS model tuning the number of PCs to
# use as predictors
set.seed(123)
cv_model_pls <- train(
  Attrition ~ .,
  data = churn_train,
  method = "pls",
  family = "binomial",
  trControl = trainControl(method = "cv", number = 10),
  preProcess = c("zv", "center", "scale"),
  tuneLength = 16
)
```

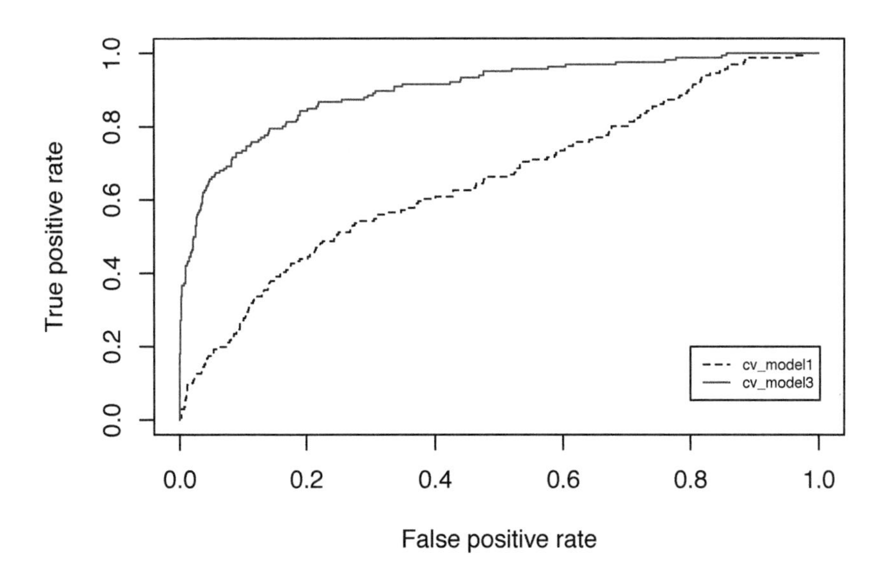

FIGURE 5.4: ROC curve for cross-validated models 1 and 3. The increase in the AUC represents the 'lift' that we achieve with model 3.

```
# Model with lowest RMSE
cv_model_pls$bestTune
##    ncomp
## 14    14

# Plot cross-validated RMSE
ggplot(cv_model_pls)
```

5.6 Model concerns

As with linear models, it is important to check the adequacy of the logistic regression model (in fact, this should be done for all parametric models). This was discussed for linear models in Section 4.5 where the residuals played an important role. Although not as common, residual analysis and diagnostics are

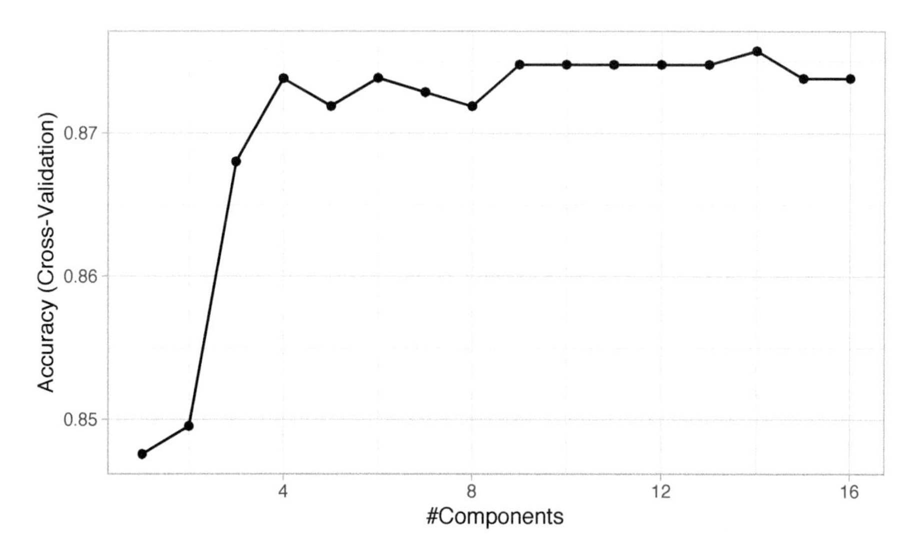

FIGURE 5.5: The 10-fold cross validation RMSE obtained using PLS with 1–16 principal components.

equally important to generalized linear models. The problem is that there is no obvious way to define what a residual is for more general models. For instance, how might we define a residual in logistic regression when the outcome is either 0 or 1? Nonetheless attempts have been made and a number of useful diagnostics can be constructed based on the idea of a *pseudo residual*; see, for example, Harrell (2015), Section 10.4.

More recently, Liu and Zhang (2018) introduced the concept of *surrogate residuals* that allows for residual-based diagnostic procedures and plots not unlike those in traditional linear regression (e.g., checking for outliers and misspecified link functions). For an overview with examples in R using the **sure** package, see Greenwell et al. (2018c).

5.7 Feature interpretation

Similar to linear regression, once our preferred logistic regression model is identified, we need to interpret how the features are influencing the results. As with normal linear regression models, variable importance for logistic regression models can be computed using the absolute value of the z-statistic for each coefficient (albeit with the same issues previously discussed). Using `vip::vip()` we can extract our top 20 influential variables. Figure 5.6 illustrates that `OverTime` is the most influential followed by `JobSatisfaction`, and `EnvironmentSatisfaction`.

```
vip(cv_model3, num_features = 20)
```

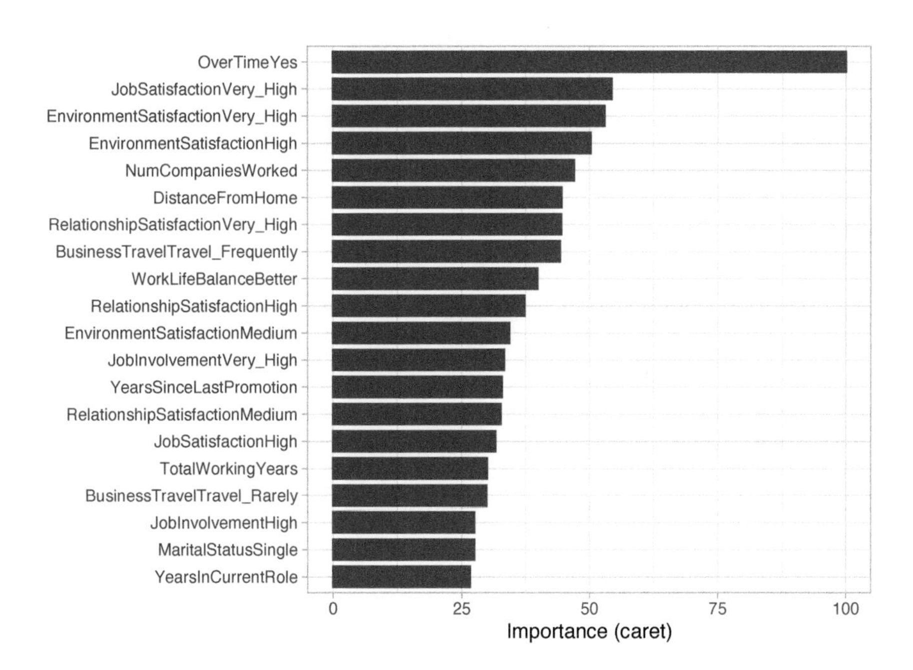

FIGURE 5.6: Top 20 most important variables for the PLS model.

Similar to linear regression, logistic regression assumes a monotonic linear relationship. However, the linear relationship occurs on the logit scale; on the probability scale, the relationship will be nonlinear. This is illustrated by the PDP in Figure 5.7 which illustrates the functional relationship between the predicted probability of attrition and the number of companies an employee has worked for (`NumCompaniesWorked`) while taking into account the average effect of all the other predictors in the model. Employees who've experienced more employment changes tend to have a high probability of making another change in the future.

Furthermore, the PDPs for the top three categorical predictors (`OverTime`, `JobSatisfaction`, and `EnvironmentSatisfaction`) illustrate the change in predicted probability of attrition based on the employee's status for each predictor.

See the online supplemental material for the code to reproduce the plots in Figure 5.7.

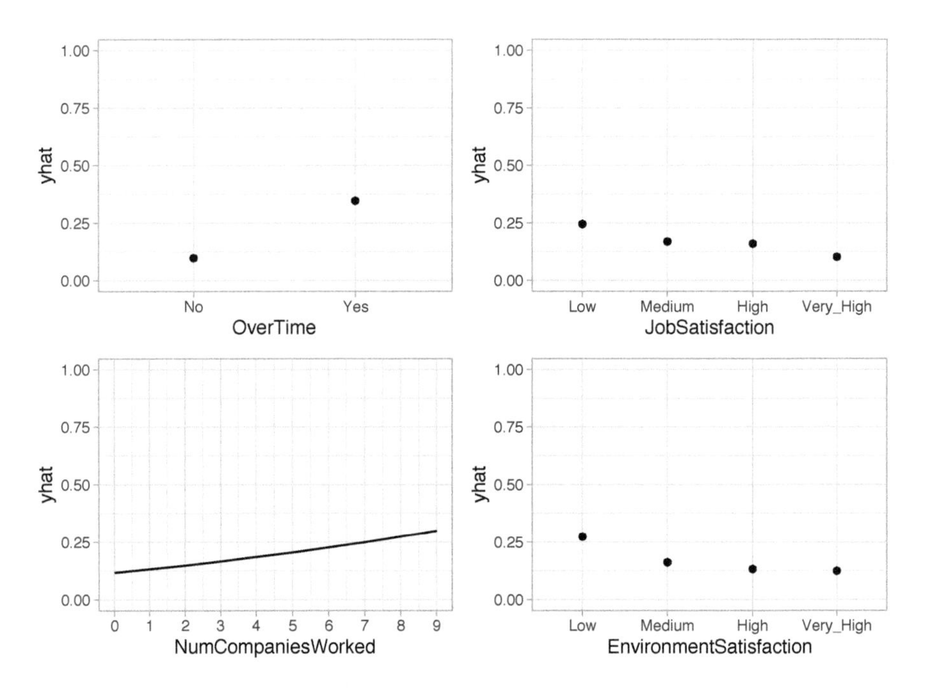

FIGURE 5.7: Partial dependence plots for the first four most important variables. We can see how the predicted probability of attrition changes for each value of the influential predictors.

5.8 Final thoughts

Logistic regression provides an alternative to linear regression for binary classification problems. However, similar to linear regression, logistic regression suffers from the many assumptions involved in the algorithm (i.e. linear relationship of the coefficient, multicollinearity). Moreover, often we have more than two classes to predict which is commonly referred to as multinomial classification. Although multinomial extensions of logistic regression exist, the assumptions made only increase and, often, the stability of the coefficient estimates (and therefore the accuracy) decrease. Future chapters will discuss more advanced algorithms that provide a more natural and trustworthy approach to binary and multinomial classification prediction.

6

Regularized Regression

Linear models (LMs) provide a simple, yet effective, approach to predictive modeling. Moreover, when certain assumptions required by LMs are met (e.g., constant variance), the estimated coefficients are unbiased and, of all linear unbiased estimates, have the lowest variance. However, in today's world, data sets being analyzed typically contain a large number of features. As the number of features grow, certain assumptions typically break down and these models tend to overfit the training data, causing our out of sample error to increase. **Regularization** methods provide a means to constrain or *regularize* the estimated coefficients, which can reduce the variance and decrease out of sample error.

6.1 Prerequisites

This chapter leverages the following packages. Most of these packages are playing a supporting role while the main emphasis will be on the **glmnet** package (Friedman et al., 2018).

```
# Helper packages
library(recipes)  # for feature engineering

# Modeling packages
library(glmnet)   # for implementing regularized regression
library(caret)    # for automating the tuning process

# Model interpretability packages
library(vip)      # for variable importance
```

To illustrate various regularization concepts we'll continue working with the `ames_train` and `ames_test` data sets created in Section 2.7; however, at the end of the chapter we'll also apply regularized regression to the employee attrition data.

6.2 Why regularize?

The easiest way to understand regularized regression is to explain how and why it is applied to ordinary least squares (OLS). The objective in OLS regression is to find the *hyperplane*[1] (e.g., a straight line in two dimensions) that minimizes the sum of squared errors (SSE) between the observed and predicted response values (see Figure 6.1 below). This means identifying the hyperplane that minimizes the grey lines, which measure the vertical distance between the observed (red dots) and predicted (blue line) response values.

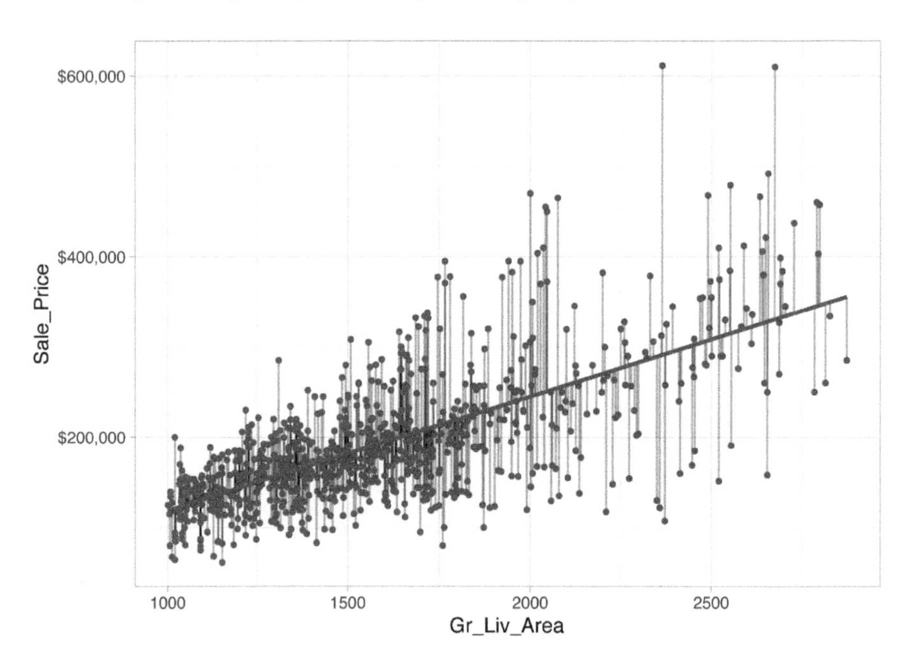

FIGURE 6.1: Fitted regression line using Ordinary Least Squares.

More formally, the objective function being minimized can be written as:

$$\text{minimize}\left(SSE = \sum_{i=1}^{n}(y_i - \hat{y}_i)^2\right) \tag{6.1}$$

As we discussed in Chapter 4, the OLS objective function performs quite well when our data adhere to a few key assumptions:

- Linear relationship;

[1]See Section 14.2 for more discussion on hyperplanes.

- There are more observations (n) than features (p) ($n > p$);
- No or little multicollinearity.

For classicial statistical inference procedures (e.g., confidence intervals based on the classic t-statistic) to be valid, we also need to make stronger assumptions regarding normality (of the errors) and homoscedasticity (i.e., constant error variance).

Many real-life data sets, like those common to *text mining* and *genomic studies* are *wide*, meaning they contain a larger number of features ($p > n$). As p increases, we're more likely to violate some of the OLS assumptions and alternative approaches should be considered. This was briefly illustrated in Chapter 4 where the presence of multicollinearity was diminishing the interpretability of our estimated coefficients due to inflated variance. By reducing multicollinearity, we were able to increase our model's accuracy. Of course, multicollinearity can also occur when $n > p$.

Having a large number of features invites additional issues in using classic regression models. For one, having a large number of features makes the model much less interpretable. Additionally, when $p > n$, there are many (in fact infinite) solutions to the OLS problem! In such cases, it is useful (and practical) to assume that a smaller subset of the features exhibit the strongest effects (something called the *bet on sparsity principal* (see Hastie et al., 2015, p. 2).). For this reason, we sometimes prefer estimation techniques that incorporate *feature selection*. One approach to this is called *hard thresholding* feature selection, which includes many of the traditional linear model selection approaches like *forward selection* and *backward elimination*. These procedures, however, can be computationally inefficient, do not scale well, and treat a feature as either in or out of the model (hence the name hard thresholding). In contrast, a more modern approach, called *soft thresholding*, slowly pushes the effects of irrelevant features toward zero, and in some cases, will zero out entire coefficients. As will be demonstrated, this can result in more accurate models that are also easier to interpret.

With wide data (or data that exhibits multicollinearity), one alternative to OLS regression is to use regularized regression (also commonly referred to as *penalized* models or *shrinkage* methods as in Friedman et al. (2001) and Kuhn and Johnson (2013)) to constrain the total size of all the coefficient estimates. This constraint helps to reduce the magnitude and fluctuations of the coefficients and will reduce the variance of our model (at the expense of no longer being unbiased—a reasonable compromise).

The objective function of a regularized regression model is similar to OLS, albeit with a penalty term P.

$$\text{minimize}\,(SSE + P) \tag{6.2}$$

This penalty parameter constrains the size of the coefficients such that the only way the coefficients can increase is if we experience a comparable decrease in the sum of squared errors (SSE).

This concept generalizes to all GLM models (e.g., logistic and Poisson regression) and even some *survival models*. So far, we have been discussing OLS and the sum of squared errors loss function. However, different models within the GLM family have different loss functions (see Chapter 4 of Friedman et al. (2001)). Yet we can think of the penalty parameter all the same—it constrains the size of the coefficients such that the only way the coefficients can increase is if we experience a comparable decrease in the model's loss function.

There are three common penalty parameters we can implement:

1. Ridge;
2. Lasso (or LASSO);
3. Elastic net (or ENET), which is a combination of ridge and lasso.

6.2.1 Ridge penalty

Ridge regression (Hoerl and Kennard, 1970) controls the estimated coefficients by adding $\lambda \sum_{j=1}^{p} \beta_j^2$ to the objective function.

$$\text{minimize}\left(SSE + \lambda \sum_{j=1}^{p} \beta_j^2\right) \tag{6.3}$$

The size of this penalty, referred to as L^2 (or Euclidean) norm, can take on a wide range of values, which is controlled by the *tuning parameter* λ. When $\lambda = 0$ there is no effect and our objective function equals the normal OLS regression objective function of simply minimizing SSE. However, as $\lambda \to \infty$, the penalty becomes large and forces the coefficients toward zero (but not all the way). This is illustrated in Figure 6.2 where exemplar coefficients have been regularized with λ ranging from 0 to over 8,000.

Although these coefficients were scaled and centered prior to the analysis, you will notice that some are quite large when λ is near zero. Furthermore, you'll notice that feature `x1` has a large negative parameter that fluctuates until $\lambda \approx 7$ where it then continuously shrinks toward zero. This is indicative of multicollinearity and likely illustrates that constraining our coefficients with $\lambda > 7$ may reduce the variance, and therefore the error, in our predictions.

In essence, the ridge regression model pushes many of the correlated features toward each other rather than allowing for one to be wildly positive and the

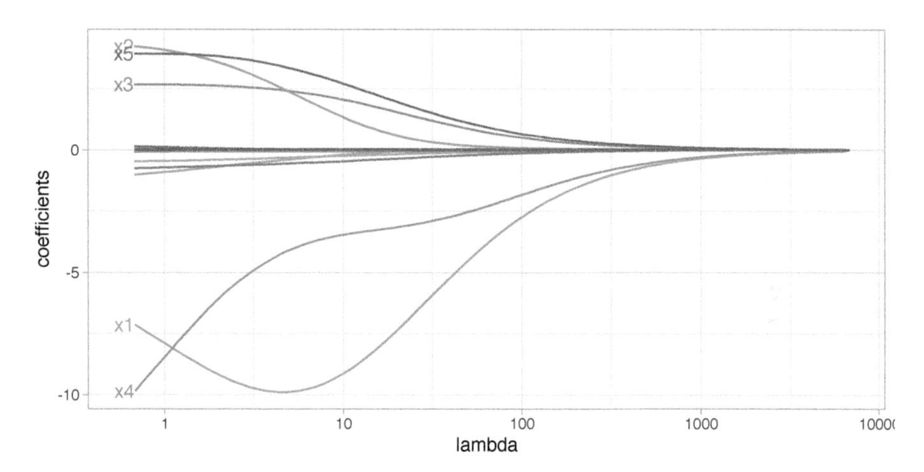

FIGURE 6.2: Ridge regression coefficients for 15 exemplar predictor variables as λ grows from $0 \rightarrow \infty$. As λ grows larger, our coefficient magnitudes are more constrained.

other wildly negative. In addition, many of the less-important features also get pushed toward zero. This helps to provide clarity in identifying the important signals in our data (i.e., the labeled features in Figure 6.2).

However, ridge regression does not perform feature selection and will retain **all** available features in the final model. Therefore, a ridge model is good if you believe there is a need to retain all features in your model yet reduce the noise that less influential variables may create (e.g., in smaller data sets with severe multicollinearity). If greater interpretation is necessary and many of the features are redundant or irrelevant then a lasso or elastic net penalty may be preferable.

6.2.2 Lasso penalty

The lasso (*least absolute shrinkage and selection operator*) penalty (Tibshirani, 1996) is an alternative to the ridge penalty that requires only a small modification. The only difference is that we swap out the L^2 norm for an L^1 norm: $\lambda \sum_{j=1}^{p} |\beta_j|$:

$$\text{minimize} \left(SSE + \lambda \sum_{j=1}^{p} |\beta_j| \right) \tag{6.4}$$

Whereas the ridge penalty pushes variables to *approximately but not equal to zero*, the lasso penalty will actually push coefficients all the way to zero as illustrated in Figure 6.3. Switching to the lasso penalty not only improves the model but it also conducts automated feature selection.

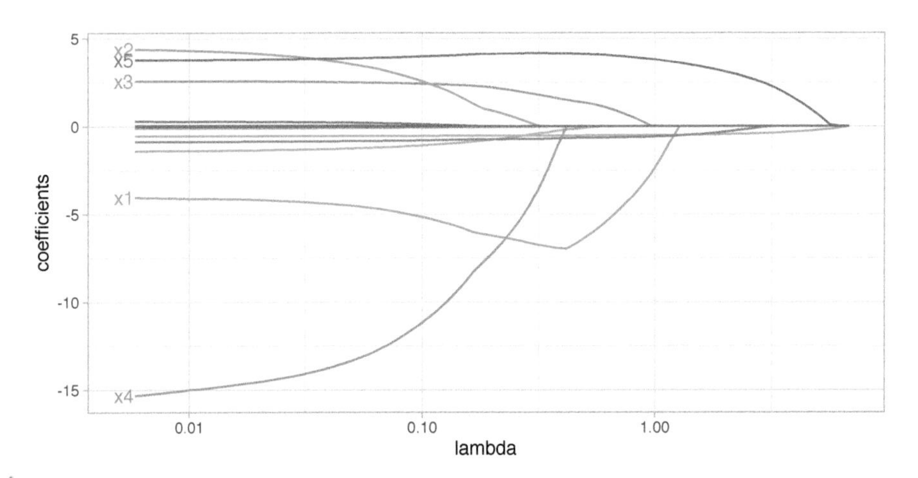

FIGURE 6.3: Lasso regression coefficients as λ grows from $0 \to \infty$.

In the figure above we see that when $\lambda < 0.01$ all 15 variables are included in the model, when $\lambda \approx 0.5$ 9 variables are retained, and when $log(\lambda) = 1$ only 5 variables are retained. Consequently, when a data set has many features, lasso can be used to identify and extract those features with the largest (and most consistent) signal.

6.2.3 Elastic nets

A generalization of the ridge and lasso penalties, called the *elastic net* (Zou and Hastie, 2005), combines the two penalties:

$$\text{minimize}\left(SSE + \lambda_1 \sum_{j=1}^{p} \beta_j^2 + \lambda_2 \sum_{j=1}^{p} |\beta_j|\right) \tag{6.5}$$

Although lasso models perform feature selection, when two strongly correlated features are pushed towards zero, one may be pushed fully to zero while the other remains in the model. Furthermore, the process of one being in and one being out is not very systematic. In contrast, the ridge regression penalty is a little more effective in systematically handling correlated features together. Consequently, the advantage of the elastic net penalty is that it enables effective regularization via the ridge penalty with the feature selection characteristics of the lasso penalty.

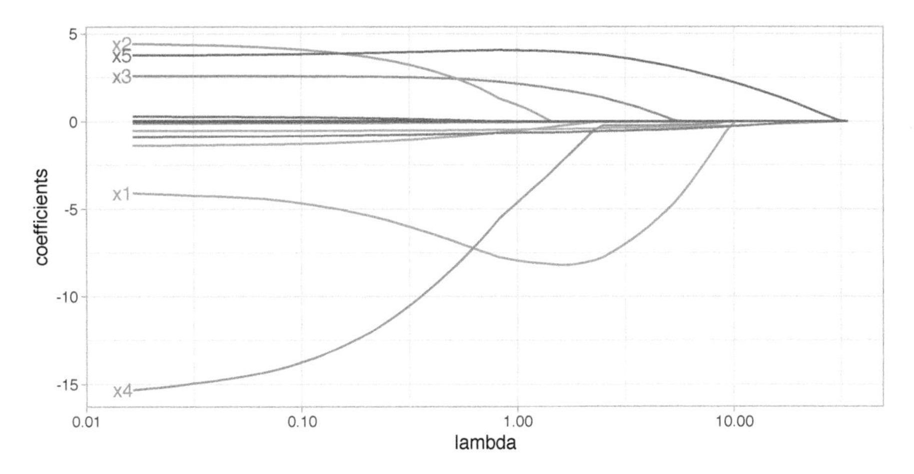

FIGURE 6.4: Elastic net coefficients as λ grows from $0 \to \infty$.

6.3 Implementation

First, we illustrate an implementation of regularized regression using the direct engine **glmnet**. This will provide you with a strong sense of what is happening with a regularized model. Realize there are other implementations available (e.g., **h2o**, **elasticnet**, **penalized**). Then, in Section 6.4, we'll demonstrate how to apply a regularized model so we can properly compare it with our previous predictive models.

The **glmnet** package is extremely efficient and fast, even on very large data sets (mostly due to its use of Fortran to solve the lasso problem via *coordinate descent*); note, however, that it only accepts the non-formula XY interface (2.3.1) so prior to modeling we need to separate our feature and target sets.

The following uses `model.matrix` to dummy encode our feature set (see `Matrix::sparse.model.matrix` for increased efficiency on larger data sets). We also log transform the response variable which is not required; however, parametric models such as regularized regression are sensitive to skewed response values so transforming can often improve predictive performance.

```
# Create training  feature matrices
# we use model.matrix(...)[, -1] to discard the intercept
```

```
X <- model.matrix(Sale_Price ~ ., ames_train)[, -1]

# transform y with log transformation
Y <- log(ames_train$Sale_Price)
```

To apply a regularized model we can use the `glmnet::glmnet()` function. The `alpha` parameter tells **glmnet** to perform a ridge (`alpha = 0`), lasso (`alpha = 1`), or elastic net (`0 < alpha < 1`) model. By default, **glmnet** will do two things that you should be aware of:

1. Since regularized methods apply a penalty to the coefficients, we need to ensure our coefficients are on a common scale. If not, then predictors with naturally larger values (e.g., total square footage) will be penalized more than predictors with naturally smaller values (e.g., total number of rooms). By default, **glmnet** automatically standardizes your features. If you standardize your predictors prior to **glmnet** you can turn this argument off with `standardize = FALSE`.
2. **glmnet** will fit ridge models across a wide range of λ values, which is illustrated in Figure 6.5.

```
# Apply ridge regression to attrition data
ridge <- glmnet(
  x = X,
  y = Y,
  alpha = 0
)

plot(ridge, xvar = "lambda")
```

We can see the exact λ values applied with `ridge$lambda`. Although you can specify your own λ values, by default **glmnet** applies 100 λ values that are data derived.

glmnet can auto-generate the appropriate λ values based on the data; the vast majority of the time you will have little need to adjust this default.

We can also access the coefficients for a particular model using `coef()`. **glmnet** stores all the coefficients for each model in order of largest to smallest λ. Here we just peak at the two largest coefficients (which correspond to `Latitude` & `Overall_QualVery_Excellent`) for the largest (289.0010) and smallest

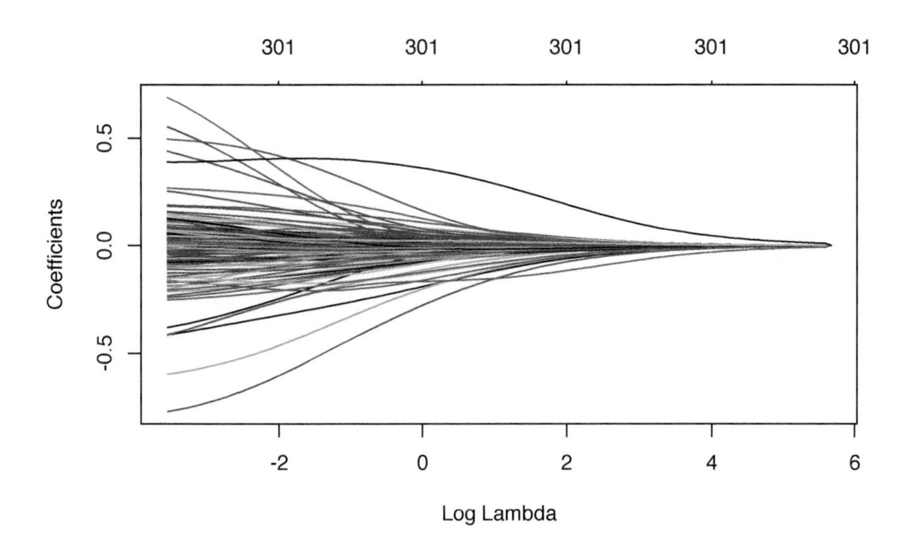

FIGURE 6.5: Coefficients for our ridge regression model as λ grows from $0 \rightarrow \infty$.

(0.02791035) λ values. You can see how the largest λ value has pushed most of these coefficients to nearly 0.

```
# lambdas applied to penalty parameter
ridge$lambda %>% head()
## [1] 289 263 240 219 199 182

# small lambda results in large coefficients
coef(ridge)[c("Latitude", "Overall_QualVery_Excellent"), 100]
##                   Latitude Overall_QualVery_Excellent
##                      0.389                      0.127

# large lambda results in small coefficients
coef(ridge)[c("Latitude", "Overall_QualVery_Excellent"), 1]
##                   Latitude Overall_QualVery_Excellent
##                   6.78e-36                   9.72e-37
```

At this point, we do not understand how much improvement we are experiencing in our loss function across various λ values.

6.4 Tuning

Recall that λ is a tuning parameter that helps to control our model from over-fitting to the training data. To identify the optimal λ value we can use k-fold cross-validation (CV). `glmnet::cv.glmnet()` can perform k-fold CV, and by default, performs 10-fold CV. Below we perform a CV **glmnet** model with both a ridge and lasso penalty separately:

By default, `glmnet::cv.glmnet()` uses MSE as the loss function but you can also use mean absolute error (MAE) for continuous outcomes by changing the `type.measure` argument; see `?glmnet::cv.glmnet()` for more details.

```
# Apply CV ridge regression to Ames data
ridge <- cv.glmnet(
  x = X,
  y = Y,
  alpha = 0
)

# Apply CV lasso regression to Ames data
lasso <- cv.glmnet(
  x = X,
  y = Y,
  alpha = 1
)

# plot results
par(mfrow = c(1, 2))
plot(ridge, main = "Ridge penalty\n\n")
plot(lasso, main = "Lasso penalty\n\n")
```

Figure 6.6 illustrates the 10-fold CV MSE across all the λ values. In both models we see a slight improvement in the MSE as our penalty $log(\lambda)$ gets larger, suggesting that a regular OLS model likely overfits the training data. But as we constrain it further (i.e., continue to increase the penalty), our MSE starts to decrease. The numbers across the top of the plot refer to the number of features in the model. Ridge regression does not force any variables to exactly zero so all features will remain in the model but we see the number of variables retained in the lasso model decrease as the penalty increases.

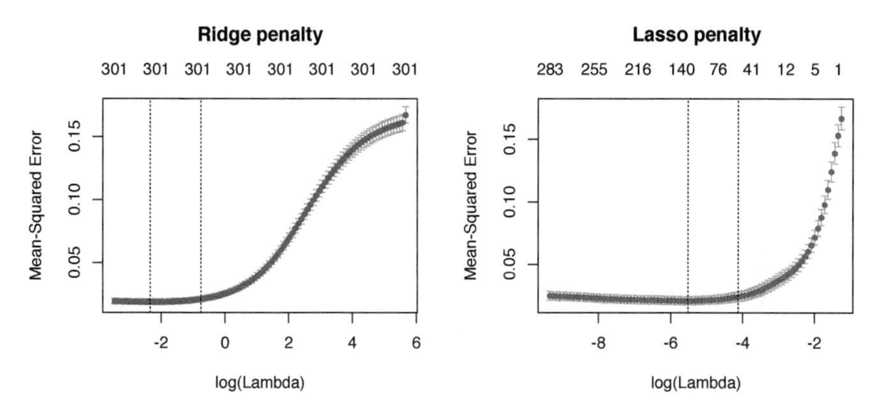

FIGURE 6.6: 10-fold CV MSE for a ridge and lasso model. First dotted vertical line in each plot represents the λ with the smallest MSE and the second represents the λ with an MSE within one standard error of the minimum MSE.

The first and second vertical dashed lines represent the λ value with the minimum MSE and the largest λ value within one standard error of it. The minimum MSE for our ridge model is 0.01899 (produced when $\lambda = 0.09686$ whereas the minimum MSE for our lasso model is 0.02090 (produced when $\lambda = 0.00400$).

```
# Ridge model
min(ridge$cvm)       # minimum MSE
## [1] 0.019
ridge$lambda.min     # lambda for this min MSE
## [1] 0.0969

ridge$cvm[ridge$lambda == ridge$lambda.1se]  # 1-SE rule
## [1] 0.0212
ridge$lambda.1se  # lambda for this MSE
## [1] 0.471

# Lasso model
min(lasso$cvm)       # minimum MSE
## [1] 0.0209
lasso$lambda.min     # lambda for this min MSE
## [1] 0.004

lasso$cvm[lasso$lambda == lasso$lambda.1se]  # 1-SE rule
## [1] 0.0241
```

```
lasso$lambda.1se  # lambda for this MSE
## [1] 0.0162
```

We can assess this visually. Figure 6.7 plots the estimated coefficients across the range of λ values. The dashed red line represents the λ value with the smallest MSE and the dashed blue line represents largest λ value that falls within one standard error of the minimum MSE. This shows you how much we can constrain the coefficients while still maximizing predictive accuracy.

Above, we saw that both ridge and lasso penalties provide similar MSEs; however, these plots illustrate that ridge is still using all 299 features whereas the lasso model can get a similar MSE while reducing the feature set from 299 down to 131. However, there will be some variability with this MSE and we can reasonably assume that we can achieve a similar MSE with a slightly more constrained model that uses only 63 features. Although this lasso model does not offer significant improvement over the ridge model, we get approximately the same accuracy by using only 63 features! If describing and interpreting the predictors is an important component of your analysis, this may significantly aid your endeavor.

```
# Ridge model
ridge_min <- glmnet(
  x = X,
  y = Y,
  alpha = 0
)

# Lasso model
lasso_min <- glmnet(
  x = X,
  y = Y,
  alpha = 1
)

par(mfrow = c(1, 2))
# plot ridge model
plot(ridge_min, xvar = "lambda", main = "Ridge penalty\n\n")
abline(v = log(ridge$lambda.min), col = "red", lty = "dashed")
abline(v = log(ridge$lambda.1se), col = "blue", lty = "dashed")
```

```
# plot lasso model
plot(lasso_min, xvar = "lambda", main = "Lasso penalty\n\n")
abline(v = log(lasso$lambda.min), col = "red", lty = "dashed")
abline(v = log(lasso$lambda.1se), col = "blue", lty = "dashed")
```

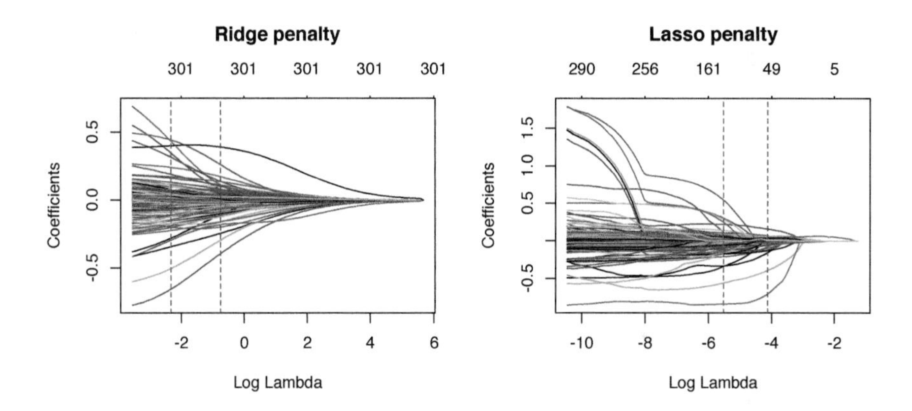

FIGURE 6.7: Coefficients for our ridge and lasso models. First dotted vertical line in each plot represents the λ with the smallest MSE and the second represents the λ with an MSE within one standard error of the minimum MSE.

So far we've implemented a pure ridge and pure lasso model. However, we can implement an elastic net the same way as the ridge and lasso models, by adjusting the `alpha` parameter. Any `alpha` value between 0–1 will perform an elastic net. When `alpha = 0.5` we perform an equal combination of penalties whereas `alpha` < 0.5 will have a heavier ridge penalty applied and `alpha` > 0.5 will have a heavier lasso penalty.

Often, the optimal model contains an `alpha` somewhere between 0–1, thus we want to tune both the λ and the `alpha` parameters. As in Chapters 4 and 5, we can use the **caret** package to automate the tuning process. This ensures that any feature engineering is appropriately applied within each resample. The following performs a grid search over 10 values of the alpha parameter between 0–1 and ten values of the lambda parameter from the lowest to highest lambda values identified by **glmnet**.

This grid search took roughly **71 seconds** to compute.

The following snippet of code shows that the model that minimized RMSE

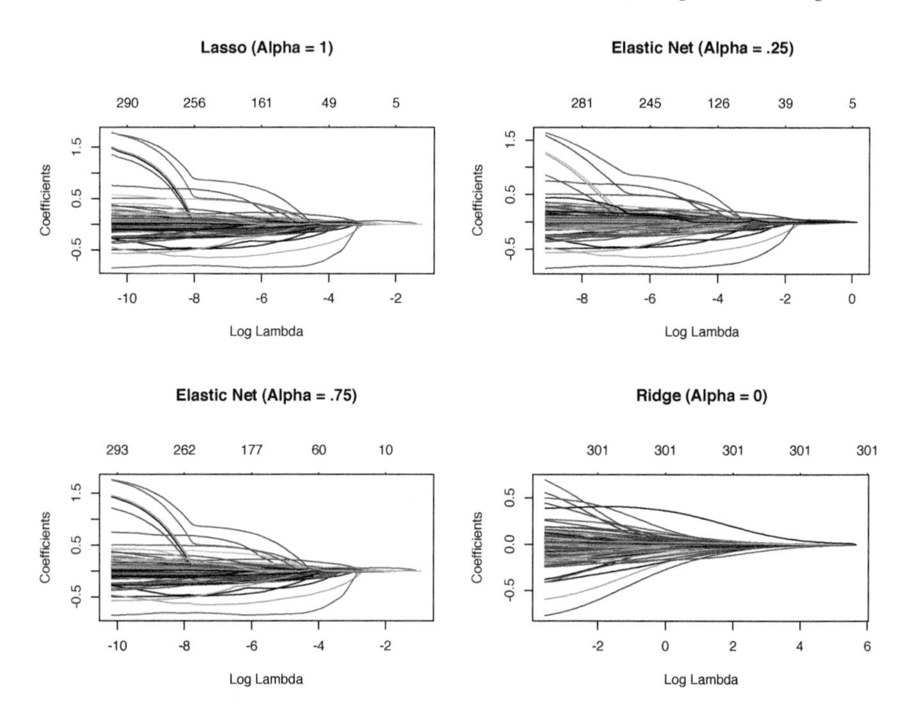

FIGURE 6.8: Coefficients for various penalty parameters.

used an alpha of 0.1 and λ of 0.04688. The minimum RMSE of 0.1419461 ($MSE = 0.1419461^2 = 0.0201487$) is in line with the full ridge model produced earlier. Figure 6.9 illustrates how the combination of alpha values (x-axis) and λ values (line color) influence the RMSE.

```
# for reproducibility
set.seed(123)

# grid search across
cv_glmnet <- train(
  x = X,
  y = Y,
  method = "glmnet",
  preProc = c("zv", "center", "scale"),
  trControl = trainControl(method = "cv", number = 10),
  tuneLength = 10
)
```

```r
# model with lowest RMSE
cv_glmnet$bestTune
##    alpha lambda
## 8    0.1 0.0469

# plot cross-validated RMSE
ggplot(cv_glmnet)
```

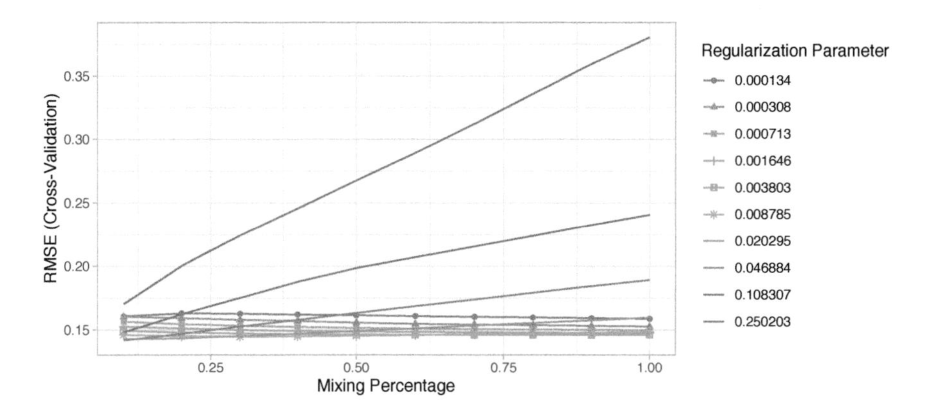

FIGURE 6.9: The 10-fold cross valdation RMSE across 10 alpha values (x-axis) and 10 lambda values (line color).

So how does this compare to our previous best model for the Ames data set? Keep in mind that for this chapter we log transformed the response variable (`Sale_Price`). Consequently, to provide a fair comparison to our previously obtained PLS model's RMSE of $29,970, we need to re-transform our predicted values. The following illustrates that our optimal regularized model achieved an RMSE of $23,503. Introducing a penalty parameter to constrain the coefficients provided quite an improvement over our previously obtained dimension reduction approach.

```r
# predict sales price on training data
pred <- predict(cv_glmnet, X)

# compute RMSE of transformed predicted
RMSE(exp(pred), exp(Y))
## [1] 23503
```

6.5 Feature interpretation

Variable importance for regularized models provides a similar interpretation as in linear (or logistic) regression. Importance is determined by magnitude of the standardized coefficients and we can see in Figure 6.10 some of the same features that were considered highly influential in our PLS model, albeit in differing order (i.e. `Gr_Liv_Area`, `Overall_Qual`, `Total_Bsmt_SF`, `First_Flr_SF`).

```
vip(cv_glmnet, num_features = 20, bar = FALSE)
```

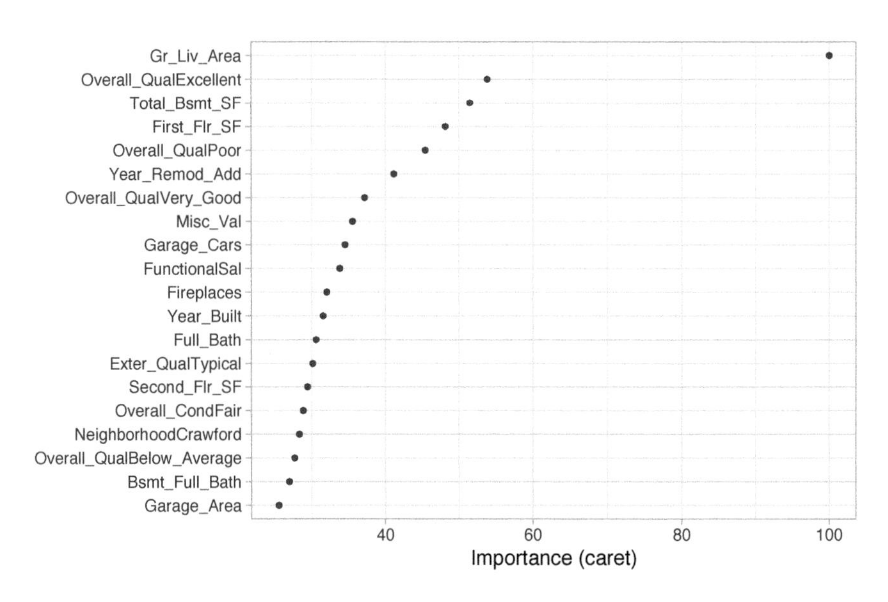

FIGURE 6.10: Top 20 most important variables for the optimal regularized regression model.

Similar to linear and logistic regression, the relationship between the features and response is monotonic linear. However, since we modeled our response with a log transformation, the estimated relationships will still be monotonic but non-linear on the original response scale. Figure 6.11 illustrates the relationship between the top four most influential variables (i.e., largest absolute coefficients) and the non-transformed sales price. All relationships are positive in nature, as the values in these features increase (or for `Overall_QualExcellent` if it exists) the average predicted sales price increases.

However, we see the 5th most influential variable is `Overall_QualPoor`. When

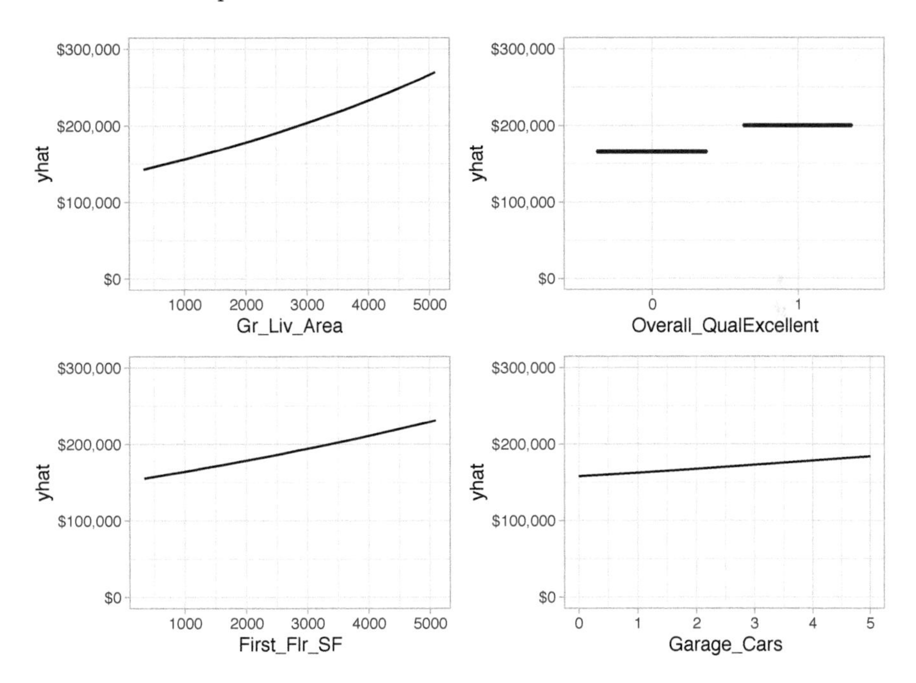

FIGURE 6.11: Partial dependence plots for the first four most important variables.

a home has an overall quality rating of poor we see that the average predicted sales price decreases versus when it has some other overall quality rating. Consequently, its important to not only look at the variable importance ranking, but also observe the positive or negative nature of the relationship.

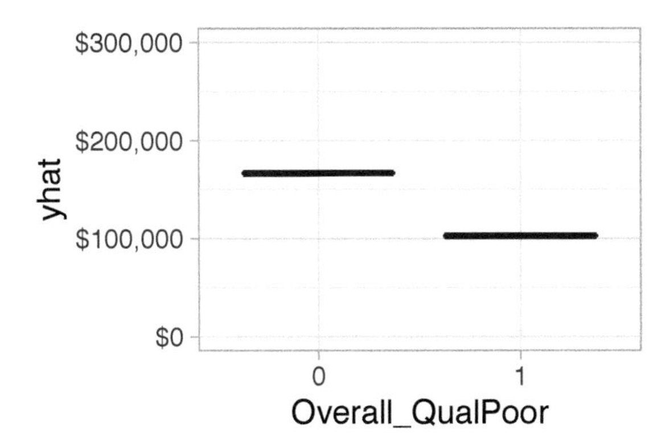

FIGURE 6.12: Partial dependence plot for when overall quality of a home is (1) versus is not poor (0).

6.6 Attrition data

We saw that regularization significantly improved our predictive accuracy for the Ames data set, but how about for the employee attrition example? In Chapter 5 we saw a maximum CV accuracy of 86.3% for our logistic regression model. We see a little improvement in the following with some preprocessing; however, performing a regularized logistic regression model provides us with an additional 0.8% improvement in accuracy (likely within the margin of error).

```
df <- attrition %>% mutate_if(is.ordered, factor, ordered = FALSE)

# Create training (70%) and test (30%) sets for the
# rsample::attrition data. Use set.seed for reproducibility
set.seed(123)
churn_split <- initial_split(df, prop = .7, strata = "Attrition")
train <- training(churn_split)
test  <- testing(churn_split)

# train logistic regression model
set.seed(123)
glm_mod <- train(
  Attrition ~ .,
  data = train,
  method = "glm",
  family = "binomial",
  preProc = c("zv", "center", "scale"),
  trControl = trainControl(method = "cv", number = 10)
  )

# train regularized logistic regression model
set.seed(123)
penalized_mod <- train(
  Attrition ~ .,
  data = train,
  method = "glmnet",
  family = "binomial",
  preProc = c("zv", "center", "scale"),
  trControl = trainControl(method = "cv", number = 10),
  tuneLength = 10
  )
```

```
# extract out of sample performance measures
summary(resamples(list(
  logistic_model = glm_mod,
  penalized_model = penalized_mod
  )))$statistics$Accuracy
##                  Min. 1st Qu. Median  Mean 3rd Qu.
## logistic_model  0.837   0.850  0.879 0.876   0.891
## penalized_model 0.845   0.876  0.883 0.884   0.892
##                  Max. NA's
## logistic_model  0.931    0
## penalized_model 0.941    0
```

6.7 Final thoughts

Regularized regression provides many great benefits over traditional GLMs when applied to large data sets with lots of features. It provides a great option for handling the $n > p$ problem, helps minimize the impact of multicollinearity, and can perform automated feature selection. It also has relatively few hyperparameters which makes them easy to tune, computationally efficient compared to other algorithms discussed in later chapters, and memory efficient.

However, regularized regression does require some feature preprocessing. Notably, all inputs must be numeric; however, some packages (e.g., **caret** and **h2o**) automate this process. They cannot automatically handle missing data, which requires you to remove or impute them prior to modeling. Similar to GLMs, they are also not robust to outliers in both the feature and target. Lastly, regularized regression models still assume a monotonic linear relationship (always increasing or decreasing in a linear fashion). It is also up to the analyst whether or not to include specific interaction effects.

7

Multivariate Adaptive Regression Splines

The previous chapters discussed algorithms that are intrinsically linear. Many of these models can be adapted to nonlinear patterns in the data by manually adding nonlinear model terms (e.g., squared terms, interaction effects, and other transformations of the original features); however, to do so you the analyst must know the specific nature of the nonlinearities and interactions *a priori*. Alternatively, there are numerous algorithms that are inherently nonlinear. When using these models, the exact form of the nonlinearity does not need to be known explicitly or specified prior to model training. Rather, these algorithms will search for, and discover, nonlinearities and interactions in the data that help maximize predictive accuracy.

This chapter discusses *multivariate adaptive regression splines* (MARS) (Friedman, 1991), an algorithm that automatically creates a piecewise linear model which provides an intuitive stepping block into nonlinearity after grasping the concept of multiple linear regression. Future chapters will focus on other nonlinear algorithms.

7.1 Prerequisites

For this chapter we will use the following packages:

```
# Helper packages
library(dplyr)     # for data wrangling
library(ggplot2)   # for awesome plotting
library(rsample)   # for data splitting

# Modeling packages
library(earth)     # for fitting MARS models
library(caret)     # for automating the tuning process

# Model interpretability packages
```

```
library(vip)       # for variable importance
library(pdp)       # for variable relationships
```

To illustrate various concepts we'll continue with the `ames_train` and `ames_test` data sets created in Section 2.7.

7.2 The basic idea

In the previous chapters, we focused on linear models (where the analyst has to explicitly specify any nonlinear relationships and interaction effects). We illustrated some of the advantages of linear models such as their ease and speed of computation and also the intuitive nature of interpreting their coefficients. However, linear models make a strong assumption about linearity, and this assumption is often a poor one, which can affect predictive accuracy.

We can extend linear models to capture any non-linear relationship. Typically, this is done by explicitly including polynomial terms (e.g., X_1^2) or step functions. Polynomial regression is a form of regression in which the relationship between X and Y is modeled as a dth degree polynomial in X. For example, Equation (7.1) represents a polynomial regression function where Y is modeled as a d-th degree polynomial in X. Generally speaking, it is unusual to use d greater than 3 or 4 as the larger d becomes, the easier the function fit becomes overly flexible and oddly shaped...especially near the boundaries of the range of X values. Increasing d also tends to increase the presence of multicollinearity.

$$y_i = \beta_0 + \beta_1 x_i + \beta_2 x_i^2 + \beta_3 x_i^3 \cdots + \beta_d x_i^d + \epsilon_i, \tag{7.1}$$

An alternative to polynomials is to use step functions. Whereas polynomial functions impose a global non-linear relationship, step functions break the range of X into bins, and fit a simple constant (e.g., the mean response) in each. This amounts to converting a continuous feature into an ordered categorical variable such that our linear regression function is converted to Equation (7.2)

$$y_i = \beta_0 + \beta_1 C_1(x_i) + \beta_2 C_2(x_i) + \beta_3 C_3(x_i) \cdots + \beta_d C_d(x_i) + \epsilon_i, \tag{7.2}$$

where $C_1(x)$ represents X values ranging from $c_1 \leq X < c_2$, $C_2(X)$ represents X values ranging from $c_2 \leq X < c_3$, ..., $C_d(X)$ represents X values ranging from $c_{d-1} \leq X < c_d$. Figure 7.1 contrasts linear, polynomial, and step function fits for non-linear, non-monotonic simulated data.

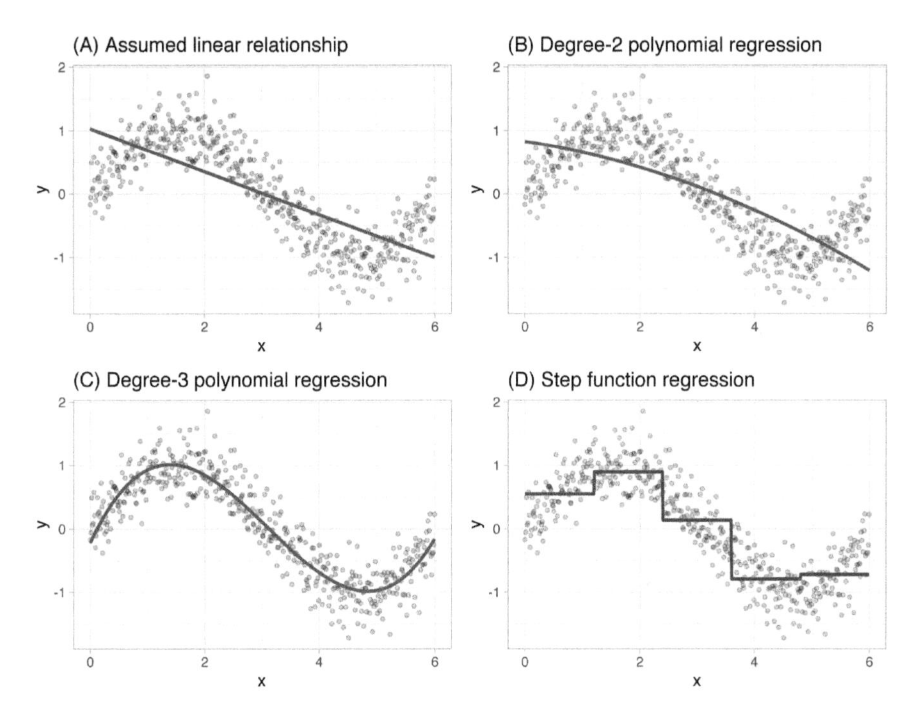

FIGURE 7.1: Blue line represents predicted ('y') values as a function of 'x' for alternative approaches to modeling explicit nonlinear regression patterns. (A) Traditional linear regression approach does not capture any nonlinearity unless the predictor or response is transformed (i.e. log transformation). (B) Degree-2 polynomial, (C) Degree-3 polynomial, (D) Step function fitting cutting 'x' into six categorical levels.

Although useful, the typical implementation of polynomial regression and step functions require the user to explicitly identify and incorporate which variables should have what specific degree of interaction or at what points of a variable X should cut points be made for the step functions. Considering many data sets today can easily contain 50, 100, or more features, this would require an enormous and unnecessary time commitment from an analyst to determine these explicit non-linear settings.

7.2.1 Multivariate adaptive regression splines

Multivariate adaptive regression splines (MARS) provide a convenient approach to capture the nonlinearity relationships in the data by assessing cutpoints (*knots*) similar to step functions. The procedure assesses each data point for each predictor as a knot and creates a linear regression model with the candidate feature(s). For example, consider our non-linear, non-monotonic data above where $Y = f(X)$. The MARS procedure will first look for the single point across the range of `X` values where two different linear relationships between `Y` and `X` achieve the smallest error (e.g., smallest SSE). What results is known as a hinge function $h(x-a)$, where a is the cutpoint value. For a single knot (Figure 7.2 (A)), our hinge function is $h(\text{x} - 1.183606)$ such that our two linear models for `Y` are

$$\text{y} = \begin{cases} \beta_0 + \beta_1(1.183606 - \text{x}) & \text{x} < 1.183606, \\ \beta_0 + \beta_1(\text{x} - 1.183606) & \text{x} > 1.183606 \end{cases} \tag{7.3}$$

Once the first knot has been found, the search continues for a second knot which is found at $x = 4.898114$ (Figure 7.2 (B)). This results in three linear models for `y`:

$$\text{y} = \begin{cases} \beta_0 + \beta_1(1.183606 - \text{x}) & \text{x} < 1.183606, \\ \beta_0 + \beta_1(\text{x} - 1.183606) & \text{x} > 1.183606 \quad \& \quad \text{x} < 4.898114, \\ \beta_0 + \beta_1(4.898114 - \text{x}) & \text{x} > 4.898114 \end{cases} \tag{7.4}$$

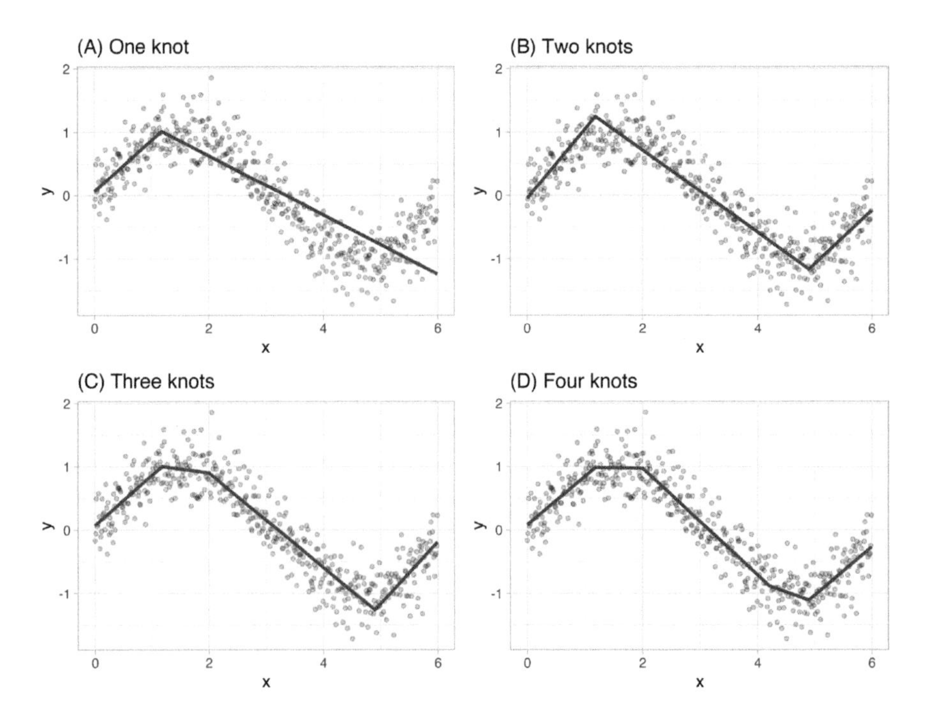

FIGURE 7.2: Examples of fitted regression splines of one (A), two (B), three (C), and four (D) knots.

This procedure continues until many knots are found, producing a (potentially) highly non-linear prediction equation. Although including many knots may allow us to fit a really good relationship with our training data, it may not generalize very well to new, unseen data. Consequently, once the full set of knots has been identified, we can sequentially remove knots that do not contribute significantly to predictive accuracy. This process is known as "pruning" and we can use cross-validation, as we have with the previous models, to find the optimal number of knots.

7.3 Fitting a basic MARS model

We can fit a direct engine MARS model with the **earth** package (from mda:mars by Trevor Hastie and utilities with Thomas Lumley's leaps wrapper., 2019). By default, `earth::earth()` will assess all potential knots across all supplied features and then will prune to the optimal number of knots based on an expected change in R^2 (for the training data) of less than 0.001. This calculation is performed by the Generalized cross-validation (GCV) procedure, which is a computational shortcut for linear models that produces an approximate leave-one-out cross-validation error metric (Golub et al., 1979).

The term "MARS" is trademarked and licensed exclusively to Salford Systems: `https://www.salford-systems.com`. We can use MARS as an abbreviation; however, it cannot be used for competing software solutions. This is why the R package uses the name **earth**. Although, according to the package documentation, a backronym for "earth" is "Enhanced Adaptive Regression Through Hinges".

The following applies a basic MARS model to our **ames** example. The results show us the final models GCV statistic, generalized R^2 (GRSq), and more.

```
# Fit a basic MARS model
mars1 <- earth(
  Sale_Price ~ .,
  data = ames_train
)

# Print model summary
print(mars1)
```

```
## Selected 36 of 41 terms, and 24 of 307 predictors
## Termination condition: RSq changed by less than 0.001 at 41 terms
## Importance: First_Flr_SF, Second_Flr_SF, ...
## Number of terms at each degree of interaction: 1 35 (additive model)
## GCV 5.11e+08     RSS 9.79e+11     GRSq 0.921     RSq 0.926
```

It also shows us that 36 of 41 terms were used from 24 of the 307 original predictors. But what does this mean? If we were to look at all the coefficients, we would see that there are 36 terms in our model (including the intercept). These terms include hinge functions produced from the original 307 predictors (307 predictors because the model automatically dummy encodes categorical features). Looking at the first 10 terms in our model, we see that `Gr_Liv_Area` is included with a knot at 2790 (the coefficient for $h\left(2790 - \text{Gr\_Liv\_Area}\right)$ is -55.26), `Year_Built` is included with a knot at 2002, etc.

You can check out all the coefficients with `summary(mars1)` or `coef(mars1)`.

```
summary(mars1) %>% .$coefficients %>% head(10)
##                          Sale_Price
## (Intercept)                289316.2
## h(Gr_Liv_Area-2790)           -55.3
## h(Year_Built-2002)           3040.6
## h(2002-Year_Built)           -410.9
## h(2220-Total_Bsmt_SF)         -30.7
## h(Bsmt_Unf_SF-543)            -25.4
## h(543-Bsmt_Unf_SF)             13.8
## h(Total_Bsmt_SF-1550)          39.2
## h(Garage_Cars-2)            12480.7
## h(2-Garage_Cars)            -4834.4
```

The plot method for MARS model objects provides useful performance and residual plots. Figure 7.3 illustrates the model selection plot that graphs the GCV R^2 (left-hand y-axis and solid black line) based on the number of terms retained in the model (x-axis) which are constructed from a certain number of original predictors (right-hand y-axis). The vertical dashed lined at 36 tells us the optimal number of non-intercept terms retained where marginal increases in GCV R^2 are less than 0.001.

```
plot(mars1, which = 1)
```

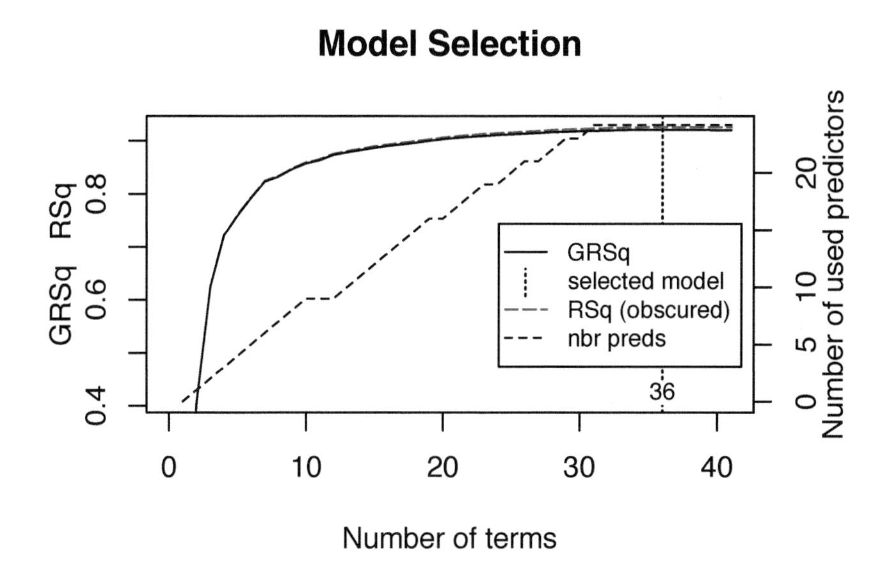

FIGURE 7.3: Model summary capturing GCV R^2 (left-hand y-axis and solid black line) based on the number of terms retained (x-axis) which is based on the number of predictors used to make those terms (right-hand side y-axis). For this model, 35 non-intercept terms were retained which are based on 26 predictors. Any additional terms retained in the model, over and above these 35, result in less than 0.001 improvement in the GCV R^2.

In addition to pruning the number of knots, `earth::earth()` allows us to also assess potential interactions between different hinge functions. The following illustrates this by including a `degree = 2` argument. You can see that now our model includes interaction terms between a maximum of two hinge functions (e.g., `h(Year_Built-2002)*h(2362-Gr_Liv_Area)` represents an interaction effect for those houses built prior to 2002 and had less than 2,362 square feet of living space above ground).

```
# Fit a basic MARS model
mars2 <- earth(
  Sale_Price ~ .,
  data = ames_train,
```

```
  degree = 2
)

# check out the first 10 coefficient terms
summary(mars2) %>% .$coefficients %>% head(10)
##                                              Sale_Price
## (Intercept)                                   230780.52
## h(Gr_Liv_Area-2790)                               94.79
## h(2790-Gr_Liv_Area)                              -50.98
## h(Year_Built-2002)                              9111.80
## h(2002-Year_Built)                              -689.25
## h(Year_Built-2002)*h(2362-Gr_Liv_Area)            -7.72
## h(Total_Bsmt_SF-1136)                             63.63
## h(1136-Total_Bsmt_SF)                            -32.70
## h(Bsmt_Unf_SF-504)                               -25.09
## h(504-Bsmt_Unf_SF)                                12.96
```

7.4 Tuning

There are two important tuning parameters associated with our MARS model: the maximum degree of interactions and the number of terms retained in the final model. We need to perform a grid search to identify the optimal combination of these hyperparameters that minimize prediction error (the above pruning process was based only on an approximation of CV model performance on the training data rather than an exact k-fold CV process). As in previous chapters, we'll perform a CV grid search to identify the optimal hyperparameter mix. Below, we set up a grid that assesses 30 different combinations of interaction complexity (`degree`) and the number of terms to retain in the final model (`nprune`).

Rarely is there any benefit in assessing greater than 3-rd degree interactions and we suggest starting out with 10 evenly spaced values for `nprune` and then you can always zoom in to a region once you find an approximate optimal solution.

```
# create a tuning grid
hyper_grid <- expand.grid(
  degree = 1:3,
  nprune = seq(2, 100, length.out = 10) %>% floor()
)

head(hyper_grid)
##   degree nprune
## 1      1      2
## 2      2      2
## 3      3      2
## 4      1     12
## 5      2     12
## 6      3     12
```

As in the previous chapters, we can use **caret** to perform a grid search using 10-fold CV. The model that provides the optimal combination includes third degree interaction effects and retains 45 terms. The cross-validated RMSE for these models is displayed in Figure 7.4; the optimal model's cross-validated RMSE was $22,888.

This grid search took roughly five minutes to complete.

```
# Cross-validated model
set.seed(123)  # for reproducibility
cv_mars <- train(
  x = subset(ames_train, select = -Sale_Price),
  y = ames_train$Sale_Price,
  method = "earth",
  metric = "RMSE",
  trControl = trainControl(method = "cv", number = 10),
  tuneGrid = hyper_grid
)

# View results
cv_mars$bestTune
##    nprune degree
## 25     45      3
ggplot(cv_mars)
```

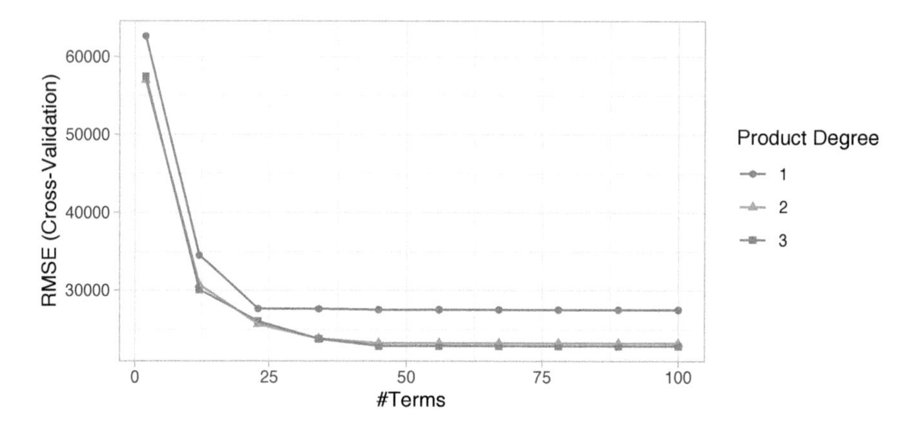

FIGURE 7.4: Cross-validated RMSE for the 30 different hyperparameter combinations in our grid search. The optimal model retains 45 terms and includes up to 3rd degree interactions.

The above grid search helps to focus where we can further refine our model tuning. As a next step, we could perform a grid search that focuses in on a refined grid space for `nprune` (e.g., comparing 35–45 terms retained). However, for brevity we'll leave this as an exercise for the reader.

So how does this compare to our previously built models for the Ames housing data? The following table compares the cross-validated RMSE for our tuned MARS model to an ordinary multiple regression model along with tuned principal component regression (PCR), partial least squares (PLS), and regularized regression (elastic net) models. By incorporating non-linear relationships and interaction effects, the MARS model provides a substantial improvement over the previous linear models that we have explored.

Notice that our elastic net model is higher than in the last chapter. This table compares these 5 modeling approaches without performing any logarithmic transformation on the target variable. However, even considering the best tuned regularized regression results from last chapter (RMSE = 23503), our optimal MARS model performs better.

TABLE 7.1: Cross-validated RMSE results for tuned MARS and regression models.

	Min.	1st Qu.	Median	Mean	3rd Qu.	Max.	NA's
LM	20945	25674	33769	37304	42967	80339	0
PCR	26925	29987	33891	33659	37180	40713	0
PLS	21147	25951	28903	29970	35261	41414	0
ENET	19938	24062	25261	29785	33829	53333	0
MARS	18967	21792	22833	22888	23786	27190	0

7.5 Feature interpretation

MARS models via `earth::earth()` include a backwards elimination feature selection routine that looks at reductions in the GCV estimate of error as each predictor is added to the model. This total reduction is used as the variable importance measure (`value = "gcv"`). Since MARS will automatically include and exclude terms during the pruning process, it essentially performs automated feature selection. If a predictor was never used in any of the MARS basis functions in the final model (after pruning), it has an importance value of zero. This is illustrated in Figure 16.1 where 27 features have > 0 importance values while the rest of the features have an importance value of zero since they were not included in the final model. Alternatively, you can also monitor the change in the residual sums of squares (RSS) as terms are added (`value = "rss"`); however, you will see very little difference between these methods.

```
# variable importance plots
p1 <- vip(cv_mars, num_features = 40, bar = FALSE, value = "gcv") +
  ggtitle("GCV")

p2 <- vip(cv_mars, num_features = 40, bar = FALSE, value = "rss") +
  ggtitle("RSS")

gridExtra::grid.arrange(p1, p2, ncol = 2)
```

Its important to realize that variable importance will only measure the impact of the prediction error as features are included; however, it does not measure the impact for particular hinge functions created for a given feature. For example, in Figure 16.1 we see that `Gr_Liv_Area` and `Year_Built` are the two most influential variables; however, variable importance does not tell us how our model is treating the non-linear patterns for each feature. Also, if we

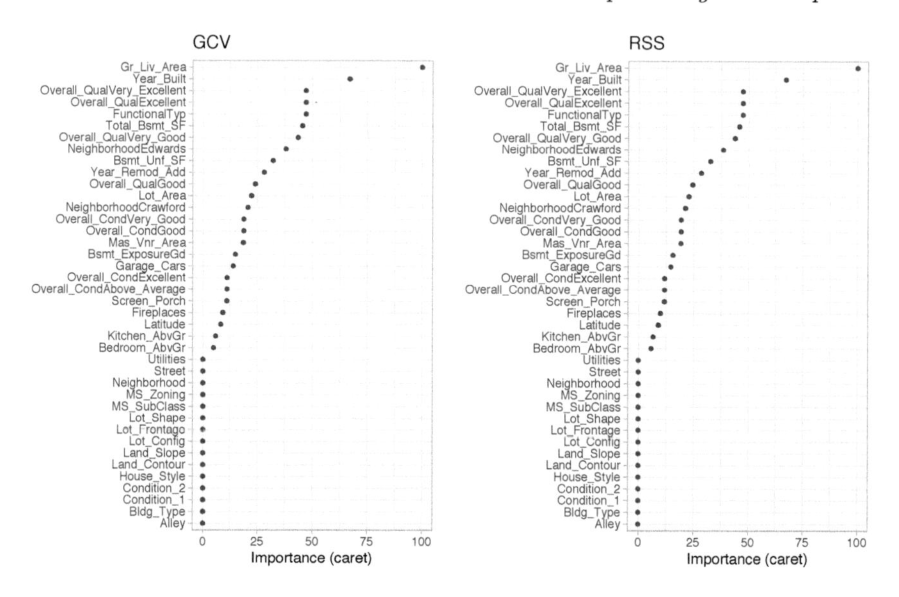

FIGURE 7.5: Variable importance based on impact to GCV (left) and RSS (right) values as predictors are added to the model. Both variable importance measures will usually give you very similar results.

look at the interaction terms our model retained, we see interactions between different hinge functions for `Gr_Liv_Area` and `Year_Built`.

```
# extract coefficients, convert to tidy data frame, and
# filter for interaction terms
cv_mars$finalModel %>%
  coef() %>%
  broom::tidy() %>%
  filter(stringr::str_detect(names, ”\\*”))
## # A tibble: 20 x 2
##    names                                         x
##    <chr>                                     <dbl>
##  1 h(Year_Built-2002) * h(2362-Gr_Liv_Area)   -6.98e+0
##  2 h(Year_Remod_Add-2007) * h(Total_Bsmt_SF~   8.93e+0
##  3 h(2007-Year_Remod_Add) * h(Total_Bsmt_SF~  -1.22e+0
##  4 NeighborhoodEdwards * h(Year_Built-2002)~  -6.71e+1
##  5 h(Lot_Area-3874) * h(3-Garage_Cars)        -1.17e+0
##  6 Bsmt_ExposureGd * h(Total_Bsmt_SF-1136)     3.12e+1
##  7 NeighborhoodCrawford * h(2002-Year_Built)   4.25e+2
##  8 h(2002-Year_Built) * h(Year_Remod_Add-19~   7.90e+0
##  9 h(2002-Year_Built) * h(1974-Year_Remod_A~   5.42e+0
```

```
## 10 h(Kitchen_AbvGr-1) * FunctionalTyp              -1.55e+4
## 11 Overall_QualVery_Excellent * FunctionalT~        9.68e+4
## 12 Overall_QualGood * FunctionalTyp                 1.32e+4
## 13 h(Lot_Area-3874) * h(Latitude-42.0014)           7.65e+0
## 14 h(Lot_Area-3874) * h(42.0014-Latitude)          -1.23e+2
## 15 h(Total_Bsmt_SF-1136) * h(115-Screen_Por~       -3.04e-1
## 16 h(Lot_Area-3874) * h(Gr_Liv_Area-2411) *~       -2.99e-3
## 17 h(Lot_Area-3874) * h(2411-Gr_Liv_Area) *~        5.29e-4
## 18 Overall_CondGood * h(2002-Year_Built)            3.33e+2
## 19 Overall_CondVery_Good * h(2002-Year_Buil~        3.68e+2
## 20 Overall_CondAbove_Average * h(2790-Gr_Li~        6.20e+0
```

To better understand the relationship between these features and `Sale_Price`, we can create partial dependence plots (PDPs) for each feature individually and also together. The individual PDPs illustrate the knots for each feature that our model found provides the best fit. For `Gr_Liv_Area`, as homes exceed 2,790 square feet, each additional square foot demands a higher marginal increase in sale price than homes with less than 2,790 square feet. Similarly, for homes built after 2002, there is a greater marginal effect on sales price based on the age of the home than for homes built prior to 2002. The interaction plot (far right figure) illustrates the stronger effect these two features have when combined.

```
# Construct partial dependence plots
p1 <- partial(cv_mars, pred.var = "Gr_Liv_Area") %>% autoplot()
p2 <- partial(cv_mars, pred.var = "Year_Built") %>% autoplot()
p3 <- partial(cv_mars, pred.var = c("Gr_Liv_Area", "Year_Built"),
              chull = TRUE) %>%
  plotPartial(palette = "inferno", contour = TRUE) %>%
  ggplotify::as.grob() # convert to grob to plot with cowplot

# Display plots in a grid
top_row <- cowplot::plot_grid(p1, p2)
cowplot::plot_grid(top_row, p3, nrow = 2, rel_heights = c(1, 2))
```

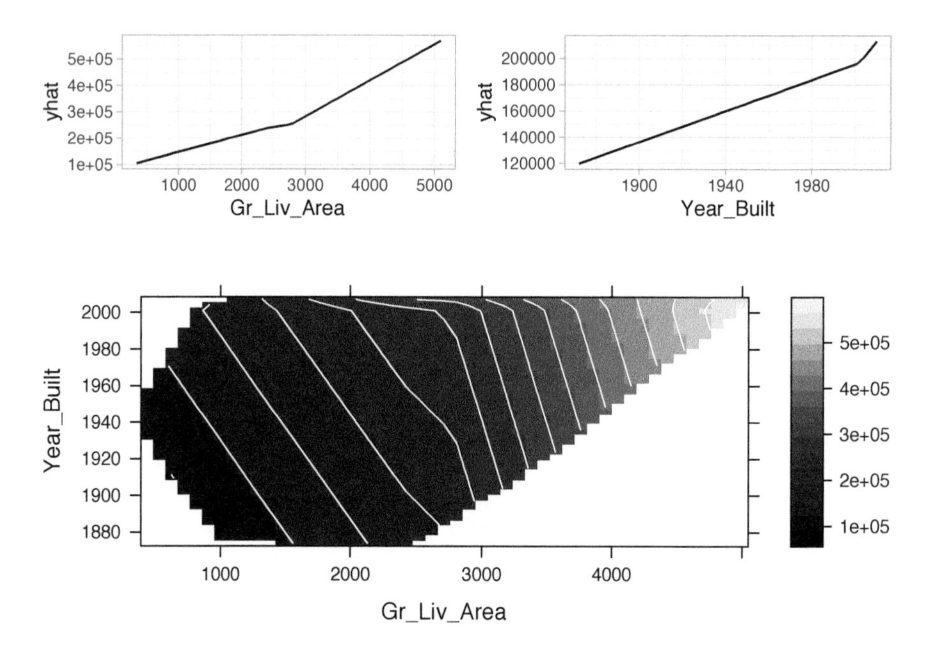

FIGURE 7.6: Partial dependence plots to understand the relationship between sale price and the living space and year built features. The PDPs tell us that as living space increases and for newer homes, predicted sale price increases dramatically.

7.6 Attrition data

The MARS method and algorithm can be extended to handle classification problems and GLMs in general.[1] We saw significant improvement to our predictive accuracy on the Ames data with a MARS model, but how about the employee attrition example? In Chapter 5 we saw a slight improvement in our cross-validated accuracy rate using regularized regression. Here, we tune a MARS model using the same search grid as we did above. We see our best models include no interaction effects and the optimal model retained 12 terms.

```
# get attrition data
df <- attrition %>% mutate_if(is.ordered, factor, ordered = FALSE)
```

[1]See Friedman et al. (2001) and Stone et al. (1997) for technical details regarding various alternative encodings for binary and mulinomial classification approaches.

TABLE 7.2: Cross-validated accuracy results for tuned MARS and regression models.

	Min.	1st Qu.	Median	Mean	3rd Qu.	Max.	NA's
GLM (logit)	0.837	0.850	0.879	0.876	0.891	0.931	0
ENET	0.845	0.876	0.883	0.884	0.892	0.941	0
MARS	0.816	0.858	0.878	0.871	0.891	0.903	0

```
# Create training (70%) and test (30%) sets for the
# rsample::attrition data.
set.seed(123)
churn_split <- initial_split(df, prop = 0.7, strata = "Attrition")
churn_train <- training(churn_split)
churn_test  <- testing(churn_split)

# for reproducibiity
set.seed(123)

# cross validated model
tuned_mars <- train(
  x = subset(churn_train, select = -Attrition),
  y = churn_train$Attrition,
  method = "earth",
  trControl = trainControl(method = "cv", number = 10),
  tuneGrid = hyper_grid
)

# best model
tuned_mars$bestTune
##    nprune degree
## 2      12      1

# plot results
ggplot(tuned_mars)
```

However, comparing our MARS model to the previous linear models (logistic regression and regularized regression), we do not see any improvement in our overall accuracy rate.

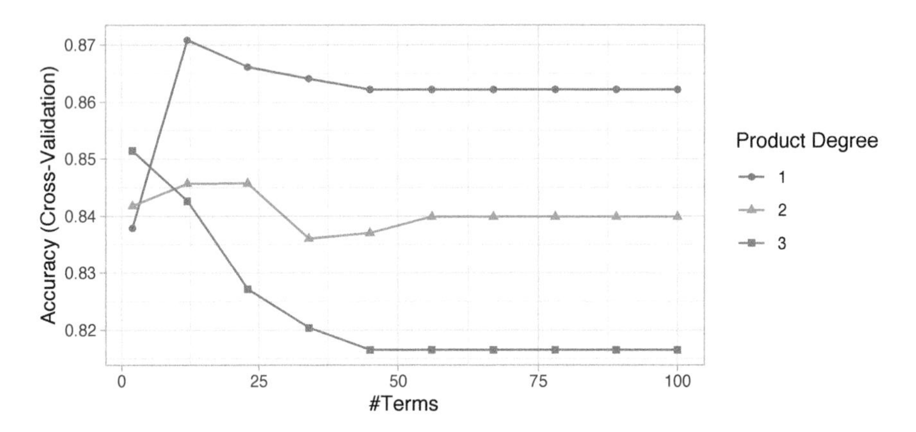

FIGURE 7.7: Cross-validated accuracy rate for the 30 different hyperparameter combinations in our grid search. The optimal model retains 12 terms and includes no interaction effects.

7.7 Final thoughts

There are several advantages to MARS. First, MARS naturally handles mixed types of predictors (quantitative and qualitative). MARS considers all possible binary partitions of the categories for a qualitative predictor into two groups.[2] Each group then generates a pair of piecewise indicator functions for the two categories. MARS also requires minimal feature engineering (e.g., feature scaling) and performs automated feature selection. For example, since MARS scans each predictor to identify a split that improves predictive accuracy, non-informative features will not be chosen. Furthermore, highly correlated predictors do not impede predictive accuracy as much as they do with OLS models.

However, one disadvantage to MARS models is that they're typically slower to train. Since the algorithm scans each value of each predictor for potential cutpoints, computational performance can suffer as both n and p increase. Also, although correlated predictors do not necessarily impede model performance, they can make model interpretation difficult. When two features are nearly perfectly correlated, the algorithm will essentially select the first one it happens to come across when scanning the features. Then, since it randomly selected one, the correlated feature will likely not be included as it adds no additional explanatory power.

[2]This is very similar to CART-like decision trees which you'll be exposed to in Chapter 9.

8

K-Nearest Neighbors

K-nearest neighbor (KNN) is a very simple algorithm in which each observation is predicted based on its "similarity" to other observations. Unlike most methods in this book, KNN is a *memory-based* algorithm and cannot be summarized by a closed-form model. This means the training samples are required at run-time and predictions are made directly from the sample relationships. Consequently, KNNs are also known as *lazy learners* (Cunningham and Delany, 2007) and can be computationally inefficient. However, KNNs have been successful in a large number of business problems (see, for example, Jiang et al. (2012) and Mccord and Chuah (2011)) and are useful for preprocessing purposes as well (as was discussed in Section 3.3.2).

8.1 Prerequisites

For this chapter we'll use the following packages:

```
# Helper packages
library(dplyr)       # for data wrangling
library(ggplot2)     # for awesome graphics
library(rsample)     # for data splitting
library(recipes)     # for feature engineering

# Modeling packages
library(caret)       # for training KNN models
```

To illustrate various concepts we'll continue working with the `ames_train` and `ames_test` data sets created in Section 2.7; however, we'll also illustrate the performance of KNNs on the employee attrition and MNIST data sets.

```
# Create training (70%) set for the rsample::attrition data
attrit <- attrition %>%
  mutate_if(is.ordered, factor, ordered = FALSE)
set.seed(123)  # for reproducibility
churn_split <- initial_split(attrit, prop = 0.7,
                             strata = "Attrition")
churn_train <- training(churn_split)

# Import MNIST training data
mnist <- dslabs::read_mnist()
names(mnist)
## [1] "train" "test"
```

8.2 Measuring similarity

The KNN algorithm identifies k observations that are "similar" or nearest to the new record being predicted and then uses the average response value (regression) or the most common class (classification) of those k observations as the predicted output.

For illustration, consider our Ames housing data. In real estate, Realtors determine what price they will list (or market) a home for based on "comps" (comparable homes). To identify comps, they look for homes that have very similar attributes to the one being sold. This can include similar features (e.g., square footage, number of rooms, and style of the home), location (e.g., neighborhood and school district), and many other attributes. The Realtor will look at the typical sale price of these comps and will usually list the new home at a very similar price to the prices these comps sold for.

As an example, Figure 8.1 maps 10 homes (blue) that are most similar to the home of interest (red). These homes are all relatively close to the target home and likely have similar characteristics (e.g., home style, size, and school district). Consequently, the Realtor would likely list the target home around the average price that these comps sold for. In essence, this is what the KNN algorithm will do.

8.2.1 Distance measures

How do we determine the similarity between observations (or homes as in Figure 8.1)? We use distance (or dissimilarity) metrics to compute the pairwise

FIGURE 8.1: The 10 nearest neighbors (blue) whose home attributes most closely resemble the house of interest (red).

differences between observations. The most common distance measures are the Euclidean (8.1) and Manhattan (8.2) distance metrics; both of which measure the distance between observation x_a and x_b for all j features.

$$\sqrt{\sum_{j=1}^{P}(x_{aj} - x_{bj})^2} \tag{8.1}$$

$$\sum_{j=1}^{P}|x_{aj} - x_{bj}| \tag{8.2}$$

Euclidean distance is the most common and measures the straight-line distance between two samples (i.e., how the crow flies). Manhattan measures the point-to-point travel time (i.e., city block) and is commonly used for binary predictors (e.g., one-hot encoded 0/1 indicator variables). A simplified example is presented below and illustrated in Figure 8.2 where the distance measures are computed for the first two homes in `ames_train` and for only two features (`Gr_Liv_Area` & `Year_Built`).

```
two_houses <- ames_train %>%
  select(Gr_Liv_Area, Year_Built) %>%
  sample_n(2)

two_houses
## # A tibble: 2 x 2
##   Gr_Liv_Area Year_Built
##         <int>      <int>
## 1         896       1961
## 2        1511       2002

# Euclidean
dist(two_houses, method = "euclidean")
##     1
## 2 616

# Manhattan
dist(two_houses, method = "manhattan")
##     1
## 2 656
```

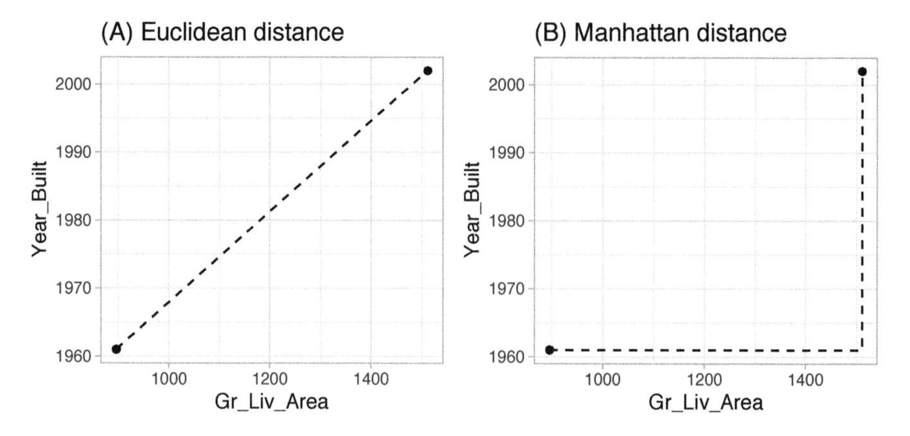

FIGURE 8.2: Euclidean (A) versus Manhattan (B) distance.

There are other metrics to measure the distance between observations. For example, the Minkowski distance is a generalization of the Euclidean and Manhattan distances and is defined as

$$\left(\sum_{j=1}^{P} |x_{aj} - x_{bj}|^q \right)^{\frac{1}{q}} \tag{8.3}$$

where $q > 0$ (Han et al., 2011). When $q = 2$ the Minkowski distance equals the Euclidean distance and when $q = 1$ it is equal to the Manhattan distance. The Mahalanobis distance is also an attractive measure to use since it accounts for the correlation between two variables (De Maesschalck et al., 2000).

8.2.2 Preprocessing

Due to the squaring in Equation (8.1), the Euclidean distance is more sensitive to outliers. Furthermore, most distance measures are sensitive to the scale of the features. Data with features that have different scales will bias the distance measures as those predictors with the largest values will contribute most to the distance between two samples. For example, consider the three home below: `home1` is a four bedroom built in 2008, `home2` is a two bedroom built in the same year, and `home3` is a three bedroom built a decade earlier.

```
home1
## # A tibble: 1 x 4
##    home  Bedroom_AbvGr Year_Built    id
##    <chr>         <int>      <int> <int>
## 1 home1             4       2008   423
home2
## # A tibble: 1 x 4
##    home  Bedroom_AbvGr Year_Built    id
##    <chr>         <int>      <int> <int>
## 1 home2             2       2008   424
home3
## # A tibble: 1 x 4
##    home  Bedroom_AbvGr Year_Built    id
##    <chr>         <int>      <int> <int>
## 1 home3             3       1998     6
```

The Euclidean distance between `home1` and `home3` is larger due to the larger difference in `Year_Built` with `home2`.

```
features <- c("Bedroom_AbvGr", "Year_Built")

# distance between home 1 and 2
dist(rbind(home1[,features], home2[,features]))
##   1
## 2 2
```

```
# distance between home 1 and 3
dist(rbind(home1[,features], home3[,features]))
##    1
## 2 10
```

However, `Year_Built` has a much larger range (1875–2010) than `Bedroom_AbvGr` (0–8). And if you ask most people, especially families with kids, the difference between 2 and 4 bedrooms is much more significant than a 10 year difference in the age of a home. If we standardize these features, we see that the difference between `home1` and `home2`'s standardized value for `Bedroom_AbvGr` is larger than the difference between `home1` and `home3`'s `Year_Built`. And if we compute the Euclidean distance between these standardized home features, we see that now `home1` and `home3` are more similar than `home1` and `home2`.

```
home1_std
## # A tibble: 1 x 4
##   home  Bedroom_AbvGr Year_Built    id
##   <chr>         <dbl>      <dbl> <int>
## 1 home1          1.38       1.21   423
home2_std
## # A tibble: 1 x 4
##   home  Bedroom_AbvGr Year_Built    id
##   <chr>         <dbl>      <dbl> <int>
## 1 home2         -1.03       1.21   424
home3_std
## # A tibble: 1 x 4
##   home  Bedroom_AbvGr Year_Built    id
##   <chr>         <dbl>      <dbl> <int>
## 1 home3         0.176      0.881     6

# distance between home 1 and 2
dist(rbind(home1_std[,features], home2_std[,features]))
##      1
## 2 2.42

# distance between home 1 and 3
dist(rbind(home1_std[,features], home3_std[,features]))
##      1
## 2 1.25
```

In addition to standardizing numeric features, all categorical features must be one-hot encoded or encoded using another method (e.g., ordinal encoding) so that all categorical features are represented numerically. Furthermore, the KNN method is very sensitive to noisy predictors since they cause similar samples to have larger magnitudes and variability in distance values. Consequently, removing irrelevant, noisy features often leads to significant improvement.

8.3 Choosing k

The performance of KNNs is very sensitive to the choice of k. This was illustrated in Section 2.5.3 where low values of k typically overfit and large values often underfit. At the extremes, when $k = 1$, we base our prediction on a single observation that has the closest distance measure. In contrast, when $k = n$, we are simply using the average (regression) or most common class (classification) across all training samples as our predicted value.

There is no general rule about the best k as it depends greatly on the nature of the data. For high signal data with very few noisy (irrelevant) features, smaller values of k tend to work best. As more irrelevant features are involved, larger values of k are required to smooth out the noise. To illustrate, we saw in Section 3.8.3 that we optimized the RMSE for the `ames_train` data with $k = 12$. The `ames_train` data has 2054 observations, so such a small k likely suggests a strong signal exists. In contrast, the `churn_train` data has 1030 observations and Figure 8.3 illustrates that our loss function is not optimized until $k = 271$. Moreover, the max ROC value is 0.8078 and the overall proportion of attriting employees to non-attriting is 0.839. This suggest there is likely not a very strong signal in the Attrition data.

When using KNN for classification, it is best to assess odd numbers for k to avoid ties in the event there is equal proportion of response levels (i.e. when $k = 2$ one of the neighbors could have class “0” while the other neighbor has class “1”).

```
# Create blueprint
blueprint <- recipe(Attrition ~ ., data = churn_train) %>%
  step_nzv(all_nominal()) %>%
  step_integer(contains("Satisfaction")) %>%
  step_integer(WorkLifeBalance) %>%
  step_integer(JobInvolvement) %>%
```

```
  step_dummy(all_nominal(), -all_outcomes(), one_hot = TRUE) %>%
  step_center(all_numeric(), -all_outcomes()) %>%
  step_scale(all_numeric(), -all_outcomes())

# Create a resampling method
cv <- trainControl(
  method = "repeatedcv",
  number = 10,
  repeats = 5,
  classProbs = TRUE,
  summaryFunction = twoClassSummary
)

# Create a hyperparameter grid search
hyper_grid <- expand.grid(
  k = floor(seq(1, nrow(churn_train)/3, length.out = 20))
)

# Fit knn model and perform grid search
knn_grid <- train(
  blueprint,
  data = churn_train,
  method = "knn",
  trControl = cv,
  tuneGrid = hyper_grid,
  metric = "ROC"
)

ggplot(knn_grid)
```

8.4 MNIST example

The MNIST data set is significantly larger than the Ames housing and attrition data sets. Because we want this example to run locally and in a reasonable amount of time (< 1 hour), we will train our initial models on a random sample of 10,000 rows from the training set.

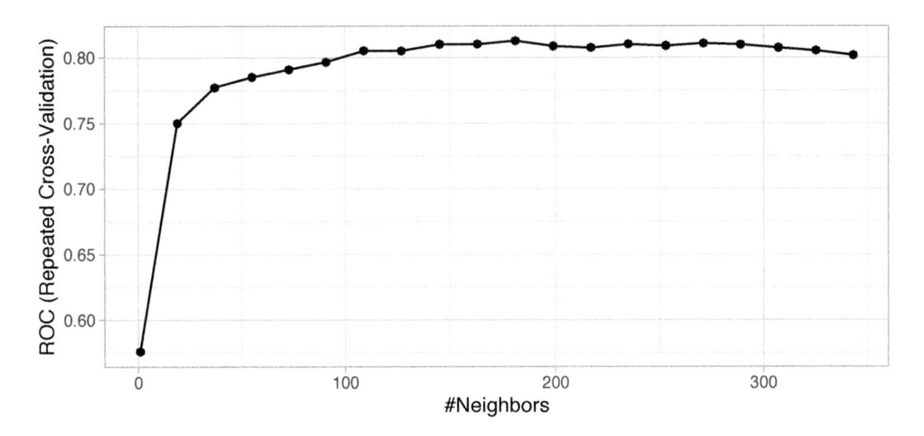

FIGURE 8.3: Cross validated search grid results for Attrition training data where 20 values between 1 and 343 are assessed for k. When k = 1, the predicted value is based on a single observation that is closest to the target sample and when k = 343, the predicted value is based on the response with the largest proportion for 1/3 of the training sample.

```
set.seed(123)
index <- sample(nrow(mnist$train$images), size = 10000)
mnist_x <- mnist$train$images[index, ]
mnist_y <- factor(mnist$train$labels[index])
```

Recall that the MNIST data contains 784 features representing the darkness (0–255) of pixels in images of handwritten numbers (0–9). As stated in Section 8.2.2, KNN models can be severely impacted by irrelevant features. One culprit of this is zero, or near-zero variance features (see Section 3.4). Figure 8.4 illustrates that there are nearly 125 features that have zero variance and many more that have very little variation.

```
mnist_x %>%
  as.data.frame() %>%
  map_df(sd) %>%
  gather(feature, sd) %>%
  ggplot(aes(sd)) +
  geom_histogram(binwidth = 1)
```

Figure 8.5 shows which features are driving this concern. Images (A)–(C) illustrate typical handwritten numbers from the test set. Image (D) illustrates which features in our images have variability. The white in the center shows

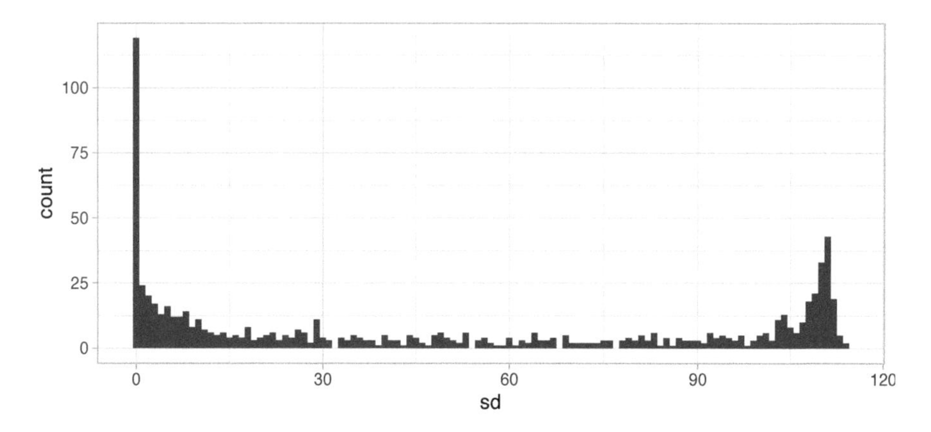

FIGURE 8.4: Distribution of variability across the MNIST features. We see a significant number of zero variance features that should be removed.

that the features that represent the center pixels have regular variability whereas the black exterior highlights that the features representing the edge pixels in our images have zero or near-zero variability. These features have low variability in pixel values because they are rarely drawn on.

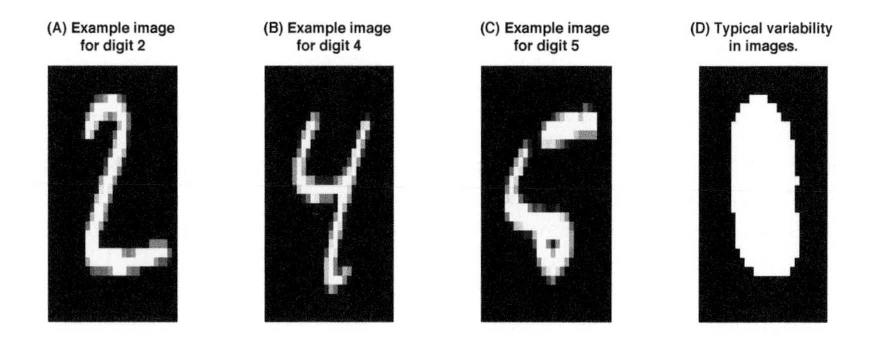

FIGURE 8.5: Example images (A)-(C) from our data set and (D) highlights near-zero variance features around the edges of our images.

By identifying and removing these zero (or near-zero) variance features, we end up keeping 249 of the original 784 predictors. This can cause dramatic improvements to both the accuracy and speed of our algorithm. Furthermore, by removing these upfront we can remove some of the overhead experienced by `caret::train()`. Furthermore, we need to add column names to the feature matrices as these are required by **caret**.

```
# Rename features
colnames(mnist_x) <- paste0("V", 1:ncol(mnist_x))

# Remove near zero variance features manually
nzv <- nearZeroVar(mnist_x)
index <- setdiff(1:ncol(mnist_x), nzv)
mnist_x <- mnist_x[, index]
```

Next we perform our search grid. Since we are working with a larger data set, using resampling (e.g., k-fold cross validation) becomes costly. Moreover, as we have more data, our estimated error rate produced by a simple train vs. validation set becomes less biased and variable. Consequently, the following CV procedure (`cv`) uses 70% of our data to train and the remaining 30% for validation. We can adjust the `number` of times we do this which becomes similar to the bootstrap procedure discussed in Section 2.4.

Our hyperparameter grid search assesses 13 k values between 1–25 and takes approximately 3 minutes.

```
# Use train/validate resampling method
cv <- trainControl(
  method = "LGOCV",
  p = 0.7,
  number = 1,
  savePredictions = TRUE
)

# Create a hyperparameter grid search
hyper_grid <- expand.grid(k = seq(3, 25, by = 2))

# Execute grid search
knn_mnist <- train(
  mnist_x,
  mnist_y,
  method = "knn",
  tuneGrid = hyper_grid,
  preProc = c("center", "scale"),
  trControl = cv
)

ggplot(knn_mnist)
```

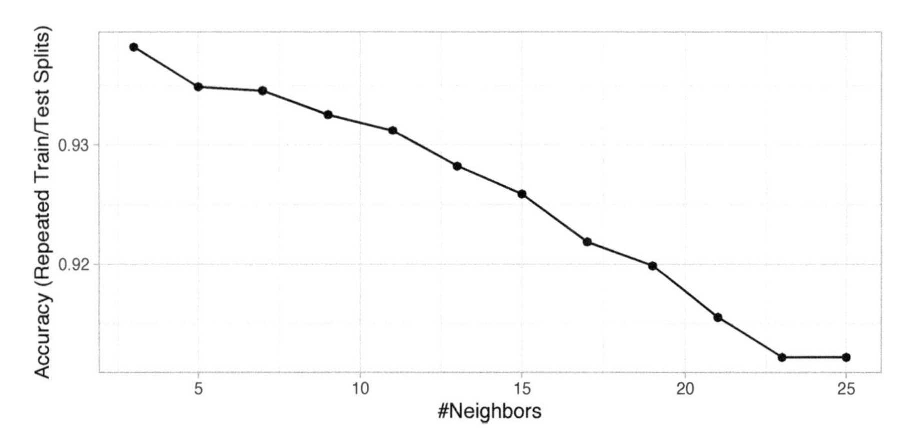

FIGURE 8.6: KNN search grid results for the MNIST data

Figure 8.6 illustrates the grid search results and our best model used 3 nearest neighbors and provided an accuracy of 93.8%. Looking at the results for each class, we can see that 8s were the hardest to detect followed by 2s, 3s, and 4s (based on sensitivity). The most common incorrectly predicted digit is 1 (specificity).

```
# Create confusion matrix
cm <- confusionMatrix(knn_mnist$pred$pred, knn_mnist$pred$obs)
cm$byClass[, c(1:2, 11)]  # sensitivity, specificity, & accuracy
##          Sensitivity Specificity Balanced Accuracy
## Class: 0       0.964       0.996             0.980
## Class: 1       0.992       0.984             0.988
## Class: 2       0.916       0.996             0.956
## Class: 3       0.916       0.992             0.954
## Class: 4       0.870       0.996             0.933
## Class: 5       0.915       0.991             0.953
## Class: 6       0.980       0.989             0.984
## Class: 7       0.933       0.990             0.961
## Class: 8       0.822       0.998             0.910
## Class: 9       0.933       0.985             0.959
```

Feature importance for KNNs is computed by finding the features with the smallest distance measure (see Equation (8.1)). Since the response variable in the MNIST data is multiclass, the variable importance scores below sort the features by maximum importance across the classes.

```
# Top 20 most important features
vi <- varImp(knn_mnist)
vi
## ROC curve variable importance
##
##   variables are sorted by maximum importance across the classes
##   only 20 most important variables shown (out of 249)
##
##          X0    X1    X2    X3    X4    X5    X6    X7
## V435 100.0 100.0 100.0 100.0 100.0 100.0 100.0 100.0
## V407  99.4  99.4  99.4  99.4  99.4  99.4  99.4  99.4
## V463  97.9  97.9  97.9  97.9  97.9  97.9  97.9  97.9
## V379  97.4  97.4  97.4  97.4  97.4  97.4  97.4  97.4
## V434  95.9  95.9  95.9  95.9  95.9  95.9  96.7  95.9
## V380  96.1  96.1  96.1  96.1  96.1  96.1  96.1  96.1
## V462  95.6  95.6  95.6  95.6  95.6  95.6  95.6  95.6
## V408  95.4  95.4  95.4  95.4  95.4  95.4  95.4  95.4
## V352  93.5  93.5  93.5  93.5  93.5  93.5  93.5  93.5
## V490  93.1  93.1  93.1  93.1  93.1  93.1  93.1  93.1
## V406  92.9  92.9  92.9  92.9  92.9  92.9  92.9  92.9
## V437  70.8  60.4  92.8  52.0  71.1  83.4  75.5  91.1
## V351  92.4  92.4  92.4  92.4  92.4  92.4  92.4  92.4
## V409  70.5  76.1  88.1  54.5  79.9  77.7  84.9  91.9
## V436  90.0  90.0  90.9  90.0  90.0  90.0  91.4  90.0
## V464  76.7  76.5  90.2  76.5  76.5  76.6  77.7  82.0
## V491  89.5  89.5  89.5  89.5  89.5  89.5  89.5  89.5
## V598  68.0  68.0  88.4  68.0  68.0  84.9  68.0  88.2
## V465  63.1  36.6  87.7  38.2  50.7  80.6  59.9  84.3
## V433  63.7  55.7  76.7  55.7  57.4  55.7  87.6  68.4
##          X8   X9
## V435 100.0 80.6
## V407  99.4 75.2
## V463  97.9 83.3
## V379  97.4 86.6
## V434  95.9 76.2
## V380  96.1 88.0
## V462  95.6 83.4
## V408  95.4 75.0
## V352  93.5 87.1
## V490  93.1 81.9
## V406  92.9 74.6
## V437  52.0 70.8
## V351  92.4 82.1
## V409  52.7 76.1
```

```
## V436  90.0 78.8
## V464  76.5 76.7
## V491  89.5 77.4
## V598  68.0 38.8
## V465  57.1 63.1
## V433  55.7 63.7
```

We can plot these results to get an understanding of what pixel features are driving our results. The image shows that the most influential features lie around the edges of numbers (outer white circle) and along the very center. This makes intuitive sense as many key differences between numbers lie in these areas. For example, the main difference between a 3 and an 8 is whether the left side of the number is enclosed.

```
# Get median value for feature importance
imp <- vi$importance %>%
  rownames_to_column(var = "feature") %>%
  gather(response, imp, -feature) %>%
  group_by(feature) %>%
  summarize(imp = median(imp))

# Create tibble for all edge pixels
edges <- tibble(
  feature = paste0("V", nzv),
  imp = 0
)

# Combine and plot
imp <- rbind(imp, edges) %>%
  mutate(ID  = as.numeric(str_extract(feature, "\\d+"))) %>%
  arrange(ID)
image(matrix(imp$imp, 28, 28), col = gray(seq(0, 1, 0.05)),
      xaxt="n", yaxt="n")
```

We can look at a few of our correct (left) and incorrect (right) predictions in Figure 8.8. When looking at the incorrect predictions, we can rationalize some of the errors (e.g., the actual 4 where we predicted a 1 has a strong vertical stroke compared to the rest of the number's features, the actual 2 where we predicted a 0 is blurry and not well defined.)

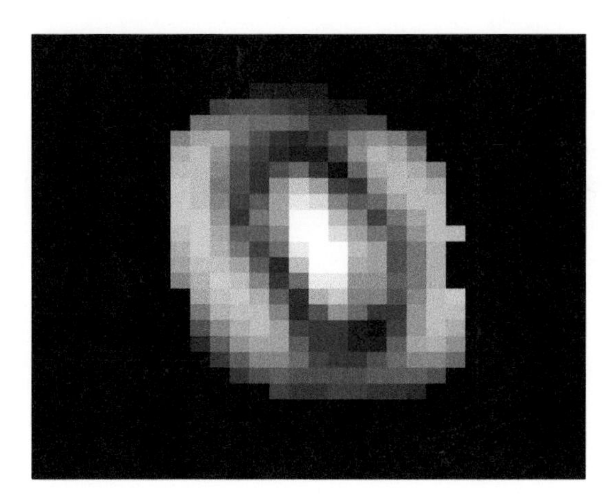

FIGURE 8.7: Image heat map showing which features, on average, are most influential across all response classes for our KNN model.

```
# Get a few accurate predictions
set.seed(9)
good <- knn_mnist$pred %>%
  filter(pred == obs) %>%
  sample_n(4)

# Get a few inaccurate predictions
set.seed(9)
bad <- knn_mnist$pred %>%
  filter(pred != obs) %>%
  sample_n(4)

combine <- bind_rows(good, bad)

# Get original feature set with all pixel features
```

```
set.seed(123)
index <- sample(nrow(mnist$train$images), 10000)
X <- mnist$train$images[index,]

# Plot results
par(mfrow = c(4, 2), mar=c(1, 1, 1, 1))
layout(matrix(seq_len(nrow(combine)), 4, 2, byrow = FALSE))
for(i in seq_len(nrow(combine))) {
  image(matrix(X[combine$rowIndex[i],], 28, 28)[, 28:1],
        col = gray(seq(0, 1, 0.05)),
        main = paste("Actual:", combine$obs[i], "  ",
                     "Predicted:", combine$pred[i]),
        xaxt="n", yaxt="n")
}
```

8.5 Final thoughts

KNNs are a very simplistic, and intuitive, algorithm that can provide average to decent predictive power, especially when the response is dependent on the local structure of the features. However, a major drawback of KNNs is their computation time, which increases by $n \times p$ for each observation. Furthermore, since KNNs are a lazy learner, they require the model be run at prediction time which limits their use for real-time modeling. Some work has been done to minimize this effect; for example the **FNN** package (Beygelzimer et al., 2019) provides a collection of fast k-nearest neighbor search algorithms and applications such as cover-tree (Beygelzimer et al., 2006) and kd-tree (Robinson, 1981).

Although KNNs rarely provide the best predictive performance, they have many benefits, for example, in feature engineering and in data cleaning and preprocessing. We discussed KNN for imputation in Section 3.3.2. Bruce and Bruce (2017) discuss another approach that uses KNNs to add a *local knowledge* feature. This includes running a KNN to estimate the predicted output or class and using this predicted value as a new feature for downstream modeling. However, this approach also invites more opportunities for target leakage.

Other alternatives to traditional KNNs such as using invariant metrics, tangent distance metrics, and adaptive nearest neighbor methods are also discussed in Friedman et al. (2001) and are worth exploring.

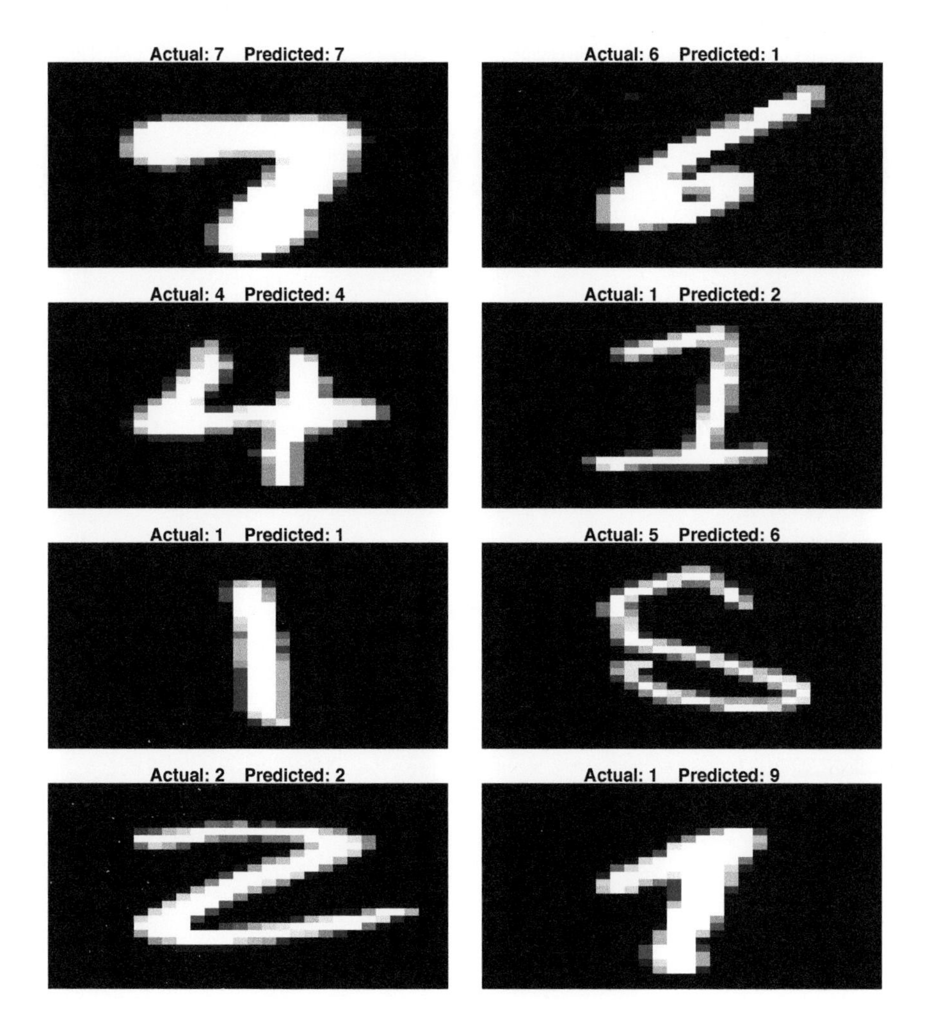

FIGURE 8.8: Actual images from the MNIST data set along with our KNN model's predictions. Left column illustrates a few accurate predictions and the right column illustrates a few inaccurate predictions.

9

Decision Trees

Tree-based models are a class of nonparametric algorithms that work by partitioning the feature space into a number of smaller (non-overlapping) regions with similar response values using a set of *splitting rules*. Predictions are obtained by fitting a simpler model (e.g., a constant like the average response value) in each region. Such *divide-and-conquer* methods can produce simple rules that are easy to interpret and visualize with *tree diagrams*. As we'll see, decision trees offer many benefits; however, they typically lack in predictive performance compared to more complex algorithms like neural networks and MARS. However, future chapters will discuss powerful ensemble algorithms—like random forests and gradient boosting machines—which are constructed by combining together many decision trees in a clever way. This chapter will provide you with a strong foundation in decision trees.

9.1 Prerequisites

In this chapter we'll use the following packages:

```
# Helper packages
library(dplyr)       # for data wrangling
library(ggplot2)     # for awesome plotting

# Modeling packages
library(rpart)       # direct engine for decision tree application
library(caret)       # meta engine for decision tree application

# Model interpretability packages
library(rpart.plot)  # for plotting decision trees
library(vip)         # for feature importance
library(pdp)         # for feature effects
```

We'll continue to illustrate the main concepts using the Ames housing example from Section 2.7.

9.2 Structure

There are many methodologies for constructing decision trees but the most well-known is the **c**lassification **a**nd **r**egression **t**ree (CART) algorithm proposed in Breiman (1984).[1] A basic decision tree partitions the training data into homogeneous subgroups (i.e., groups with similar response values) and then fits a simple *constant* in each subgroup (e.g., the mean of the within group response values for regression). The subgroups (also called nodes) are formed recursively using binary partitions formed by asking simple yes-or-no questions about each feature (e.g., is `age < 18`?). This is done a number of times until a suitable stopping criteria is satisfied (e.g., a maximum depth of the tree is reached). After all the partitioning has been done, the model predicts the output based on (1) the average response values for all observations that fall in that subgroup (regression problem), or (2) the class that has majority representation (classification problem). For classification, predicted probabilities can be obtained using the proportion of each class within the subgroup.

What results is an inverted tree-like structure such as that in Figure 9.1. In essence, our tree is a set of rules that allows us to make predictions by asking simple yes-or-no questions about each feature. For example, if the customer is loyal, has household income greater than $150,000, and is shopping in a store, the exemplar tree diagram in Figure 9.1 would predict that the customer will redeem a coupon.

We refer to the first subgroup at the top of the tree as the *root node* (this node contains all of the training data). The final subgroups at the bottom of the tree are called the *terminal nodes* or *leaves*. Every subgroup in between is referred to as an internal node. The connections between nodes are called *branches*.

[1] Other decision tree algorithms include the Iterative Dichotomiser 3 (Quinlan, 1986), C4.5 (Quinlan et al., 1996), Chi-square automatic interaction detection (Kass, 1980), Conditional inference trees (Hothorn et al., 2006), and more.

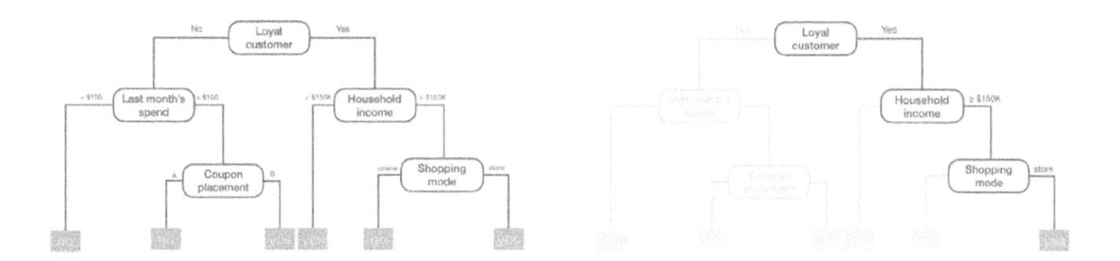

FIGURE 9.1: Exemplar decision tree predicting whether or not a customer will redeem a coupon (yes or no) based on the customer's loyalty, household income, last month's spend, coupon placement, and shopping mode.

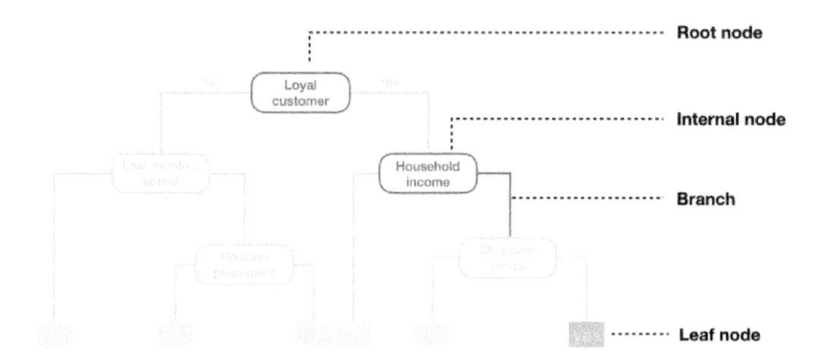

FIGURE 9.2: Terminology of a decision tree.

9.3 Partitioning

As illustrated above, CART uses *binary recursive partitioning* (it's recursive because each split or rule depends on the the splits above it). The objective at each node is to find the "best" feature (x_i) to partition the remaining data into one of two regions (R_1 and R_2) such that the overall error between the actual response (y_i) and the predicted constant (c_i) is minimized. For regression problems, the objective function to minimize is the total SSE as defined in Equation (9.1) below:

$$SSE = \sum_{i \in R_1} (y_i - c_1)^2 + \sum_{i \in R_2} (y_i - c_2)^2 \tag{9.1}$$

For classification problems, the partitioning is usually made to maximize the reduction in cross-entropy or the Gini index (see Section 2.6).[2]

[2]Gini index and cross-entropy are the two most commonly applied loss functions used

In both regression and classification trees, the objective of partitioning is to minimize dissimilarity in the terminal nodes. However, we suggest Therneau et al. (1997) for a more thorough discussion regarding binary recursive partitioning.

Having found the best feature/split combination, the data are partitioned into two regions and the splitting process is repeated on each of the two regions (hence the name binary recursive partitioning). This process is continued until a suitable stopping criterion is reached (e.g., a maximum depth is reached or the tree becomes "too complex").

It's important to note that a single feature can be used multiple times in a tree. For example, say we have data generated from a simple sin function with Gaussian noise: $Y_i \overset{iid}{\sim} N(\sin(X_i), \sigma^2)$, for $i = 1, 2, \dots, 500$. A regression tree built with a single root node (often referred to as a decision stump) leads to a split occurring at $x = 3.1$.

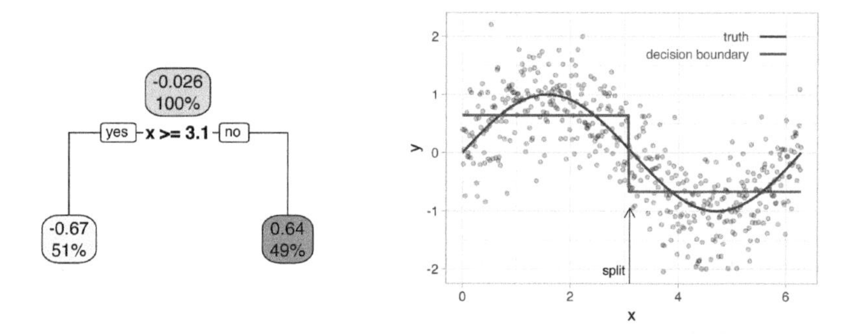

FIGURE 9.3: Decision tree illustrating the single split on feature x (left). The resulting decision boundary illustrates the predicted value when x < 3.1 (0.64), and when x > 3.1 (-0.67) (right).

If we build a deeper tree, we'll continue to split on the same feature (x) as illustrated in Figure 9.4. This is because x is the only feature available to split on so it will continue finding the optimal splits along this feature's values until a pre-determined stopping criteria is reached.

However, even when many features are available, a single feature may still dominate if it continues to provide the best split after each successive partition. For example, a decision tree applied to the iris data set (Fisher, 1936) where the species of the flower (setosa, versicolor, and virginica) is predicted based on two features (sepal width and sepal length) results in an optimal decision

for decision trees. Classification error is rarely used to determine partitions as they are less sensitive to poor performing splits (Friedman et al., 2001).

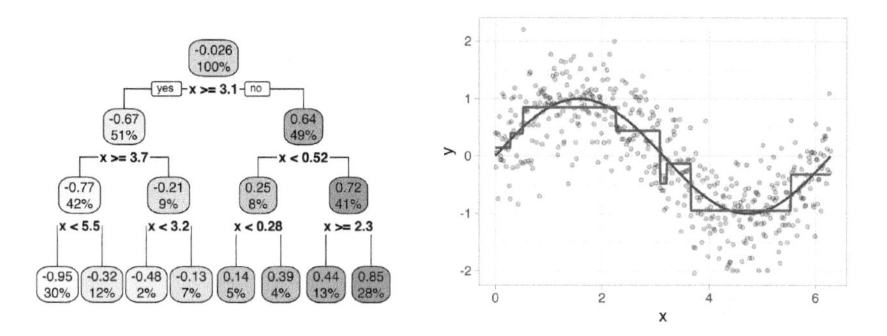

FIGURE 9.4: Decision tree illustrating with depth = 3, resulting in 7 decision splits along values of feature x and 8 prediction regions (left). The resulting decision boundary (right).

tree with two splits on each feature. Also, note how the decision boundary in a classification problem results in rectangular regions enclosing the observations. The predicted value is the response class with the greatest proportion within the enclosed region.

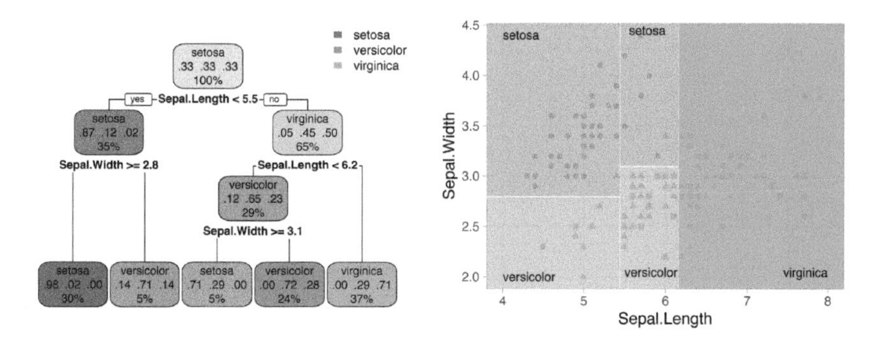

FIGURE 9.5: Decision tree for the iris classification problem (left). The decision boundary results in rectangular regions that enclose the observations. The class with the highest proportion in each region is the predicted value (right).

9.4 How deep?

This leads to an important question: how deep (i.e., complex) should we make the tree? If we grow an overly complex tree as in Figure 9.6, we tend to overfit to our training data resulting in poor generalization performance.

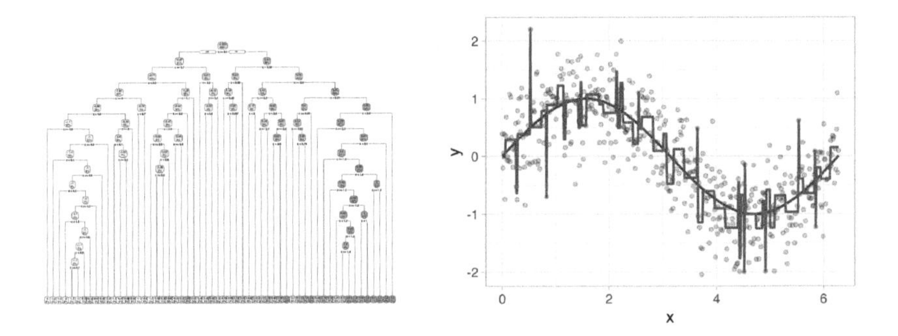

FIGURE 9.6: Overfit decision tree with 56 splits.

Consequently, there is a balance to be achieved in the depth and complexity of the tree to optimize predictive performance on future unseen data. To find this balance, we have two primary approaches: (1) early stopping and (2) pruning.

9.4.1 Early stopping

Early stopping explicitly restricts the growth of the tree. There are several ways we can restrict tree growth but two of the most common approaches are to restrict the tree depth to a certain level or to restrict the minimum number of observations allowed in any terminal node. When limiting tree depth we stop splitting after a certain depth (e.g., only grow a tree that has a depth of 5 levels). The shallower the tree the less variance we have in our predictions; however, at some point we can start to inject too much bias as shallow trees (e.g., stumps) are not able to capture interactions and complex patterns in our data.

When restricting minimum terminal node size (e.g., leaf nodes must contain at least 10 observations for predictions) we are deciding to not split intermediate nodes which contain too few data points. At the far end of the spectrum, a terminal node's size of one allows for a single observation to be captured in the leaf node and used as a prediction (in this case, we're interpolating the training data). This results in high variance and poor generalizability. On the other hand, large values restrict further splits therefore reducing variance.

These two approaches can be implemented independently of one another; however, they do have interaction effects as illustrated by Figure 9.7.

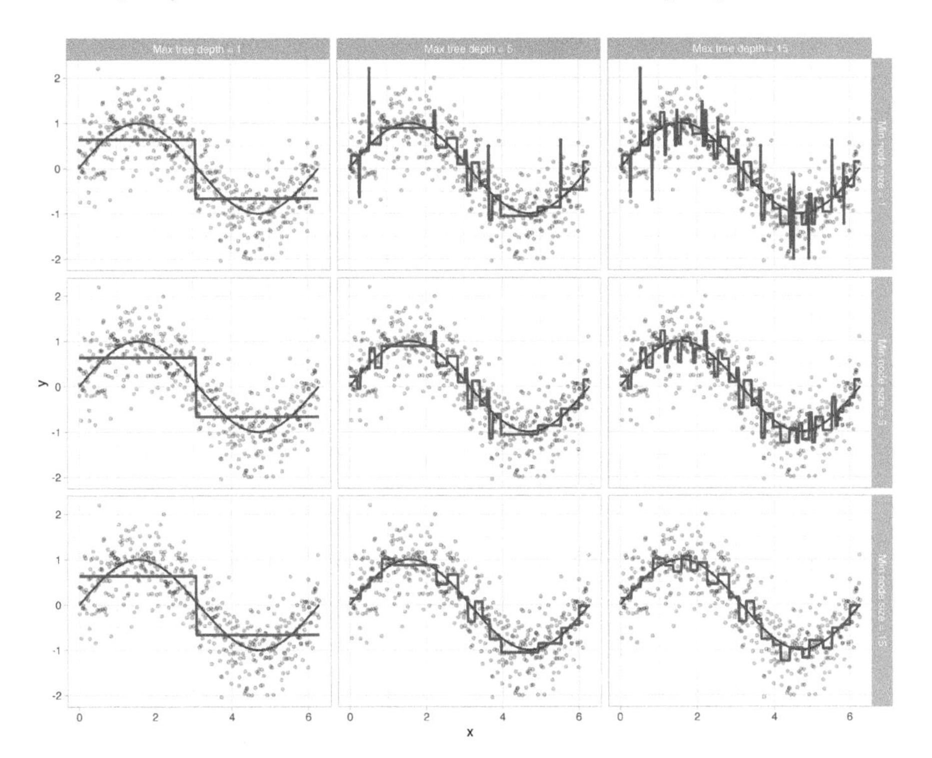

FIGURE 9.7: Illustration of how early stopping affects the decision boundary of a regression decision tree. The columns illustrate how tree depth impacts the decision boundary and the rows illustrate how the minimum number of observations in the terminal node influences the decision boundary.

9.4.2 Pruning

An alternative to explicitly specifying the depth of a decision tree is to grow a very large, complex tree and then *prune* it back to find an optimal subtree. We find the optimal subtree by using a *cost complexity parameter* (α) that penalizes our objective function in Equation (9.1) for the number of terminal nodes of the tree (T) as in Equation (9.2).

$$\texttt{minimize}\left\{SSE + \alpha|T|\right\} \tag{9.2}$$

For a given value of α we find the smallest pruned tree that has the lowest penalized error. You may recognize the close association to the lasso penalty discussed in Chapter 6. As with the regularization methods, smaller penalties tend to produce more complex models, which result in larger trees. Whereas larger penalties result in much smaller trees. Consequently, as a tree grows larger, the reduction in the SSE must be greater than the cost complexity penalty. Typically, we evaluate multiple models across a spectrum of α and use CV to identify the optimal value and, therefore, the optimal subtree that generalizes best to unseen data.

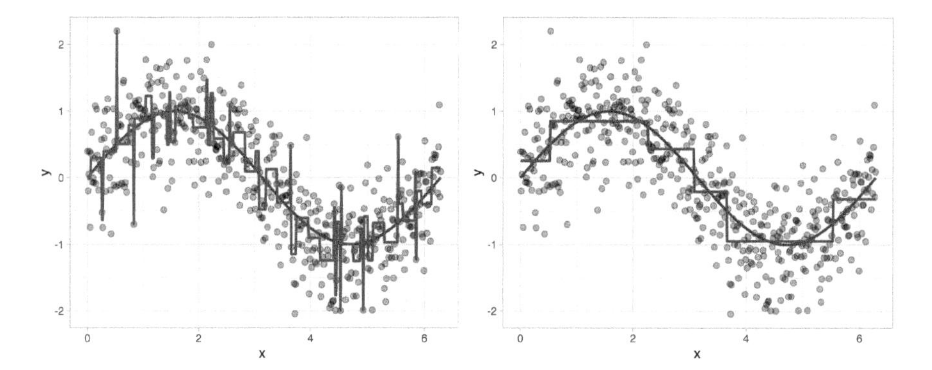

FIGURE 9.8: To prune a tree, we grow an overly complex tree (left) and then use a cost complexity parameter to identify the optimal subtree (right).

9.5 Ames housing example

We can fit a regression tree using `rpart` and then visualize it using `rpart.plot`. The fitting process and the visual output of regression trees and classification trees are very similar. Both use the formula method for expressing the model (similar to `lm()`). However, when fitting a regression tree, we need to set `method = "anova"`. By default, `rpart()` will make an intelligent guess as to what method to use based on the data type of your response column, but it's good practice to set this explicitly.

```
ames_dt1 <- rpart(
  formula = Sale_Price ~ .,
  data    = ames_train,
  method  = "anova"
)
```

Once we've fit our model we can take a peak at the decision tree output. This prints various information about the different splits. For example, we start with `2054` observations at the root node and the first variable we split on (i.e., the first variable gave the largest reduction in SSE) is `Overall_Qual`. We see that at the first node all observations with `Overall_Qual` $\in$ {`Very_Poor`, `Poor`, `Fair`, `Below_Average`, `Average`, `Above_Average`, `Good`} go to the 2nd (`2)`) branch. The total number of observations that follow this branch (`1721`), their average sales price (`156200`) and SSE (`3.964e+12`) are listed. If you look for the 3rd branch (`3)`) you will see that `333` observations with `Overall_Qual` $\in$ {`Very_Good`, `Excellent`, `Very_Excellent`} follow this branch and their average sales prices is `304600` and the SEE in this region is `2.917e+12`. Basically, this is telling us that `Overall_Qual` is an important predictor on sales price with those homes on the upper end of the quality spectrum having almost double the average sales price.

```
ames_dt1
## n= 2054
##
## node), split, n, deviance, yval
##       * denotes terminal node
##
##  1) root 2054 1.32e+13 181000
##    2) Overall_Qual=Very_Poor,Poor,Fair,Below_Average,Average,Abo...
##      4) Neighborhood=North_Ames,Old_Town,Edwards,Sawyer,Mitchell...
##        8) Overall_Qual=Very_Poor,Poor,Fair,Below_Average 195 1.6...
##        9) Overall_Qual=Average,Above_Average,Good 827 8.15e+11 1...
##         18) First_Flr_SF< 1.21e+03 631 3.84e+11 132000 *
##         19) First_Flr_SF>=1.21e+03 196 2.74e+11 165000 *
##      5) Neighborhood=College_Creek,Somerset,Northridge_Heights,G...
##       10) Gr_Liv_Area< 1.72e+03 492 5.18e+11 178000
##         20) Total_Bsmt_SF< 1.33e+03 353 2.33e+11 167000 *
##         21) Total_Bsmt_SF>=1.33e+03 139 1.37e+11 205000 *
##       11) Gr_Liv_Area>=1.72e+03 194 3.38e+11 229000 *
##    3) Overall_Qual=Very_Good,Excellent,Very_Excellent 346 2.92e+...
##      6) Overall_Qual=Very_Good 249 9.55e+11 272000
##       12) Gr_Liv_Area< 1.97e+03 152 3.13e+11 244000 *
##       13) Gr_Liv_Area>=1.97e+03 97 3.32e+11 317000 *
##      7) Overall_Qual=Excellent,Very_Excellent 97 1.04e+12 387000
##       14) Total_Bsmt_SF< 1.9e+03 65 2.32e+11 349000 *
##       15) Total_Bsmt_SF>=1.9e+03 32 5.14e+11 465000
##         30) Year_Built>=2e+03 25 2.70e+11 430000 *
##         31) Year_Built< 2e+03 7 9.74e+10 593000 *
```

We can visualize our tree model with `rpart.plot()`. The `rpart.plot()` function has many plotting options, which we'll leave to the reader to explore. However, in the default print it will show the percentage of data that fall in each node and the predicted outcome for that node. One thing you may notice is that this tree contains 10 internal nodes resulting in 11 terminal nodes. In other words, this tree is partitioning on only 10 features even though there are 80 variables in the training data. Why is that?

```
rpart.plot(ames_dt1)
```

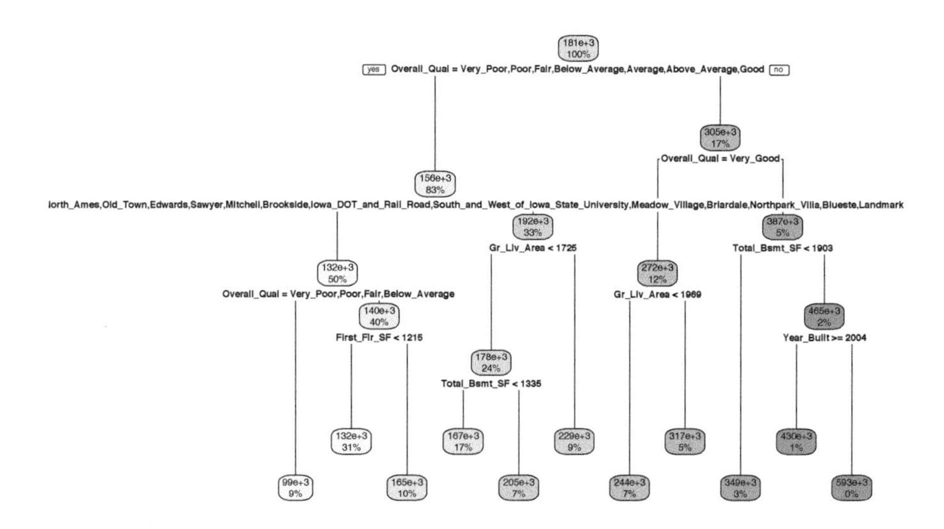

FIGURE 9.9: Diagram displaying the pruned decision tree for the Ames Housing data.

Behind the scenes `rpart()` is automatically applying a range of cost complexity (α values to prune the tree). To compare the error for each α value, `rpart()` performs a 10-fold CV (by default). In this example we find diminishing returns after 12 terminal nodes as illustrated in Figure 9.10 (y-axis is the CV error, lower x-axis is the cost complexity (α) value, upper x-axis is the number of terminal nodes (i.e., tree size $= |T|$). You may also notice the dashed line which goes through the point $|T| = 8$. Breiman (1984) suggested that in actual practice, it's common to instead use the smallest tree within 1 standard error (SE) of the minimum CV error (this is called the *1-SE rule*). Thus, we could use a tree with 8 terminal nodes and reasonably expect to experience similar results within a small margin of error.

```
plotcp(ames_dt1)
```

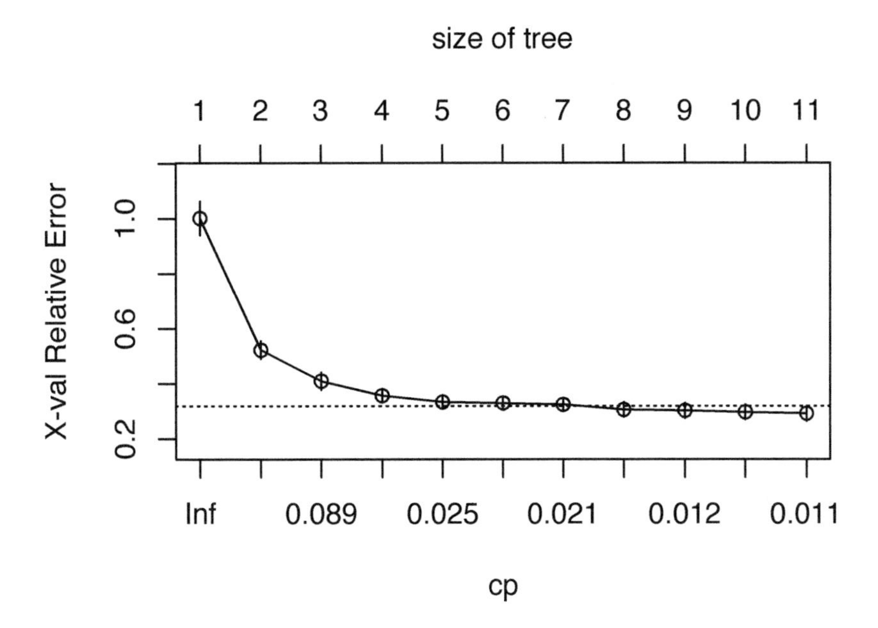

FIGURE 9.10: Pruning complexity parameter (cp) plot illustrating the relative cross validation error (y-axis) for various cp values (lower x-axis). Smaller cp values lead to larger trees (upper x-axis). Using the 1-SE rule, a tree size of 10-12 provides optimal cross validation results.

To illustrate the point of selecting a tree with 11 terminal nodes (or 8 if you go by the 1-SE rule), we can force `rpart()` to generate a full tree by setting `cp = 0` (no penalty results in a fully grown tree). Figure 9.11 shows that after 11 terminal nodes, we see diminishing returns in error reduction as the tree grows deeper. Thus, we can significantly prune our tree and still achieve minimal expected error.

```
ames_dt2 <- rpart(
    formula = Sale_Price ~ .,
    data    = ames_train,
    method  = "anova",
    control = list(cp = 0, xval = 10)
)
```

```r
plotcp(ames_dt2)
abline(v = 11, lty = "dashed")
```

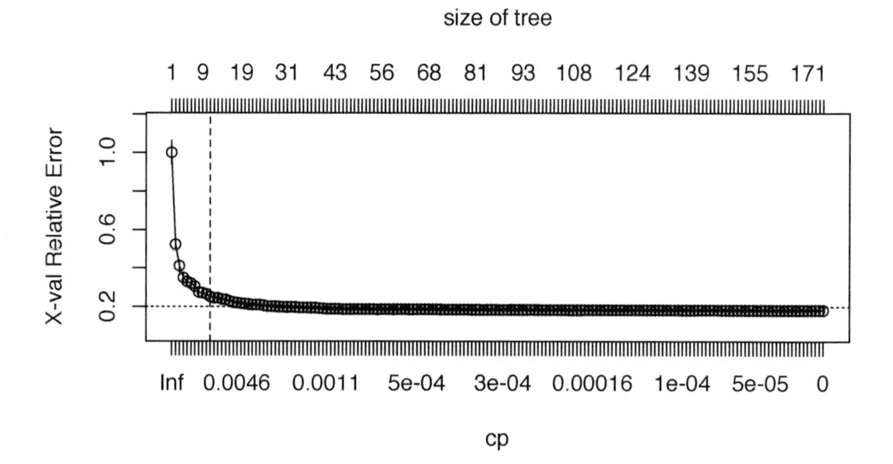

FIGURE 9.11: Pruning complexity parameter plot for a fully grown tree. Significant reduction in the cross validation error is achieved with tree sizes 6-20 and then the cross validation error levels off with minimal or no additional improvements.

So, by default, `rpart()` is performing some automated tuning, with an optimal subtree of 10 total splits, 11 terminal nodes, and a cross-validated SSE of 0.292. Although `rpart()` does not provide the RMSE or other metrics, you can use **caret**. In both cases, smaller penalties (deeper trees) are providing better CV results.

```r
# rpart cross validation results
ames_dt1$cptable
##        CP nsplit rel error xerror   xstd
## 1  0.4794      0     1.000  1.001 0.0612
## 2  0.1129      1     0.521  0.523 0.0320
## 3  0.0700      2     0.408  0.410 0.0311
## 4  0.0276      3     0.338  0.357 0.0222
## 5  0.0235      4     0.310  0.334 0.0218
## 6  0.0220      5     0.287  0.330 0.0245
## 7  0.0204      6     0.265  0.324 0.0242
## 8  0.0119      7     0.244  0.306 0.0264
## 9  0.0112      8     0.232  0.303 0.0271
```

```
## 10 0.0110        9      0.221  0.297 0.0270
## 11 0.0100       10      0.210  0.292 0.0270

# caret cross validation results
ames_dt3 <- train(
  Sale_Price ~ .,
  data = ames_train,
  method = "rpart",
  trControl = trainControl(method = "cv", number = 10),
  tuneLength = 20
)

ggplot(ames_dt3)
```

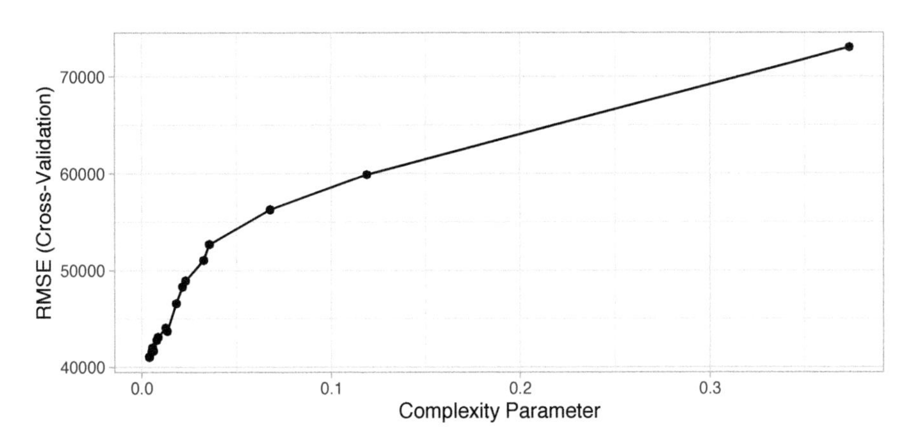

FIGURE 9.12: Cross-validated accuracy rate for the 20 different α parameter values in our grid search. Lower α values (deeper trees) help to minimize errors.

9.6 Feature interpretation

To measure feature importance, the reduction in the loss function (e.g., SSE) attributed to each variable at each split is tabulated. In some instances, a single variable could be used multiple times in a tree; consequently, the total reduction in the loss function across all splits by a variable are summed up and used as the total feature importance. When using **caret**, these values are standardized so that the most important feature has a value of 100 and the remaining features are scored based on their relative reduction in the loss function. Also, since there may be candidate variables that are important but are not used in a split, the top competing variables are also tabulated at each split.

Figure 9.13 illustrates the top 40 features in the Ames housing decision tree. Similar to MARS (Chapter 7), decision trees perform automated feature selection where uninformative features are not used in the model. We can see this in Figure 9.13 where the bottom four features in the plot have zero importance.

```
vip(ames_dt3, num_features = 40, bar = FALSE)
```

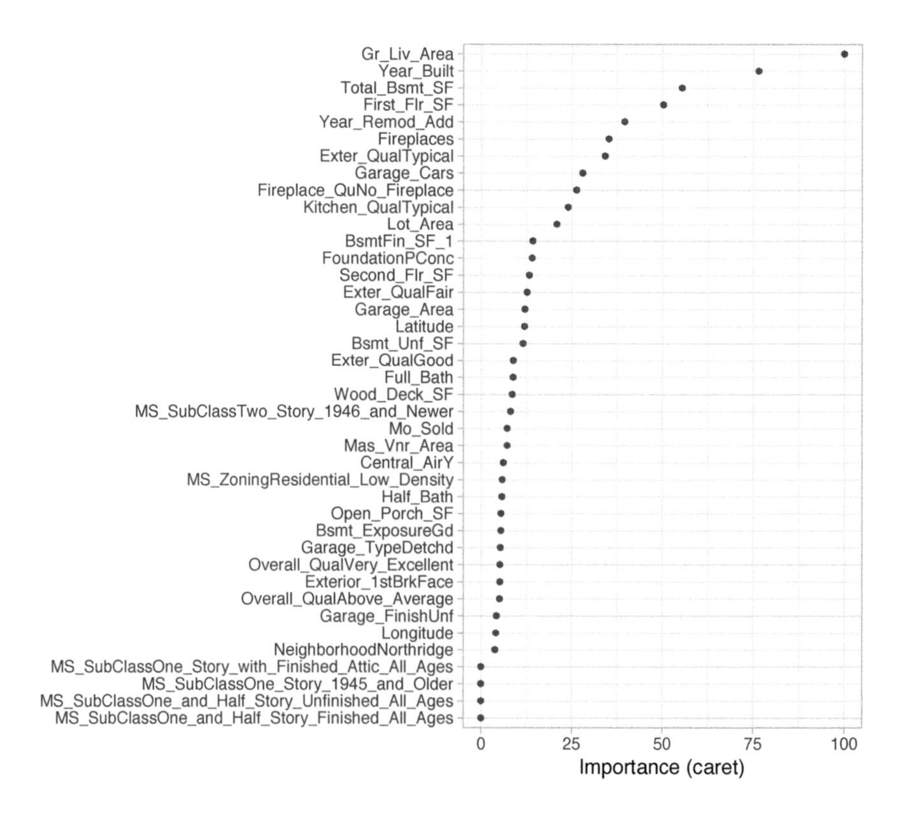

FIGURE 9.13: Variable importance based on the total reduction in MSE for the Ames Housing decision tree.

If we look at the same partial dependence plots that we created for the MARS models (Section 7.5), we can see the similarity in how decision trees are modeling the relationship between the features and target. In Figure 9.14, we see that

`Gr_Liv_Area` has a non-linear relationship such that it has increasingly stronger effects on the predicted sales price for `Gr_liv_Area` values between 1000–2500 but then has little, if any, influence when it exceeds 2500. However, the 3-D plot of the interaction effect between `Gr_Liv_Area` and `Year_Built` illustrates a key difference in how decision trees have rigid non-smooth prediction surfaces compared to MARS; in fact, MARS was developed as an improvement to CART for regression problems.

```
# Construct partial dependence plots
p1 <- partial(ames_dt3, pred.var = "Gr_Liv_Area") %>% autoplot()
p2 <- partial(ames_dt3, pred.var = "Year_Built") %>% autoplot()
p3 <- partial(ames_dt3, pred.var = c("Gr_Liv_Area", "Year_Built")) %>%
  plotPartial(levelplot = FALSE, zlab = "yhat", drape = TRUE,
              colorkey = TRUE, screen = list(z = -20, x = -60))

# Display plots side by side
gridExtra::grid.arrange(p1, p2, p3, ncol = 3)
```

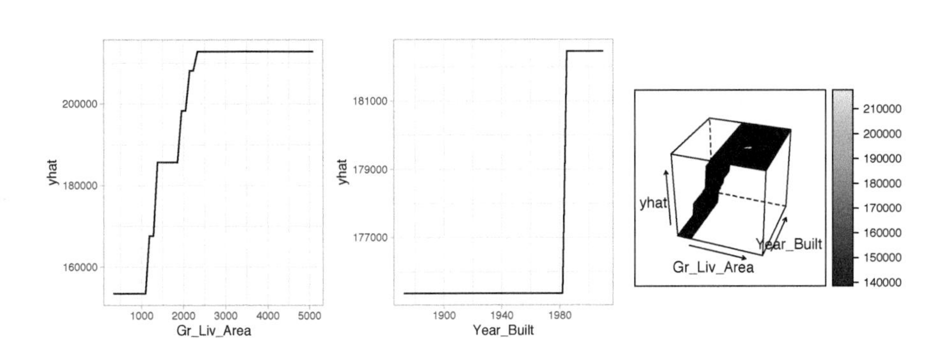

FIGURE 9.14: Partial dependence plots to understand the relationship between sale price and the living space, and year built features.

9.7 Final thoughts

Decision trees have a number of advantages. Trees require very little pre-processing. This is not to say feature engineering may not improve upon a decision tree, but rather, that there are no pre-processing requirements. Monotonic transformations (e.g., log, exp, and $\sqrt{}$) are not required to meet algorithm assumptions as in many parametric models; instead, they only shift the location of the optimal split points. Outliers typically do not bias the results as much since the binary partitioning simply looks for a single location to make a split within the distribution of each feature.

Decision trees can easily handle categorical features without preprocessing. For unordered categorical features with more than two levels, the classes are ordered based on the outcome (for regression problems, the mean of the response is used and for classification problems, the proportion of the positive outcome class is used). For more details see Friedman et al. (2001), Breiman and Ihaka (1984), Ripley (2007), Fisher (1958), and Loh and Vanichsetakul (1988).

Missing values often cause problems with statistical models and analyses. Most procedures deal with them by refusing to deal with them—incomplete observations are tossed out. However, most decision tree implementations can easily handle missing values in the features and do not require imputation. This is handled in various ways but most commonly by creating a new "missing" class for categorical variables or using surrogate splits (see Therneau et al. (1997) for details).

However, individual decision trees generally do not often achieve state-of-the-art predictive accuracy. In this chapter, we saw that the best pruned decision tree, although it performed better than linear regression (Chapter 4), had a very poor RMSE ($41,019) compared to some of the other models we've built. This is driven by the fact that decision trees are composed of simple yes-or-no rules that create rigid non-smooth decision boundaries. Furthermore, we saw that deep trees tend to have high variance (and low bias) and shallow trees tend to be overly bias (but low variance). In the chapters that follow, we'll see how we can combine multiple trees together into very powerful prediction models called *ensembles*.

10

Bagging

In Section 2.4.2 we learned about bootstrapping as a resampling procedure, which creates b new bootstrap samples by drawing samples with replacement of the original training data. This chapter illustrates how we can use bootstrapping to create an *ensemble* of predictions. Bootstrap aggregating, also called *bagging*, is one of the first ensemble algorithms[1] machine learning practitioners learn and is designed to improve the stability and accuracy of regression and classification algorithms. By model averaging, bagging helps to reduce variance and minimize overfitting. Although it is usually applied to decision tree methods, it can be used with any type of method.

10.1 Prerequisites

In this chapter we'll make use of the following packages:

```
# Helper packages
library(dplyr)       # for data wrangling
library(ggplot2)     # for awesome plotting
library(doParallel)  # for parallel backend to foreach
library(foreach)     # for parallel processing with for loops

# Modeling packages
library(caret)       # for general model fitting
library(rpart)       # for fitting decision trees
library(ipred)       # for fitting bagged decision trees
```

We'll continue to illustrate the main concepts with the `ames_train` data set created in Section 2.7.

[1] Also commonly referred to as a meta-algorithm.

10.2 Why and when bagging works

Bootstrap aggregating (bagging) prediction models is a general method for fitting multiple versions of a prediction model and then combining (or ensembling) them into an aggregated prediction (Breiman, 1996a). Bagging is a fairly straight forward algorithm in which b bootstrap copies of the original training data are created, the regression or classification algorithm (commonly referred to as the *base learner*) is applied to each bootstrap sample and, in the regression context, new predictions are made by averaging the predictions together from the individual base learners. When dealing with a classification problem, the base learner predictions are combined using plurality vote or by averaging the estimated class probabilities together. This is represented in Equation (10.1) where X is the record for which we want to generate a prediction, $\widehat{f_{bag}}$ is the bagged prediction, and $\widehat{f_1}(X), \widehat{f_2}(X), \dots, \widehat{f_b}(X)$ are the predictions from the individual base learners.

$$\widehat{f_{bag}} = \widehat{f_1}(X) + \widehat{f_2}(X) + \cdots + \widehat{f_b}(X) \tag{10.1}$$

Because of the aggregation process, bagging effectively reduces the variance of an individual base learner (i.e., averaging reduces variance); however, bagging does not always improve upon an individual base learner. As discussed in Section 2.5, some models have larger variance than others. Bagging works especially well for unstable, high variance base learners—algorithms whose predicted output undergoes major changes in response to small changes in the training data (Dietterich, 2000a,b). This includes algorithms such as decision trees and KNN (when k is sufficiently small). However, for algorithms that are more stable or have high bias, bagging offers less improvement on predicted outputs since there is less variability (e.g., bagging a linear regression model will effectively just return the original predictions for large enough b).

The general idea behind bagging is referred to as the "wisdom of the crowd" effect and was popularized by Surowiecki (2005). It essentially means that the aggregation of information in large diverse groups results in decisions that are often better than could have been made by any single member of the group. The more diverse the group members are then the more diverse their perspectives and predictions will be, which often leads to better aggregated information. Think of estimating the number of jelly beans in a jar at a carinival. While any individual guess is likely to be way off, you'll often find that the averaged guesses tends to be a lot closer to the true number.

This is illustrated in Figure 10.1, which compares bagging $b = 100$ polynomial regression models, MARS models, and CART decision trees. You can see that the low variance base learner (polynomial regression) gains very little from bagging while the higher variance learner (decision trees) gains significantly more. Not only does bagging help minimize the high variability (instability) of single trees, but it also helps to smooth out the prediction surface.

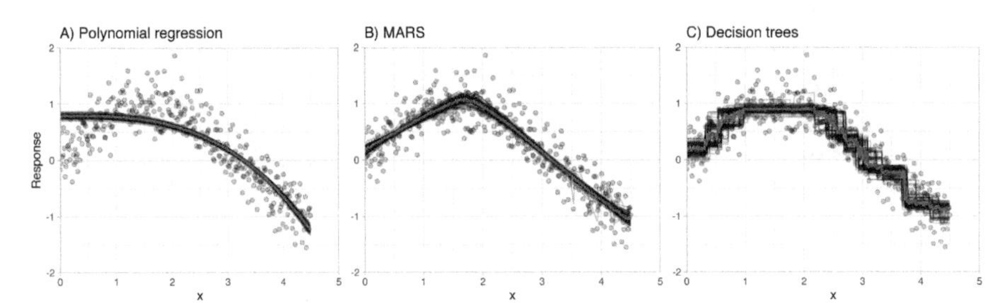

FIGURE 10.1: The effect of bagging 100 base learners. High variance models such as decision trees (C) benefit the most from the aggregation effect in bagging, whereas low variance models such as polynomial regression (A) show little improvement.

Optimal performance is often found by bagging 50–500 trees. Data sets that have a few strong predictors typically require less trees; whereas data sets with lots of noise or multiple strong predictors may need more. Using too many trees will not lead to overfitting. However, it's important to realize that since multiple models are being run, the more iterations you perform the more computational and time requirements you will have. As these demands increase, performing k-fold CV can become computationally burdensome.

A benefit to creating ensembles via bagging, which is based on resampling with replacement, is that it can provide its own internal estimate of predictive performance with the out-of-bag (OOB) sample (see Section 2.4.2). The OOB sample can be used to test predictive performance and the results usually compare well compared to k-fold CV assuming your data set is sufficiently large (say $n \geq 1,000$). Consequently, as your data sets become larger and your bagging iterations increase, it is common to use the OOB error estimate as a proxy for predictive performance.

Think of the OOB estimate of generalization performance as an unstructured, but free CV statistic.

10.3 Implementation

In Chapter 9, we saw how decision trees performed in predicting the sales price for the Ames housing data. Performance was subpar compared to the MARS (Chapter 7) and KNN (Chapter 8) models we fit, even after tuning to find the optimal pruned tree. Rather than use a single pruned decision tree, we can use, say, 100 bagged unpruned trees (by not pruning the trees we're keeping bias low and variance high which is when bagging will have the biggest effect). As the below code chunk illustrates, we gain significant improvement over our individual (pruned) decision tree (RMSE of 26,462 for bagged trees vs. 41,019 for the single decision tree).

The `bagging()` function comes from the **ipred** package and we use `nbagg` to control how many iterations to include in the bagged model and `coob = TRUE` indicates to use the OOB error rate. By default, `bagging()` uses `rpart::rpart()` for decision tree base learners but other base learners are available. Since bagging just aggregates a base learner, we can tune the base learner parameters as normal. Here, we pass parameters to `rpart()` with the `control` parameter and we build deep trees (no pruning) that require just two observations in a node to split.

```
# make bootstrapping reproducible
set.seed(123)

# train bagged model
ames_bag1 <- bagging(
  formula = Sale_Price ~ .,
  data = ames_train,
  nbagg = 100,
  coob = TRUE,
  control = rpart.control(minsplit = 2, cp = 0)
)

ames_bag1
##
## Bagging regression trees with 100 bootstrap replications
##
## Call: bagging.data.frame(formula = Sale_Price ~ ., data = ames_train,
##     nbagg = 100, coob = TRUE, control = rpart.control(minsplit = 2,
```

```
##         cp = 0))
##
## Out-of-bag estimate of root mean squared error:  26462
```

One thing to note is that typically, the more trees the better. As we add more trees we're averaging over more high variance decision trees. Early on, we see a dramatic reduction in variance (and hence our error) but eventually the error will typically flatline and stabilize signaling that a suitable number of trees has been reached. Often, we need only 50–100 trees to stabilize the error (in other cases we may need 500 or more). For the Ames data we see that the error is stabilizing with just over 100 trees so we'll likely not gain much improvement by simply bagging more trees.

Unfortunately, `bagging()` does not provide the RMSE by tree so to produce this error curve we iterated over `nbagg` values of 1–200 and applied the same `bagging()` function above.

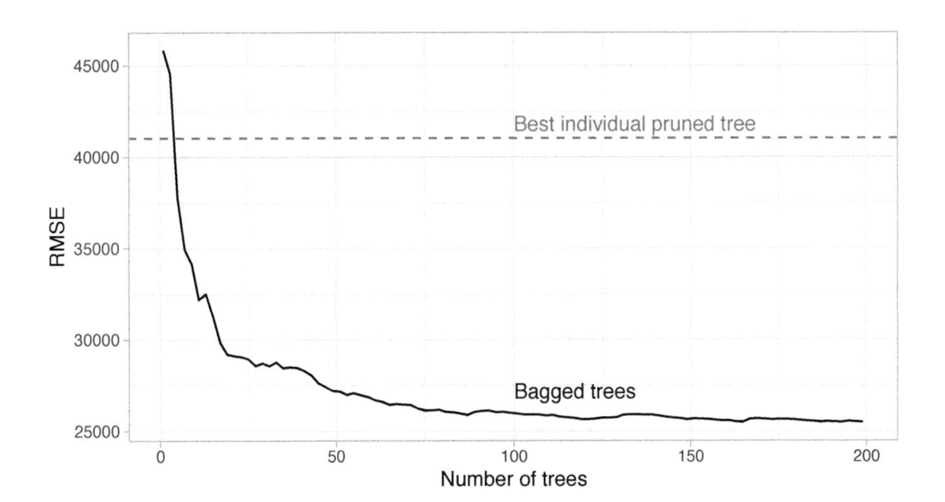

FIGURE 10.2: Error curve for bagging 1-200 deep, unpruned decision trees. The benefit of bagging is optimized at 187 trees although the majority of error reduction occurred within the first 100 trees.

We can also apply bagging within **caret** and use 10-fold CV to see how well our ensemble will generalize. We see that the cross-validated RMSE for 200 trees is similar to the OOB estimate (difference of 495). However, using the OOB error took 58 seconds to compute whereas performing the following 10-fold CV took roughly 26 minutes on our machine!

```
ames_bag2 <- train(
  Sale_Price ~ .,
  data = ames_train,
  method = "treebag",
  trControl = trainControl(method = "cv", number = 10),
  nbagg = 200,
  control = rpart.control(minsplit = 2, cp = 0)
)
ames_bag2
## Bagged CART
##
## 2054 samples
##   80 predictor
##
## No pre-processing
## Resampling: Cross-Validated (10 fold)
## Summary of sample sizes: 1849, 1848, 1848, 1849, 1849, 1847, ...
## Resampling results:
##
##   RMSE      Rsquared   MAE
##   26957.06  0.8900689  16713.14
```

10.4 Easily parallelize

As stated in Section 10.2, bagging can become computationally intense as the number of iterations increases. Fortunately, the process of bagging involves fitting models to each of the bootstrap samples which are completely independent of one another. This means that each model can be trained in parallel and the results aggregated in the end for the final model. Consequently, if you have access to a large cluster or number of cores, you can more quickly create bagged ensembles on larger data sets.

The following illustrates parallelizing the bagging algorithm (with $b = 160$ decision trees) on the Ames housing data using eight cores and returning the predictions for the test data for each of the trees.

```
# Create a parallel socket cluster
cl <- makeCluster(8) # use 8 workers
registerDoParallel(cl) # register the parallel backend
```

```
# Fit trees in parallel and compute predictions on the test set
predictions <- foreach(
  icount(160),
  .packages = "rpart",
  .combine = cbind
  ) %dopar% {
    # bootstrap copy of training data
    index <- sample(nrow(ames_train), replace = TRUE)
    ames_train_boot <- ames_train[index, ]

    # fit tree to bootstrap copy
    bagged_tree <- rpart(
      Sale_Price ~ .,
      control = rpart.control(minsplit = 2, cp = 0),
      data = ames_train_boot
      )

    predict(bagged_tree, newdata = ames_test)
}

predictions[1:5, 1:7]
##   result.1 result.2 result.3 result.4 result.5 result.6 result.7
## 1   158000   195500   187500   190500   185000   187500   183000
## 2   180000   240000   180000   180000   180000   172500   188500
## 3   163990   166000   178900   180000   176000   181755   175900
## 4   210000   174000   224000   236500   233555   215000   223000
## 5   129000   137500   143250   147000   147000   139000   144000
```

We can then do some data wrangling to compute and plot the RMSE as additional trees are added. Our results, illustrated in Figure 10.3, closely resemble the results obtained in Figure 10.2. This also illustrates how the OOB error closely approximates the test error.

```
predictions %>%
  as.data.frame() %>%
  mutate(
    observation = 1:n(),
    actual = ames_test$Sale_Price) %>%
  tidyr::gather(tree, predicted, -c(observation, actual)) %>%
  group_by(observation) %>%
  mutate(tree = stringr::str_extract(tree, '\\d+') %>% as.numeric()) %>%
  ungroup() %>%
```

```
  arrange(observation, tree) %>%
  group_by(observation) %>%
  mutate(avg_prediction = cummean(predicted)) %>%
  group_by(tree) %>%
  summarize(RMSE = RMSE(avg_prediction, actual)) %>%
  ggplot(aes(tree, RMSE)) +
  geom_line() +
  xlab('Number of trees')
```

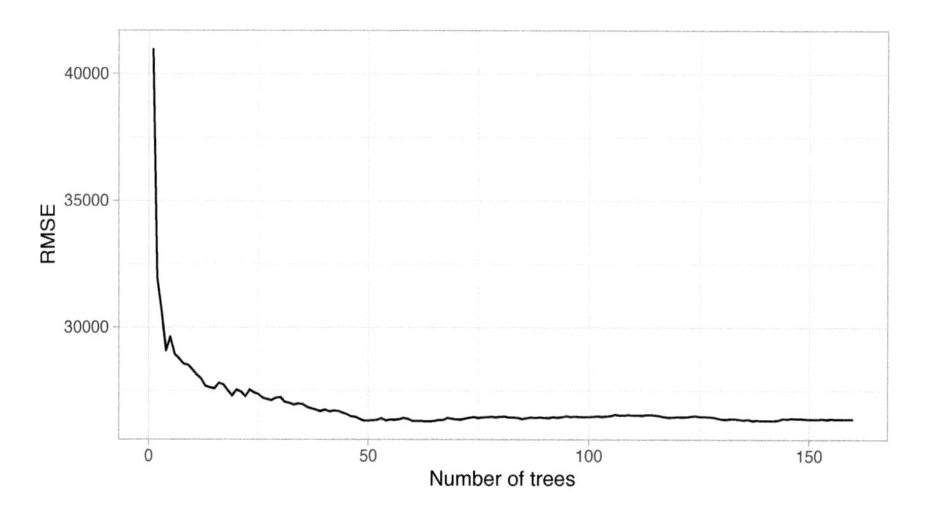

FIGURE 10.3: Error curve for custom parallel bagging of 1-160 deep, unpruned decision trees.

```
# Shutdown parallel cluster
stopCluster(cl)
```

10.5 Feature interpretation

Unfortunately, due to the bagging process, models that are normally perceived as interpretable are no longer so. However, we can still make inferences about how features are influencing our model. Recall in Section 9.6 that we measure feature importance based on the sum of the reduction in the loss function (e.g., SSE) attributed to each variable at each split in a given tree.

For bagged decision trees, this process is similar. For each tree, we compute the sum of the reduction of the loss function across all splits. We then aggregate this measure across all trees for each feature. The features with the largest average decrease in SSE (for regression) are considered most important. Unfortunately, the **ipred** package does not capture the required information for computing variable importance but the **caret** package does. In the code chunk below, we use **vip** to construct a variable importance plot (VIP) of the top 40 features in the `ames_bag2` model.

With a single decision tree, we saw that many non-informative features were not used in the tree. However, with bagging, since we use many trees built on bootstrapped samples, we are likely to see many more features used for splits. Consequently, we tend to have many more features involved but with lower levels of importance.

```
vip::vip(ames_bag2, num_features = 40, bar = FALSE)
```

Understanding the relationship between a feature and predicted response for bagged models follows the same procedure we've seen in previous chapters. PDPs tell us visually how each feature influences the predicted output, on average. Although the averaging effect of bagging diminishes the ability to interpret the final ensemble, PDPs and other interpretability methods (Chapter 16) help us to interpret any "black box" model. Figure 10.5 highlights the unique, and sometimes non-linear, non-monotonic relationships that may exist between a feature and response.

```
# Construct partial dependence plots
p1 <- pdp::partial(
  ames_bag2,
  pred.var = "Lot_Area",
  grid.resolution = 20
  ) %>%
  autoplot()

p2 <- pdp::partial(
  ames_bag2,
  pred.var = "Lot_Frontage",
  grid.resolution = 20
  ) %>%
  autoplot()
```

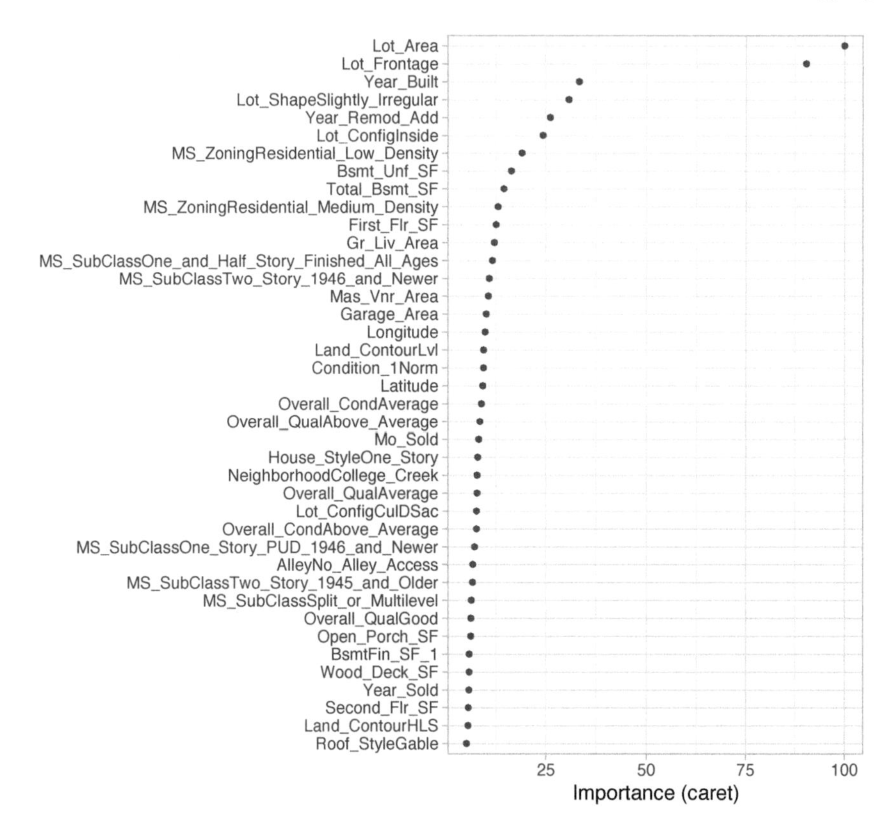

FIGURE 10.4: Variable importance for 200 bagged trees for the Ames Housing data.

```r
gridExtra::grid.arrange(p1, p2, nrow = 1)
```

10.6 Final thoughts

Bagging improves the prediction accuracy for high variance (and low bias) models at the expense of interpretability and computational speed. However, using various interpretability algorithms such as VIPs and PDPs, we can still make inferences about how our bagged model leverages feature information. Also, since bagging consists of independent processes, the algorithm is easily parallelizable.

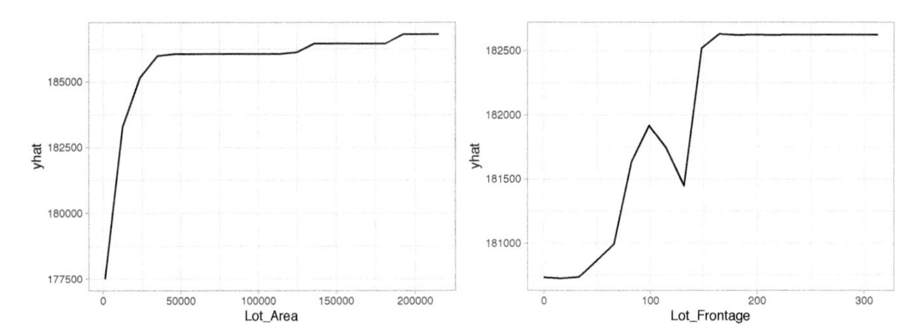

FIGURE 10.5: Partial dependence plots to understand the relationship between sales price and the lot area and frontage size features.

However, when bagging trees, a problem still exists. Although the model building steps are independent, the trees in bagging are not completely independent of each other since all the original features are considered at every split of every tree. Rather, trees from different bootstrap samples typically have similar structure to each other (especially at the top of the tree) due to any underlying strong relationships.

For example, if we create six decision trees with different bootstrapped samples of the Boston housing data (Harrison Jr and Rubinfeld, 1978), we see a similar structure as the top of the trees. Although there are 15 predictor variables to split on, all six trees have both `lstat` and `rm` variables driving the first few splits.

We use the Boston housing data in this example because it has fewer features and shorter names than the Ames housing data. Consequently, it is easier to compare multiple trees side-by-side; however, the same tree correlation problem exists in the Ames bagged model.

This characteristic is known as *tree correlation* and prevents bagging from further reducing the variance of the base learner. In the next chapter, we discuss how *random forests* extend and improve upon bagged decision trees by reducing this correlation and thereby improving the accuracy of the overall ensemble.

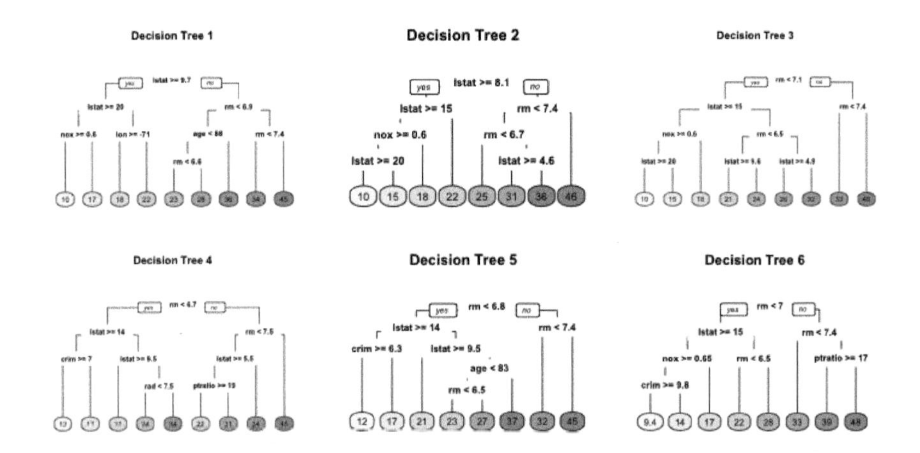

FIGURE 10.6: Six decision trees based on different bootstrap samples.

11

Random Forests

Random forests are a modification of bagged decision trees that build a large collection of *de-correlated* trees to further improve predictive performance. They have become a very popular "out-of-the-box" or "off-the-shelf" learning algorithm that enjoys good predictive performance with relatively little hyperparameter tuning. Many modern implementations of random forests exist; however, Leo Breiman's algorithm (Breiman, 2001) has largely become the authoritative procedure. This chapter will cover the fundamentals of random forests.

11.1 Prerequisites

This chapter leverages the following packages. Some of these packages play a supporting role; however, the emphasis is on how to implement random forests with the **ranger** (Wright and Ziegler, 2017) and **h2o** packages.

```
# Helper packages
library(dplyr)    # for data wrangling
library(ggplot2)  # for awesome graphics

# Modeling packages
library(ranger)   # a c++ implementation of random forest
library(h2o)      # a java-based implementation of random forest
```

We'll continue working with the ames_train data set created in Section 2.7 to illustrate the main concepts.

11.2 Extending bagging

Random forests are built using the same fundamental principles as decision trees (Chapter 9) and bagging (Chapter 10). Bagging trees introduces a random component into the tree building process by building many trees on bootstrapped copies of the training data. Bagging then aggregates the predictions across all the trees; this aggregation reduces the variance of the overall procedure and results in improved predictive performance. However, as we saw in Section 10.6, simply bagging trees results in tree correlation that limits the effect of variance reduction.

Random forests help to reduce tree correlation by injecting more randomness into the tree-growing process.[1] More specifically, while growing a decision tree during the bagging process, random forests perform *split-variable randomization* where each time a split is to be performed, the search for the split variable is limited to a random subset of m_{try} of the original p features. Typical default values are $m_{try} = \frac{p}{3}$ (regression) and $m_{try} = \sqrt{p}$ (classification) but this should be considered a tuning parameter.

The basic algorithm for a regression or classification random forest can be generalized as follows:

```
1.  Given a training data set
2.  Select number of trees to build (n_trees)
3.  for i = 1 to n_trees do
4.  |  Generate a bootstrap sample of the original data
5.  |  Grow a regression/classification tree to the bootstrapped data
6.  |  for each split do
7.  |  | Select m_try variables at random from all p variables
8.  |  | Pick the best variable/split-point among the m_try
9.  |  | Split the node into two child nodes
10. |  end
11. | Use typical tree model stopping criteria to determine when a
    | tree is complete (but do not prune)
12. end
13. Output ensemble of trees
```

When $m_{try} = p$, the algorithm is equivalent to *bagging* decision trees.

[1]See Friedman et al. (2001) for a mathematical explanation of the tree correlation phenomenon.

Since the algorithm randomly selects a bootstrap sample to train on ***and*** a random sample of features to use at each split, a more diverse set of trees is produced which tends to lessen tree correlation beyond bagged trees and often dramatically increase predictive power.

11.3 Out-of-the-box performance

Random forests have become popular because they tend to provide very good out-of-the-box performance. Although they have several hyperparameters that can be tuned, the default values tend to produce good results. Moreover, Probst et al. (2018) illustrated that among the more popular machine learning algorithms, random forests have the least variability in their prediction accuracy when tuning.

For example, if we train a random forest model[2] with all hyperparameters set to their default values, we get an OOB RMSE that is better than any model we've run thus far (without any tuning).

By default, **ranger** sets the `mtry` parameter to $\lfloor\sqrt{\texttt{\# features}}\rfloor$; however, for regression problems the preferred `mtry` to start with is $\lfloor\frac{\texttt{\# features}}{3}\rfloor$. We also set `respect.unordered.factors = "order"`. This specifies how to treat unordered factor variables and we recommend setting this to "order" (see Friedman et al. (2001) Section 9.2.4 for details).

```
# number of features
n_features <- length(setdiff(names(ames_train), "Sale_Price"))

# train a default random forest model
ames_rf1 <- ranger(
  Sale_Price ~ .,
  data = ames_train,
  mtry = floor(n_features / 3),
  respect.unordered.factors = "order",
```

[2]Here we use the **ranger** package to fit a baseline random forest. It is common for folks to first learn to implement random forests by using the original **randomForest** package (Liaw and Wiener, 2002). Although **randomForest** is a great package with many bells and whistles, **ranger** provides a much faster C++ implementation of the same algorithm.

```
  seed = 123
)

# get OOB RMSE
(default_rmse <- sqrt(ames_rf1$prediction.error))
## [1] 24859
```

11.4 Hyperparameters

Although random forests perform well out-of-the-box, there are several tunable hyperparameters that we should consider when training a model. Although we briefly discuss the main hyperparameters, Probst et al. (2019) provide a much more thorough discussion. The main hyperparameters to consider include:

(1) The number of trees in the forest
(2) The number of features to consider at any given split: m_{try}
(3) The complexity of each tree
(4) The sampling scheme
(5) The splitting rule to use during tree construction
(6) and (2) typically have the largest impact on predictive accuracy and should always be tuned. (3) and (4) tend to have marginal impact on predictive accuracy but are still worth exploring. They also have the ability to influence computational efficiency. (5) tends to have the smallest impact on predictive accuracy and is used primarily to increase computational efficiency.

11.4.1 Number of trees

The first consideration is the number of trees within your random forest. Although not technically a hyperparameter, the number of trees needs to be sufficiently large to stabilize the error rate. A good rule of thumb is to start with 10 times the number of features as illustrated in Figure 11.1; however, as you adjust other hyperparameters such as m_{try} and node size, more or fewer trees may be required. More trees provide more robust and stable error estimates and variable importance measures; however, the impact on computation time increases linearly with the number of trees.

Start with $p \times 10$ trees and adjust as necessary

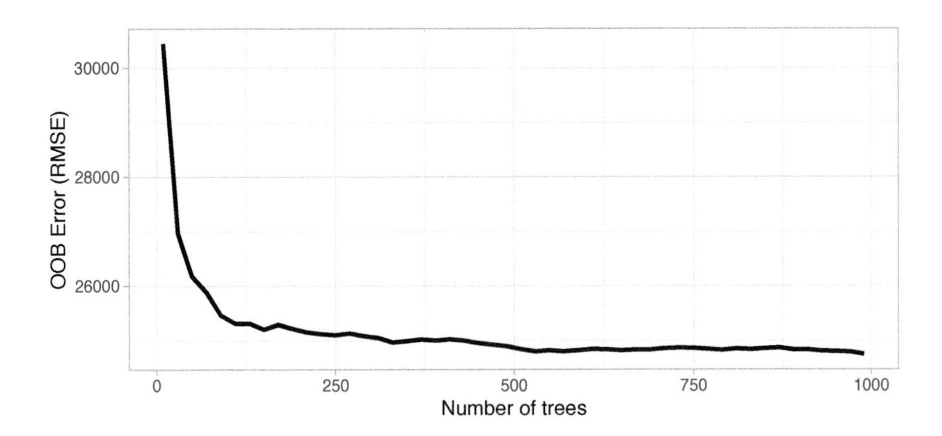

FIGURE 11.1: The Ames data has 80 features and starting with 10 times the number of features typically ensures the error estimate converges.

11.4.2 m_{try}

The hyperparameter that controls the split-variable randomization feature of random forests is often referred to as m_{try} and it helps to balance low tree correlation with reasonable predictive strength. With regression problems the default value is often $m_{try} = \frac{p}{3}$ and for classification $m_{try} = \sqrt{p}$. However, when there are fewer relevant predictors (e.g., noisy data) a higher value of m_{try} tends to perform better because it makes it more likely to select those features with the strongest signal. When there are many relevant predictors, a lower m_{try} might perform better.

Start with five evenly spaced values of m_{try} across the range 2–p centered at the recommended default as illustrated in Figure 11.2.

11.4.3 Tree complexity

Random forests are built on individual decision trees; consequently, most random forest implementations have one or more hyperparameters that allow us to control the depth and complexity of the individual trees. This will

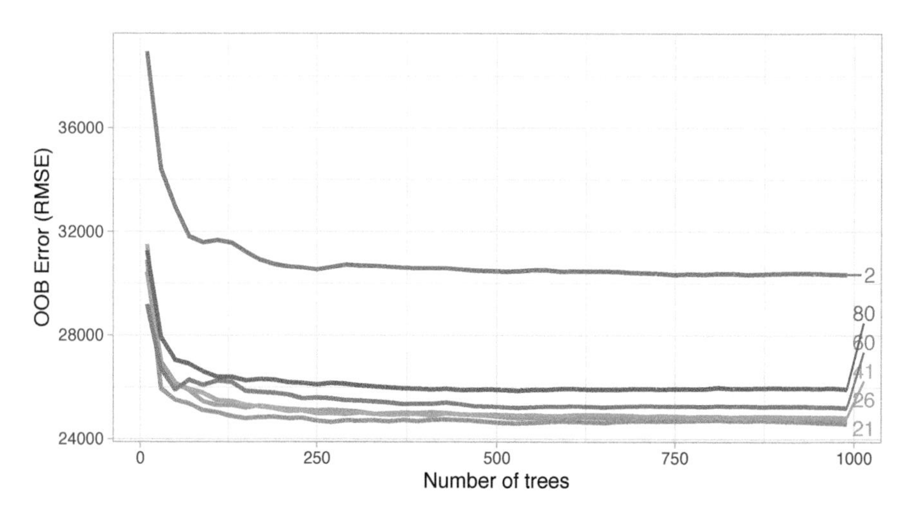

FIGURE 11.2: For the Ames data, an mtry value slightly lower (21) than the default (26) improves performance.

often include hyperparameters such as node size, max depth, max number of terminal nodes, or the required node size to allow additional splits. Node size is probably the most common hyperparameter to control tree complexity and most implementations use the default values of one for classification and five for regression as these values tend to produce good results (Díaz-Uriarte and De Andres, 2006; Goldstein et al., 2011). However, Segal (2004) showed that if your data has many noisy predictors and higher m_{try} values are performing best, then performance may improve by increasing node size (i.e., decreasing tree depth and complexity). Moreover, if computation time is a concern then you can often decrease run time substantially by increasing the node size and have only marginal impacts to your error estimate as illustrated in Figure 11.3.

When adjusting node size start with three values between 1–10 and adjust depending on impact to accuracy and run time.

11.4.4 Sampling scheme

The default sampling scheme for random forests is bootstrapping where 100% of the observations are sampled with replacement (in other words, each bootstrap copy has the same size as the original training data); however, we can adjust both the sample size and whether to sample with or without replacement. The

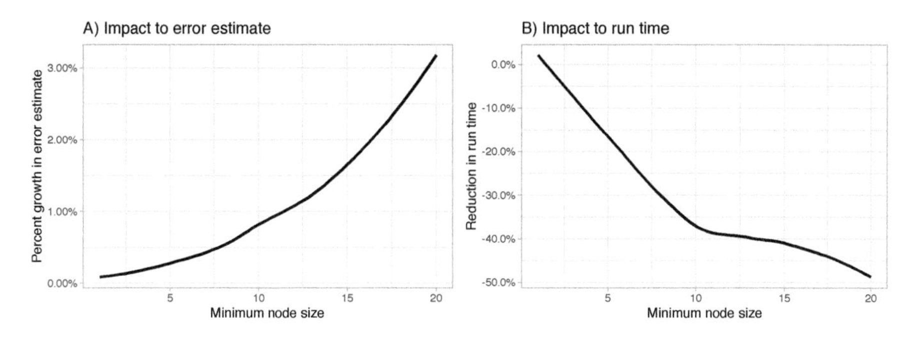

FIGURE 11.3: Increasing node size to reduce tree complexity will often have a larger impact on computation speed (right) than on your error estimate.

sample size parameter determines how many observations are drawn for the training of each tree. Decreasing the sample size leads to more diverse trees and thereby lower between-tree correlation, which can have a positive effect on the prediction accuracy. Consequently, if there are a few dominating features in your data set, reducing the sample size can also help to minimize between-tree correlation.

Also, when you have many categorical features with a varying number of levels, sampling with replacement can lead to biased variable split selection (Janitza et al., 2016; Strobl et al., 2007). Consequently, if you have categories that are not balanced, sampling without replacement provides a less biased use of all levels across the trees in the random forest.

Assess 3–4 values of sample sizes ranging from 25%–100% and if you have unbalanced categorical features try sampling without replacement.

11.4.5 Split rule

Recall the default splitting rule during random forests tree building consists of selecting, out of all splits of the (randomly selected m_{try}) candidate variables, the split that minimizes the Gini impurity (in the case of classification) and the SSE (in case of regression). However, Strobl et al. (2007) illustrated that these default splitting rules favor the selection of features with many possible splits (e.g., continuous variables or categorical variables with many categories) over variables with fewer splits (the extreme case being binary variables, which have only one possible split). *Conditional inference trees* (Hothorn et al., 2006) implement an alternative splitting mechanism that helps to reduce this variable

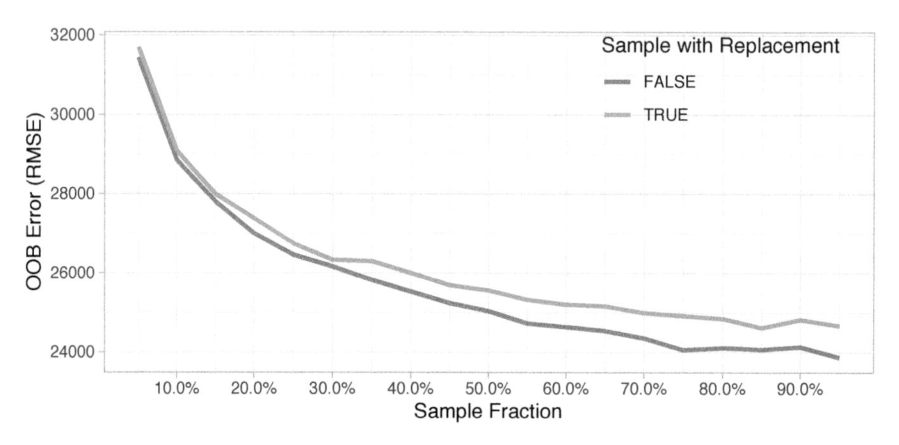

FIGURE 11.4: The Ames data has several imbalanced categorical features such as neighborhood, zoning, overall quality, and more. Consequently, sampling without replacement appears to improve performance as it leads to less biased split variable selection and more uncorrelated trees.

selection bias.[3] However, ensembling conditional inference trees has yet to be proven superior with regards to predictive accuracy and they take a lot longer to train.

To increase computational efficiency, splitting rules can be randomized where only a random subset of possible splitting values is considered for a variable (Geurts et al., 2006). If only a single random splitting value is randomly selected then we call this procedure *extremely randomized trees*. Due to the added randomness of split points, this method tends to have no improvement, or often a negative impact, on predictive accuracy.

Regarding runtime, extremely randomized trees are the fastest as the cutpoints are drawn completely randomly, followed by the classical random forest, while for conditional inference forests the runtime is the largest (Probst et al., 2019).

If you need to increase computation time significantly try completely randomized trees; however, be sure to assess predictive accuracy to traditional split rules as this approach often has a negative impact on your loss function.

[3]Conditional inference trees are available in the **partykit** (Hothorn and Zeileis, 2015) and **ranger** packages among others.

11.5 Tuning strategies

As we introduce more complex algorithms with greater number of hyperparameters, we should become more strategic with our tuning strategies. One way to become more strategic is to consider how we proceed through our grid search. Up to this point, all our grid searches have been *full Cartesian grid searches* where we assess every combination of hyperparameters of interest. We could continue to do the same; for example, the next code block searches across 120 combinations of hyperparameter settings.

This grid search takes approximately 2 minutes.

```
# create hyperparameter grid
hyper_grid <- expand.grid(
  mtry = floor(n_features * c(.05, .15, .25, .333, .4)),
  min.node.size = c(1, 3, 5, 10),
  replace = c(TRUE, FALSE),
  sample.fraction = c(.5, .63, .8),
  rmse = NA
)

# execute full cartesian grid search
for(i in seq_len(nrow(hyper_grid))) {
  # fit model for ith hyperparameter combination
  fit <- ranger(
    formula         = Sale_Price ~ .,
    data            = ames_train,
    num.trees       = n_features * 10,
    mtry            = hyper_grid$mtry[i],
    min.node.size   = hyper_grid$min.node.size[i],
    replace         = hyper_grid$replace[i],
    sample.fraction = hyper_grid$sample.fraction[i],
    verbose         = FALSE,
    seed            = 123,
    respect.unordered.factors = 'order',
  )
  # export OOB error
  hyper_grid$rmse[i] <- sqrt(fit$prediction.error)
}
```

```
# assess top 10 models
hyper_grid %>%
  arrange(rmse) %>%
  mutate(perc_gain = (default_rmse - rmse) / default_rmse * 100) %>%
  head(10)
##    mtry min.node.size replace sample.fraction  rmse
## 1    32             1   FALSE             0.8 23975
## 2    32             3   FALSE             0.8 24023
## 3    32             5   FALSE             0.8 24033
## 4    26             3   FALSE             0.8 24104
## 5    20             1   FALSE             0.8 24132
## 6    26             5   FALSE             0.8 24144
## 7    20             3   FALSE             0.8 24195
## 8    26             1   FALSE             0.8 24216
## 9    32            10   FALSE             0.8 24224
## 10   20             5   FALSE             0.8 24249
##    perc_gain
## 1       3.56
## 2       3.36
## 3       3.33
## 4       3.04
## 5       2.92
## 6       2.88
## 7       2.67
## 8       2.59
## 9       2.55
## 10      2.45
```

If we look at the results we see that the top 10 models are all near or below an RMSE of 24000 (a 2.5%–3.5% improvement over our baseline model). In these results, the default `mtry` value of $\left\lfloor \frac{\texttt{\# features}}{3} \right\rfloor = 26$ is nearly sufficient and smaller node sizes (deeper trees) perform best. What stands out the most is that taking less than 100% sample rate and sampling without replacement consistently performs best. Sampling less than 100% adds additional randomness in the procedure, which helps to further de-correlate the trees. Sampling without replacement likely improves performance because this data has a lot of high cardinality categorical features that are imbalanced.

However, as we add more hyperparameters and values to search across and as our data sets become larger, you can see how a full Cartesian search can become exhaustive and computationally expensive. In addition to full Cartesian search, the **h2o** package provides a *random grid search* that allows you to jump from one random combination to another and it also provides *early stopping* rules that allow you to stop the grid search once a certain condition is met (e.g., a certain number of models have been trained, a certain runtime has elapsed, or the accuracy has stopped improving by a certain amount). Although using a random discrete search path will likely not find the optimal model, it typically does a good job of finding a very good model.

To fit a random forest model with **h2o**, we first need to initiate our **h2o** session.

```
h2o.no_progress()
h2o.init(max_mem_size = "5g")
```

Next, we need to convert our training and test data sets to objects that **h2o** can work with.

```
# convert training data to h2o object
train_h2o <- as.h2o(ames_train)

# set the response column to Sale_Price
response <- "Sale_Price"

# set the predictor names
predictors <- setdiff(colnames(ames_train), response)
```

The following fits a default random forest model with **h2o** to illustrate that our baseline results (OOB RMSE = 24655) are very similar to the baseline **ranger** model we fit earlier.

```
h2o_rf1 <- h2o.randomForest(
    x = predictors,
    y = response,
    training_frame = train_h2o,
    ntrees = n_features * 10,
    seed = 123
)

h2o_rf1
## Model Details:
## ==============
##
## H2ORegressionModel: drf
## Model ID:  DRF_model_R_1561581756490_1
```

```
## Model Summary:
##   number_of_trees number_of_internal_trees model_size_in_bytes ...
## 1             800                      800            12401089 ...
##
##
## H2ORegressionMetrics: drf
## ** Reported on training data. **
## ** Metrics reported on Out-Of-Bag training samples **
##
## MSE:  607854463
## RMSE:  24654.7
## MAE:  15078.35
## RMSLE:  0.1371323
## Mean Residual Deviance :  607854463
```

To execute a grid search in **h2o** we need our hyperparameter grid to be a list. For example, the following code searches a larger grid space than before with a total of 240 hyperparameter combinations. We then create a random grid search strategy that will stop if none of the last 10 models have managed to have a 0.1% improvement in MSE compared to the best model before that. If we continue to find improvements then we cut the grid search off after 300 seconds (5 minutes).

```
# hyperparameter grid
hyper_grid <- list(
  mtries = floor(n_features * c(.05, .15, .25, .333, .4)),
  min_rows = c(1, 3, 5, 10),
  max_depth = c(10, 20, 30),
  sample_rate = c(.55, .632, .70, .80)
)

# random grid search strategy
search_criteria <- list(
  strategy = "RandomDiscrete",
  stopping_metric = "mse",
  stopping_tolerance = 0.001,   # stop if improvement is < 0.1%
  stopping_rounds = 10,         # over the last 10 models
  max_runtime_secs = 60*5      # or stop search after 5 min.
)
```

We can then perform the grid search with `h2o.grid()`. The following executes the grid search with early stopping turned on. The early stopping we specify below in `h2o.grid()` will stop growing an individual random forest model if we have not experienced at least a 0.05% improvement in the overall OOB error in the last 10 trees. This is very useful as we can specify to build 1000 trees for each random forest model but **h2o** may only build 200 trees if we don't experience any improvement.

This grid search takes **5** minutes.

```
# perform grid search
random_grid <- h2o.grid(
  algorithm = "randomForest",
  grid_id = "rf_random_grid",
  x = predictors,
  y = response,
  training_frame = train_h2o,
  hyper_params = hyper_grid,
  ntrees = n_features * 10,
  seed = 123,
  stopping_metric = "RMSE",
  stopping_rounds = 10,          # stop if last 10 trees added
  stopping_tolerance = 0.005,    # don't improve RMSE by 0.5%
  search_criteria = search_criteria
)
```

Our grid search assessed **129** models before stopping due to time. The best model (`max_depth = 20`, `min_rows = 1`, `mtries = 20`, and `sample_rate = 0.8`) achieved an OOB RMSE of 24091. So although our random search assessed about 30% of the number of models as a full grid search would, the more efficient random search found a near-optimal model within the specified time constraint.

In fact, we re-ran the same grid search but allowed for a full search across all 240 hyperparameter combinations and the best model achieved an OOB RMSE of 23785.

```
# collect the results and sort by our model performance metric
# of choice
```

```
random_grid_perf <- h2o.getGrid(
  grid_id = "rf_random_grid",
  sort_by = "mse",
  decreasing = FALSE
)
random_grid_perf
## H2O Grid Details
## ================
##
## Grid ID: rf_random_grid
## Used hyper parameters:
##   -  max_depth
##   -  min_rows
##   -  mtries
##   -  sample_rate
## Number of models: 76
## Number of failed models: 0
##
## Hyper-Parameter Search Summary: ordered by increasing mse
##   max_depth min_rows mtries sample_rate ...                  mse
## 1        20      1.0     20         0.8 ... 5.803817602118055E8
## 2        20      1.0     32         0.8 ... 5.979963774377276E8
## 3        20      1.0     26       0.632 ... 6.078544626686643E8
## 4        30      1.0     32       0.632 ... 6.175727047678405E8
## 5        30      1.0     26        0.55 ... 6.206862379537665E8
##
## ---
##    max_depth min_rows mtries sample_rate ...                  mse
## 71        10      1.0      4        0.55 ...  9.322703689301581E8
## 72        20      5.0      4         0.7 ...  9.331406598281059E8
## 73        20      5.0      4        0.55 ... 1.0138375847024254E9
## 74        20     10.0      4       0.632 ... 1.0978382048184035E9
## 75        30     10.0      4        0.55 ... 1.1014816203603299E9
## 76        20      5.0     32         0.8 ... 1.5919839519623117E9
```

11.6 Feature interpretation

Computing feature importance and feature effects for random forests follow the same procedure as discussed in Section 10.5. However, in addition to the impurity-based measure of feature importance where we base feature importance on the average total reduction of the loss function for a given feature across all trees, random forests also typically include a *permutation-based* importance measure. In the permutation-based approach, for each tree, the OOB sample is passed down the tree and the prediction accuracy is recorded. Then the values for each variable (one at a time) are randomly permuted and the accuracy is again computed. The decrease in accuracy as a result of this randomly shuffling of feature values is averaged over all the trees for each predictor. The variables with the largest average decrease in accuracy are considered most important.

For example, we can compute both measures of feature importance with **ranger** by setting the `importance` argument.

For **ranger**, once you've identified the optimal parameter values from the grid search, you will want to re-run your model with these hyperparameter values. You can also crank up the number of trees, which will help create more stables values of variable importance.

```
# re-run model with impurity-based variable importance
rf_impurity <- ranger(
  formula = Sale_Price ~ .,
  data = ames_train,
  num.trees = 2000,
  mtry = 32,
  min.node.size = 1,
  sample.fraction = .80,
  replace = FALSE,
  importance = "impurity",
  respect.unordered.factors = "order",
  verbose = FALSE,
  seed  = 123
)

# re-run model with permutation-based variable importance
rf_permutation <- ranger(
  formula = Sale_Price ~ .,
  data = ames_train,
  num.trees = 2000,
  mtry = 32,
  min.node.size = 1,
  sample.fraction = .80,
  replace = FALSE,
```

```
  importance = "permutation",
  respect.unordered.factors = "order",
  verbose = FALSE,
  seed  = 123
)
```

The resulting VIPs are displayed in Figure 11.5. Typically, you will not see the same variable importance order between the two options; however, you will often see similar variables at the top of the plots (and also the bottom). Consequently, in this example, we can comfortably state that there appears to be enough evidence to suggest that three variables stand out as most influential:

- `Overall_Qual`
- `Gr_Liv_Area`
- `Neighborhood`

Looking at the next ~10 variables in both plots, you will also see some commonality in influential variables (e.g., `Garage_Cars`, `Exter_Qual`, `Bsmt_Qual`, and `Year_Built`).

```
p1 <- vip::vip(rf_impurity, num_features = 25, bar = FALSE)
p2 <- vip::vip(rf_permutation, num_features = 25, bar = FALSE)

gridExtra::grid.arrange(p1, p2, nrow = 1)
```

11.7 Final thoughts

Random forests provide a very powerful out-of-the-box algorithm that often has great predictive accuracy. They come with all the benefits of decision trees (with the exception of surrogate splits) and bagging but greatly reduce instability and between-tree correlation. And due to the added split variable selection attribute, random forests are also faster than bagging as they have a smaller feature search space at each tree split. However, random forests will still suffer from slow computational speed as your data sets get larger but, similar to bagging, the algorithm is built upon independent steps, and most modern implementations (e.g., **ranger**, **h2o**) allow for parallelization to improve training time.

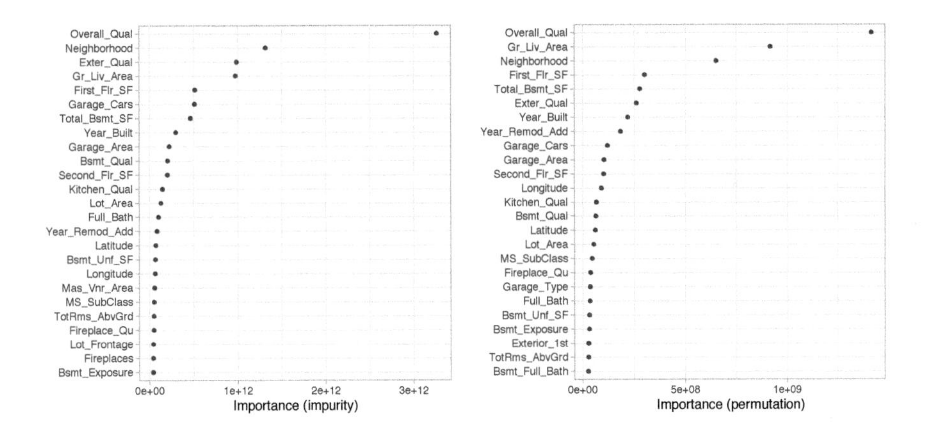

FIGURE 11.5: Top 25 most important variables based on impurity (left) and permutation (right).

12

Gradient Boosting

Gradient boosting machines (GBMs) are an extremely popular machine learning algorithm that have proven successful across many domains and is one of the leading methods for winning Kaggle competitions. Whereas random forests (Chapter 11) build an ensemble of deep independent trees, GBMs build an ensemble of shallow trees in sequence with each tree learning and improving on the previous one. Although shallow trees by themselves are rather weak predictive models, they can be "boosted" to produce a powerful "committee" that, when appropriately tuned, is often hard to beat with other algorithms. This chapter will cover the fundamentals to understanding and implementing some popular implementations of GBMs.

12.1 Prerequisites

This chapter leverages the following packages. Some of these packages play a supporting role; however, our focus is on demonstrating how to implement GBMs with the **gbm** (Greenwell et al., 2018a), **xgboost** (Chen et al., 2018), and **h2o** packages.

```
# Helper packages
library(dplyr)    # for general data wrangling needs

# Modeling packages
library(gbm)      # original implementation of regular & stochastic GBMs
library(h2o)      # for a java-based implementation of GBM variants
library(xgboost) # for fitting extreme gradient boosting
```

We'll continue working with the `ames_train` data set created in Section 2.7 to illustrate the main concepts.

We'll also demonstrate **h2o** functionality using the same setup from Section 11.5.

12.2 How boosting works

Several supervised machine learning algorithms are based on a single predictive model, for example: ordinary linear regression, penalized regression models, single decision trees, and support vector machines. Bagging and random forests, on the other hand, work by combining multiple models together into an overall ensemble. New predictions are made by combining the predictions from the individual base models that make up the ensemble (e.g., by averaging in regression). Since averaging reduces variance, bagging (and hence, random forests) are most effectively applied to models with low bias and high variance (e.g., an overgrown decision tree). While boosting is a general algorithm for building an ensemble out of simpler models (typically decision trees), it is more effectively applied to models with high bias and low variance! Although boosting, like bagging, can be applied to any type of model, it is often most effectively applied to decision trees (which we'll assume from this point on).

12.2.1 A sequential ensemble approach

The main idea of boosting is to add new models to the ensemble ***sequentially***. In essence, boosting attacks the bias-variance-tradeoff by starting with a *weak* model (e.g., a decision tree with only a few splits) and sequentially *boosts* its performance by continuing to build new trees, where each new tree in the sequence tries to fix up where the previous one made the biggest mistakes (i.e., each new tree in the sequence will focus on the training rows where the previous tree had the largest prediction errors); see Figure 12.1.

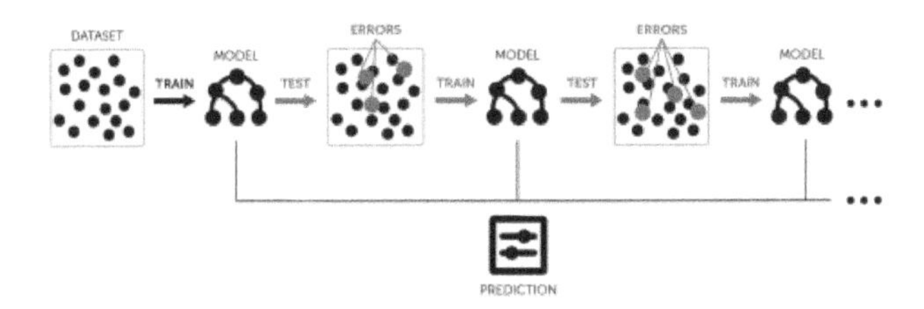

FIGURE 12.1: Sequential ensemble approach.

Let's discuss the important components of boosting in closer detail.

The base learners: Boosting is a framework that iteratively improves *any* weak learning model. Many gradient boosting applications allow you to "plug in" various classes of weak learners at your disposal. In practice however, boosted algorithms almost always use decision trees as the base-learner. Consequently, this chapter will discuss boosting in the context of decision trees.

Training weak models: A weak model is one whose error rate is only slightly better than random guessing. The idea behind boosting is that each model in the sequence slightly improves upon the performance of the previous one (essentially, by focusing on the rows of the training data where the previous tree had the largest errors or residuals). With regards to decision trees, shallow trees (i.e., trees with relatively few splits) represent a weak learner. In boosting, trees with 1–6 splits are most common.

Sequential training with respect to errors: Boosted trees are grown sequentially; each tree is grown using information from previously grown trees to improve performance. This is illustrated in the following algorithm for boosting regression trees. By fitting each tree in the sequence to the previous tree's residuals, we're allowing each new tree in the sequence to focus on the previous tree's mistakes:

1. Fit a decision tree to the data: $F_1(x) = y$,
2. We then fit the next decision tree to the residuals of the previous: $h_1(x) = y - F_1(x)$,
3. Add this new tree to our algorithm: $F_2(x) = F_1(x) + h_1(x)$,
4. Fit the next decision tree to the residuals of F_2: $h_2(x) = y - F_2(x)$,
5. Add this new tree to our algorithm: $F_3(x) = F_2(x) + h_1(x)$,
6. Continue this process until some mechanism (i.e. cross validation) tells us to stop.

The final model here is a stagewise additive model of b individual trees:

$$f(x) = \sum_{b=1}^{B} f^b(x) \tag{1}$$

Figure 12.2 illustrates with a simple example where a single predictor (x) has a true underlying sine wave relationship (blue line) with y along with some irreducible error. The first tree fit in the series is a single decision stump (i.e., a tree with a single split). Each successive decision stump thereafter is fit to the previous one's residuals. Initially there are large errors, but each additional decision stump in the sequence makes a small improvement in different areas across the feature space where errors still remain.

12.2.2 Gradient descent

Many algorithms in regression, including decision trees, focus on minimizing some function of the residuals; most typically the SSE loss function, or equivalently, the MSE or RMSE (this is accomplished through simple calculus and is

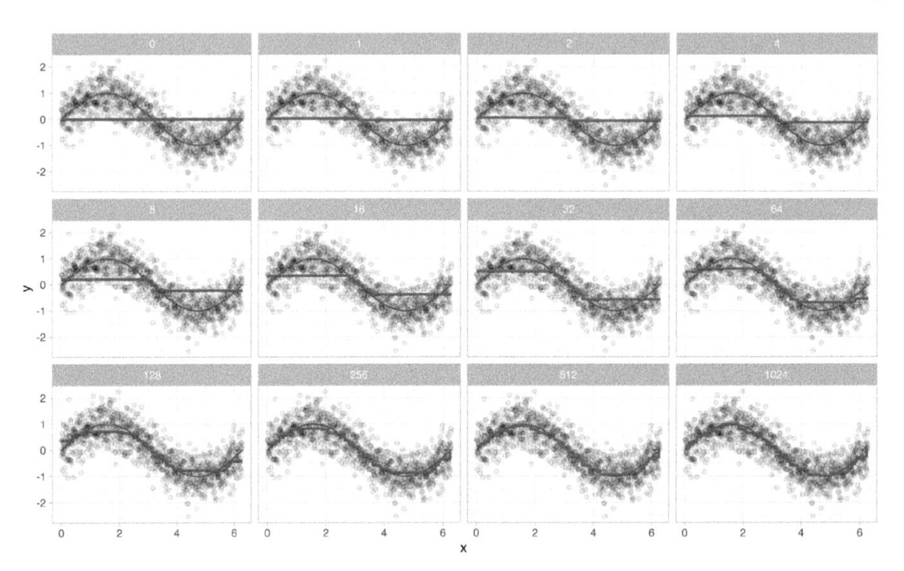

FIGURE 12.2: Boosted regression decision stumps as 0-1024 successive trees are added.

the approach taken with least squares). The boosting algorithm for regression discussed in the previous section outlines the approach of sequentially fitting regression trees to the residuals from the previous tree. This specific approach is how gradient boosting minimizes the mean squared error (SSE) loss function (for SSE loss, the gradient is nothing more than the residual error). However, we often wish to focus on other loss functions such as mean absolute error (MAE)—which is less sensitive to outliers—or to be able to apply the method to a classification problem with a loss function such as deviance, or log loss. The name ***gradient*** boosting machine comes from the fact that this procedure can be generalized to loss functions other than SSE.

Gradient boosting is considered a ***gradient descent*** algorithm. Gradient descent is a very generic optimization algorithm capable of finding optimal solutions to a wide range of problems. The general idea of gradient descent is to tweak parameter(s) iteratively in order to minimize a cost function. Suppose you are a downhill skier racing your friend. A good strategy to beat your friend to the bottom is to take the path with the steepest slope. This is exactly what gradient descent does—it measures the local gradient of the loss (cost) function for a given set of parameters (Θ) and takes steps in the direction of the descending gradient. As Figure 12.3[1] illustrates, once the gradient is zero, we have reached a minimum.

[1]Figures 12.3, 12.4, and 12.5 were inspired by Géron (2017) but completely re-written with our own code.

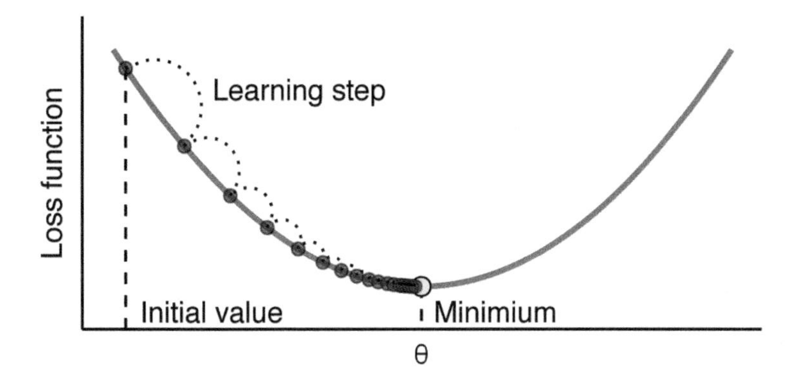

FIGURE 12.3: Gradient descent is the process of gradually decreasing the cost function (i.e. MSE) by tweaking parameter(s) iteratively until you have reached a minimum.

Gradient descent can be performed on any loss function that is differentiable. Consequently, this allows GBMs to optimize different loss functions as desired (see Friedman et al. (2001), p. 360 for common loss functions). An important parameter in gradient descent is the size of the steps which is controlled by the *learning rate*. If the learning rate is too small, then the algorithm will take many iterations (steps) to find the minimum. On the other hand, if the learning rate is too high, you might jump across the minimum and end up further away than when you started.

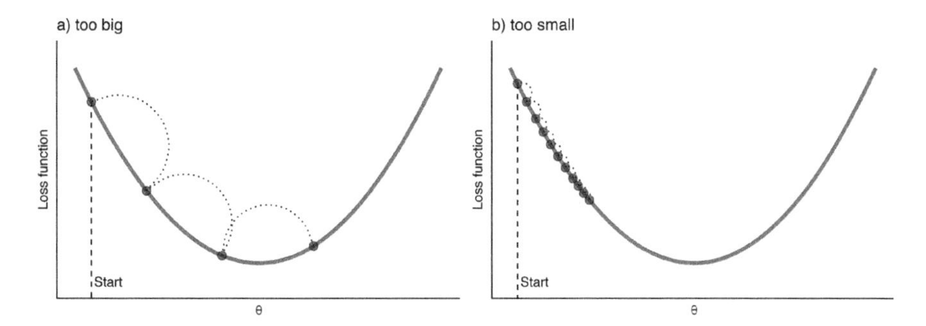

FIGURE 12.4: A learning rate that is too small will require many iterations to find the minimum. A learning rate too big may jump over the minimum.

Moreover, not all cost functions are *convex* (i.e., bowl shaped). There may be local minimas, plateaus, and other irregular terrain of the loss function that makes finding the global minimum difficult. ***Stochastic gradient descent*** can help us address this problem by sampling a fraction of the training observations (typically without replacement) and growing the next tree using that subsample.

This makes the algorithm faster but the stochastic nature of random sampling also adds some random nature in descending the loss function's gradient. Although this randomness does not allow the algorithm to find the absolute global minimum, it can actually help the algorithm jump out of local minima and off plateaus to get sufficiently near the global minimum.

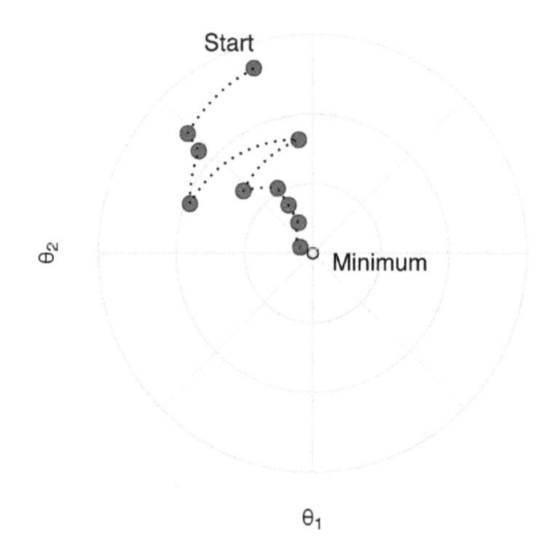

FIGURE 12.5: Stochastic gradient descent will often find a near-optimal solution by jumping out of local minimas and off plateaus.

As we'll see in the sections that follow, there are several hyperparameter tuning options available in stochastic gradient boosting (some control the gradient descent and others control the tree growing process). If properly tuned (e.g., with k-fold CV) GBMs can lead to some of the most flexible and accurate predictive models you can build!

12.3 Basic GBM

There are multiple variants of boosting algorithms with the original focused on classification problems (Kuhn and Johnson, 2013). Throughout the 1990's many approaches were developed with the most successful being the AdaBoost algorithm (Freund and Schapire, 1999). In 2000, Friedman related AdaBoost to important statistical concepts (e.g., loss functions and additive modeling), which allowed him to generalize the boosting framework to regression problems and multiple loss functions (Friedman, 2001). This led to the typical GBM model that we think of today and that most modern implementations are built on.

12.3.1 Hyperparameters

A simple GBM model contains two categories of hyperparameters: *boosting hyperparameters* and *tree-specific hyperparameters*. The two main boosting hyperparameters include:

- **Number of trees**: The total number of trees in the sequence or ensemble. The averaging of independently grown trees in bagging and random forests makes it very difficult to overfit with too many trees. However, GBMs function differently as each tree is grown in sequence to fix up the past tree's mistakes. For example, in regression, GBMs will chase residuals as long as you allow them to. Also, depending on the values of the other hyperparameters, GBMs often require many trees (it is not uncommon to have many thousands of trees) but since they can easily overfit we must find the optimal number of trees that minimize the loss function of interest with cross validation.
- **Learning rate**: Determines the contribution of each tree on the final outcome and controls how quickly the algorithm proceeds down the gradient descent (learns); see Figure 12.3. Values range from 0–1 with typical values between 0.001–0.3. Smaller values make the model robust to the specific characteristics of each individual tree, thus allowing it to generalize well. Smaller values also make it easier to stop prior to overfitting; however, they increase the risk of not reaching the optimum with a fixed number of trees and are more computationally demanding. This hyperparameter is also called *shrinkage*. Generally, the smaller this value, the more accurate the model can be but also will require more trees in the sequence.

The two main tree hyperparameters in a simple GBM model include:

- **Tree depth**: Controls the depth of the individual trees. Typical values range from a depth of 3–8 but it is not uncommon to see a tree depth of 1 (Friedman et al., 2001). Smaller depth trees such as decision stumps are computationally efficient (but require more trees); however, higher depth trees allow the algorithm to capture unique interactions but also increase the risk of over-fitting. Note that larger n or p training data sets are more tolerable to deeper trees.
- **Minimum number of observations in terminal nodes**: Also, controls the complexity of each tree. Since we tend to use shorter trees this rarely has a large impact on performance. Typical values range from 5–15 where higher values help prevent a model from learning relationships which might be highly specific to the particular sample selected for a tree (overfitting) but smaller values can help with imbalanced target classes in classification problems.

12.3.2 Implementation

There are many packages that implement GBMs and GBM variants. You can find a fairly comprehensive list at the CRAN Machine Learning Task View: `https://cran.r-project.org/web/views/MachineLearning.html`. However, the most popular original R implementation of Friedman's GBM algorithm (Friedman, 2001, 2002) is the **gbm** package.

gbm has two training functions: `gbm::gbm()` and `gbm::gbm.fit()`. The primary difference is that `gbm::gbm()` uses the formula interface to specify your model whereas `gbm::gbm.fit()` requires the separated `x` and `y` matrices; `gbm::gbm.fit()` is more efficient and recommended for advanced users.

The default settings in **gbm** include a learning rate (`shrinkage`) of 0.001. This is a very small learning rate and typically requires a large number of trees to sufficiently minimize the loss function. However, **gbm** uses a default number of trees of 100, which is rarely sufficient. Consequently, we start with a learning rate of 0.1 and increase the number of trees to train. The default depth of each tree (`interaction.depth`) is 1, which means we are ensembling a bunch of decision stumps (i.e., we are not able to capture any interaction effects). For the Ames housing data set, we increase the tree depth to 3 and use the default value for minimum number of observations required in the trees terminal nodes (`n.minobsinnode`). Lastly, we set `cv.folds = 10` to perform a 10-fold CV.

This model takes a little over 2 minutes to run.

```
# run a basic GBM model
set.seed(123)  # for reproducibility
ames_gbm1 <- gbm(
  formula = Sale_Price ~ .,
  data = ames_train,
  distribution = "gaussian",  # SSE loss function
  n.trees = 5000,
  shrinkage = 0.1,
  interaction.depth = 3,
  n.minobsinnode = 10,
  cv.folds = 10
)

# find index for number trees with minimum CV error
best <- which.min(ames_gbm1$cv.error)
```

```
# get MSE and compute RMSE
sqrt(ames_gbm1$cv.error[best])
## [1] 22252
```

Our results show a cross-validated SSE of 22252 which was achieved with 874 trees.

```
# plot error curve
gbm.perf(ames_gbm1, method = "cv")
```

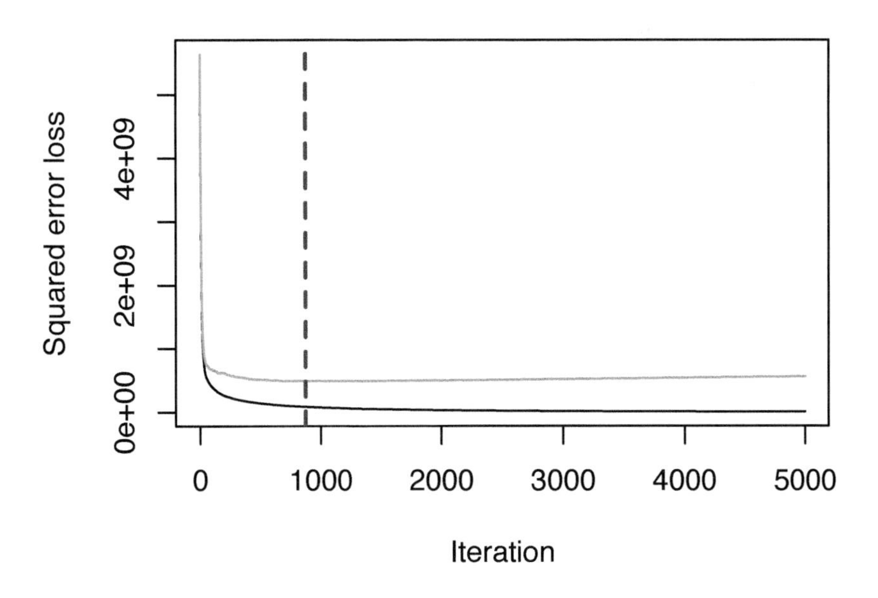

FIGURE 12.6: Training and cross-validated MSE as n trees are added to the GBM algorithm.

```
## [1] 874
```

12.3.3 General tuning strategy

Unlike random forests, GBMs can have high variability in accuracy dependent on their hyperparameter settings (Probst et al., 2018). So tuning can require much more strategy than a random forest model. Often, a good approach is to:

1. Choose a relatively high learning rate. Generally the default value of 0.1 works but somewhere between 0.05–0.2 should work across a wide range of problems.

2. Determine the optimum number of trees for this learning rate.

3. Fix tree hyperparameters and tune learning rate and assess speed vs. performance.

4. Tune tree-specific parameters for decided learning rate.

5. Once tree-specific parameters have been found, lower the learning rate to assess for any improvements in accuracy.

6. Use final hyperparameter settings and increase CV procedures to get more robust estimates. Often, the above steps are performed with a simple validation procedure or 5-fold CV due to computational constraints. If you used k-fold CV throughout steps 1–5 then this step is not necessary.

We already did (1)–(2) in the Ames example above with our first GBM model. Next, we'll do (3) and asses the performance of various learning rate values between 0.005–0.3. Our results indicate that a learning rate of 0.05 sufficiently minimizes our loss function and requires 2375 trees. All our models take a little over 2 minutes to train so we don't see any significant impacts in training time based on the learning rate.

The following grid search took us about 10 minutes.

```
# create grid search
hyper_grid <- expand.grid(
  learning_rate = c(0.3, 0.1, 0.05, 0.01, 0.005),
  RMSE = NA,
  trees = NA,
  time = NA
```

```
)

# execute grid search
for(i in seq_len(nrow(hyper_grid))) {

  # fit gbm
  set.seed(123)  # for reproducibility
  train_time <- system.time({
    m <- gbm(
      formula = Sale_Price ~ .,
      data = ames_train,
      distribution = "gaussian",
      n.trees = 5000,
      shrinkage = hyper_grid$learning_rate[i],
      interaction.depth = 3,
      n.minobsinnode = 10,
      cv.folds = 10
   )
  })

  # add SSE, trees, and training time to results
  hyper_grid$RMSE[i]  <- sqrt(min(m$cv.error))
  hyper_grid$trees[i] <- which.min(m$cv.error)
  hyper_grid$time[i]  <- train_time[["elapsed"]]

}

# results
arrange(hyper_grid, RMSE)
##   learning_rate  RMSE trees  time
## 1         0.050 21382  2375 129.5
## 2         0.010 21828  4982 126.0
## 3         0.100 22252   874 137.6
## 4         0.005 23136  5000 136.8
## 5         0.300 24454   427 139.9
```

Next, we'll set our learning rate at the optimal level (0.05) and tune the tree specific hyperparameters (`interaction.depth` and `n.minobsinnode`). Adjusting the tree-specific parameters provides us with an additional 600 reduction in RMSE.

This grid search takes about 30 minutes.

```
# search grid
hyper_grid <- expand.grid(
  n.trees = 4000,
  shrinkage = 0.05,
  interaction.depth = c(3, 5, 7),
  n.minobsinnode = c(5, 10, 15)
)

# create model fit function
model_fit <- function(n.trees, shrinkage, interaction.depth,
                      n.minobsinnode) {
  set.seed(123)
  m <- gbm(
    formula = Sale_Price ~ .,
    data = ames_train,
    distribution = "gaussian",
    n.trees = n.trees,
    shrinkage = shrinkage,
    interaction.depth = interaction.depth,
    n.minobsinnode = n.minobsinnode,
    cv.folds = 10
  )
  # compute RMSE
  sqrt(min(m$cv.error))
}

# perform search grid with functional programming
hyper_grid$rmse <- purrr::pmap_dbl(
  hyper_grid,
  ~ model_fit(
    n.trees = ..1,
    shrinkage = ..2,
    interaction.depth = ..3,
    n.minobsinnode = ..4
    )
)

# results
arrange(hyper_grid, rmse)
##   n.trees shrinkage interaction.depth n.minobsinnode  rmse
```

```
## 1     4000      0.05            5           5 20699
## 2     4000      0.05            3           5 20723
## 3     4000      0.05            7           5 21021
## 4     4000      0.05            3          10 21382
## 5     4000      0.05            5          10 21915
## 6     4000      0.05            5          15 21924
## 7     4000      0.05            3          15 21943
## 8     4000      0.05            7          10 21999
## 9     4000      0.05            7          15 22348
```

After this procedure, we took our top model's hyperparameter settings, reduced the learning rate to 0.005, and increased the number of trees (8000) to see if we got any additional improvement in accuracy. We experienced no improvement in our RMSE and our training time increased to nearly 6 minutes.

12.4 Stochastic GBMs

An important insight made by Breiman (Breiman (1996a); Breiman (2001)) in developing his bagging and random forest algorithms was that training the algorithm on a random subsample of the training data set offered additional reduction in tree correlation and, therefore, improvement in prediction accuracy. Friedman (2002) used this same logic and updated the boosting algorithm accordingly. This procedure is known as *stochastic gradient boosting* and, as illustrated in Figure 12.5, helps reduce the chances of getting stuck in local minimas, plateaus, and other irregular terrain of the loss function so that we may find a near global optimum.

12.4.1 Stochastic hyperparameters

There are a few variants of stochastic gradient boosting that can be used, all of which have additional hyperparameters:

- Subsample rows before creating each tree (available in **gbm**, **h2o**, & **xgboost**)
- Subsample columns before creating each tree (**h2o** & **xgboost**)
- Subsample columns before considering each split in each tree (**h2o** & **xgboost**)

Generally, aggressive subsampling of rows, such as selecting only 50% or less of the training data, has shown to be beneficial and typical values range between 0.5–0.8. Subsampling of columns and the impact to performance largely depends on the nature of the data and if there is strong multicollinearity or a lot of noisy features. Similar to the m_{try} parameter in random forests (Section 11.4.2), if there are fewer relevant predictors (more noisy data) higher values of column subsampling tends to perform better because it makes it more likely to select those features with the strongest signal. When there are many relevant predictors, a lower values of column subsampling tends to perform well.

When adding in a stochastic procedure, you can either include it in step 4) in the general tuning strategy above (Section 12.3.3), or once you've found the optimal basic model (after 6)). In our experience, we have not seen strong interactions between the stochastic hyperparameters and the other boosting and tree-specific hyperparameters.

12.4.2 Implementation

The following uses **h2o** to implement a stochastic GBM. We use the optimal hyperparameters found in the previous section and build onto this by assessing a range of values for subsampling rows and columns before each tree is built, and subsampling columns before each split. To speed up training we use early stopping for the individual GBM modeling process and also add a stochastic search criteria.

This grid search ran for nearly the entire 60 minutes and evaluated all 27 models.

```
# refined hyperparameter grid
# sample_rate: row subsampling
# col_sample_rate: col subsampling for each split
# col_sample_rate_per_tree: col subsampling for each tree
hyper_grid <- list(
  sample_rate = c(0.5, 0.75, 1),
  col_sample_rate = c(0.5, 0.75, 1),
  col_sample_rate_per_tree = c(0.5, 0.75, 1)
)

# random grid search strategy
search_criteria <- list(
  strategy = "RandomDiscrete",
```

```
  stopping_metric = "mse",
  stopping_tolerance = 0.001,
  stopping_rounds = 10,
  max_runtime_secs = 60*60
)

# perform grid search
grid <- h2o.grid(
  algorithm = "gbm",
  grid_id = "gbm_grid",
  x = predictors,
  y = response,
  training_frame = train_h2o,
  hyper_params = hyper_grid,
  ntrees = 5000,
  learn_rate = 0.05,
  max_depth = 5,
  min_rows = 5,
  nfolds = 10,
  stopping_rounds = 10,
  stopping_tolerance = 0,
  search_criteria = search_criteria,
  seed = 123
)

# collect the results and sort by our model performance
# metric of choice
grid_perf <- h2o.getGrid(
  grid_id = "gbm_grid",
  sort_by = "mse",
  decreasing = FALSE
)

grid_perf
## H2O Grid Details
## ================
##
## Grid ID: gbm_grid
## Used hyper parameters:
##   -  col_sample_rate
##   -  col_sample_rate_per_tree
##   -  sample_rate
## Number of models: 27
## Number of failed models: 0
```

```
##
## Hyper-Parameter Search Summary: ordered by increasing mse
##   col_sample_rate col_sample_rate_per_tree sample_rate ...      mse
## 1            0.75                      0.5        0.75 ... 4.767E8
## 2             0.5                      0.5        0.75 ... 4.872E8
## 3             0.5                     0.75        0.75 ... 4.879E8
## 4             0.5                     0.75         1.0 ... 4.919E8
## 5             0.5                      0.5         0.5 ... 4.979E8
##
## ---
##    col_sample_rate col_sample_rate_per_tree sample_rate ...      mse
## 22            0.75                     0.75         1.0 ... 5.359E8
## 23             1.0                      1.0         0.5 ... 5.409E8
## 24             1.0                      1.0        0.75 ... 5.479E8
## 25            0.75                      1.0         1.0 ... 5.517E8
## 26             1.0                     0.75         1.0 ... 5.541E8
## 27             1.0                      1.0         1.0 ... 5.856E8
```

Our grid search highlights some important results. Random sampling from the rows for each tree, randomly sampling features, and sampling features before each tree all appear to positively impact performance. Furthermore, the best sampling values are on the lower end (0.5-0.75); a further grid search may be beneficial to evaluate additional values in this lower range.

The below code chunk extracts the best performing model. In this particular case, we do not see additional improvement in our 10-fold CV RMSE over the best non-stochastic GBM model.

```
# Get model_id for the top model, chosen by cross validation error
best_model_id <- grid_perf@model_ids[[1]]
best_model <- h2o.getModel(best_model_id)

# Now let's get performance metrics on the best model
h2o.performance(model = best_model, xval = TRUE)
## H2ORegressionMetrics: gbm
## ** Reported on cross-validation data. **
## ** 10-fold cross-validation on training data (Metrics computed
##    for combined holdout predictions) **
##
## MSE:  476701099
## RMSE:  21833
## MAE:  13507
```

```
## RMSLE:  0.1234
## Mean Residual Deviance :  476701099
```

12.5 XGBoost

Extreme gradient boosting (XGBoost) is an optimized distributed gradient boosting library that is designed to be efficient, flexible, and portable across multiple languages (Chen and Guestrin, 2016). Although XGBoost provides the same boosting and tree-based hyperparameter options illustrated in the previous sections, it also provides a few advantages over traditional boosting such as:

- **Regularization**: XGBoost offers additional regularization hyperparameters, which we will discuss shortly, that provides added protection against overfitting.
- **Early stopping**: Similar to **h2o**, XGBoost implements early stopping so that we can stop model assessment when additional trees offer no improvement.
- **Parallel Processing**: Since gradient boosting is sequential in nature it is extremely difficult to parallelize. XGBoost has implemented procedures to support GPU and Spark compatibility which allows you to fit gradient boosting using powerful distributed processing engines.
- **Loss functions**: XGBoost allows users to define and optimize gradient boosting models using custom objective and evaluation criteria.
- **Continue with existing model**: A user can train an XGBoost model, save the results, and later on return to that model and continue building onto the results. Whether you shut down for the day, wanted to review intermediate results, or came up with additional hyperparameter settings to evaluate, this allows you to continue training your model without starting from scratch.
- **Different base learners**: Most GBM implementations are built with decision trees but XGBoost also provides boosted generalized linear models.
- **Multiple languages**: XGBoost offers implementations in R, Python, Julia, Scala, Java, and C++.

In addition to being offered across multiple languages, XGboost can be implemented multiple ways within R. The main R implementation is the **xgboost** package; however, as illustrated throughout many chapters one can also use **caret** as a meta engine to implement XGBoost. The **h2o** package also offers an implementation of XGBoost. In this chapter we'll demonstrate the **xgboost** package.

12.5.1 XGBoost hyperparameters

As previously mentioned, **xgboost** provides the traditional boosting and tree-based hyperparameters we discussed in Sections 12.3.1 and 12.4.1. However, **xgboost** also provides additional hyperparameters that can help reduce the chances of overfitting, leading to less prediction variability and, therefore, improved accuracy.

12.5.1.1 Regularization

xgboost provides multiple regularization parameters to help reduce model complexity and guard against overfitting. The first, `gamma`, is a pseudo-regularization hyperparameter known as a Lagrangian multiplier and controls the complexity of a given tree. `gamma` specifies a minimum loss reduction required to make a further partition on a leaf node of the tree. When `gamma` is specified, **xgboost** will grow the tree to the max depth specified but then prune the tree to find and remove splits that do not meet the specified `gamma`. `gamma` tends to be worth exploring as your trees in your GBM become deeper and when you see a significant difference between the train and test CV error. The value of `gamma` ranges from $0 - \infty$ (0 means no constraint while large numbers mean a higher regularization). What quantifies as a large `gamma` value is dependent on the loss function but generally lower values between 1–20 will do if `gamma` is influential.

Two more traditional regularization parameters include `alpha` and `lambda`. `alpha` provides an L_1 regularization (reference Section 6.2.2) and `lambda` provides an L_2 regularization (reference Section 6.2.1). Setting both of these to greater than 0 results in an elastic net regularization; similar to `gamma`, these parameters can range from $0 - \infty$. These regularization parameters limits how extreme the weights (or influence) of the leaves in a tree can become.

All three hyperparameters (`gamma`, `alpha`, `lambda`) work to constrain model complexity and reduce overfitting. Although `gamma` is more commonly implemented, your tuning strategy should explore the impact of all three. Figure 12.7 illustrates how regularization can make an overfit model more conservative on the training data which results in a slight improvement on the 10-fold CV error.

12.5.1.2 Dropout

Dropout is an alternative approach to reduce overfitting and can loosely be described as regularization. The dropout approach developed by Srivastava et al. (2014a) has been widely employed in deep learnings to prevent deep neural networks from overfitting (see Section 13.7.3). Dropout can also be used to address overfitting in GBMs. When constructing a GBM, the first

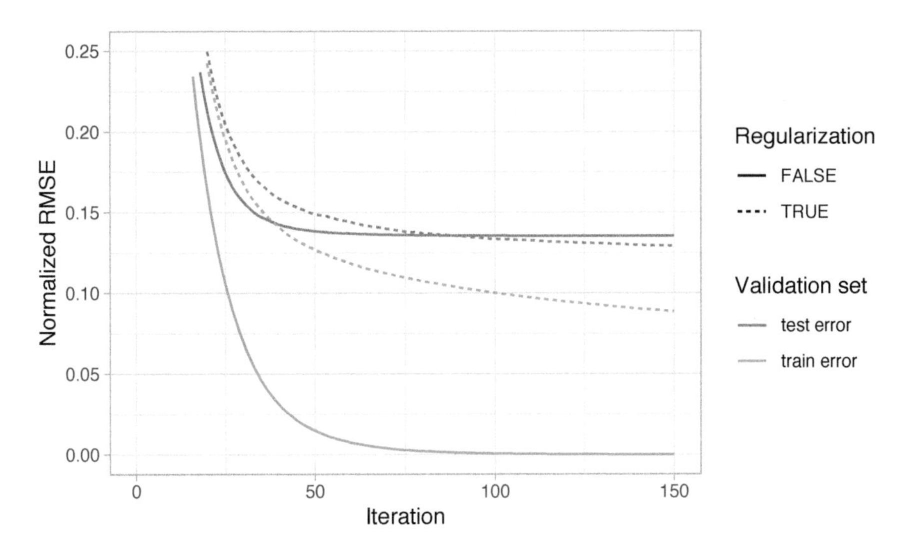

FIGURE 12.7: When a GBM model significantly overfits to the training data (blue), adding regularization (dotted line) causes the model to be more conservative on the training data, which can improve the cross-validated test error (red).

few trees added at the beginning of the ensemble typically dominate the model performance while trees added later typically improve the prediction for only a small subset of the feature space. This often increases the risk of overfitting and the idea of dropout is to build an ensemble by randomly dropping trees in the boosting sequence. This is commonly referred to as DART (Rashmi and Gilad-Bachrach, 2015) since it was initially explored in the context of *Mutliple Additive Regression Trees* (MART); DART refers to *Dropout Additive Regression Trees*. The percentage of dropouts is another regularization parameter.

Typically, when `gamma`, `alpha`, or `lambda` cannot help to control overfitting, exploring DART hyperparameters would be the next best option.[2]

12.5.2 Tuning strategy

The general tuning strategy for exploring **xgboost** hyperparameters builds onto the basic and stochastic GBM tuning strategies:

[2]See the DART booster documentation on the XGBoost website for details: `https://xgboost.readthedocs.io/en/latest/tutorials/dart.html`.

1. Crank up the number of trees and tune learning rate with early stopping
2. Tune tree-specific hyperparameters
3. Explore stochastic GBM attributes
4. If substantial overfitting occurs (e.g., large differences between train and CV error) explore regularization hyperparameters
5. If you find hyperparameter values that are substantially different from default settings, be sure to retune the learning rate
6. Obtain final "optimal" model

Running an XGBoost model with **xgboost** requires some additional data preparation. **xgboost** requires a matrix input for the features and the response to be a vector. Consequently, to provide a matrix input of the features we need to encode our categorical variables numerically (i.e. one-hot encoding, label encoding). The following numerically label encodes all categorical features and converts the training data frame to a matrix.

```
library(recipes)
xgb_prep <- recipe(Sale_Price ~ ., data = ames_train) %>%
  step_integer(all_nominal()) %>%
  prep(training = ames_train, retain = TRUE) %>%
  juice()

X <- as.matrix(xgb_prep[setdiff(names(xgb_prep), "Sale_Price")])
Y <- xgb_prep$Sale_Price
```

xgboost will accept three different kinds of matrices for the features: ordinary R matrix, sparse matrices from the **Matrix** package, or **xgboost**'s internal `xgb.DMatrix` objects. See `?xgboost::xgboost` for details.

Next, we went through a series of grid searches similar to the previous sections and found the below model hyperparameters (provided via the `params` argument) to perform quite well. Our RMSE is slightly lower than the best regular and stochastic GBM models thus far.

```
set.seed(123)
ames_xgb <- xgb.cv(
  data = X,
  label = Y,
  nrounds = 6000,
```

```
  objective = "reg:linear",
  early_stopping_rounds = 50,
  nfold = 10,
  params = list(
    eta = 0.01,
    max_depth = 3,
    min_child_weight = 3,
    subsample = 0.5,
    colsample_bytree = 0.5),
  verbose = 0
)

# minimum test CV RMSE
min(ames_xgb$evaluation_log$test_rmse_mean)
## [1] 20488
```

Next, we assess if overfitting is limiting our model's performance by performing a grid search that examines various regularization parameters (`gamma`, `lambda`, and `alpha`). Our results indicate that the best performing models use `lambda` equal to 1 and it doesn't appear that `alpha` or `gamma` have any consistent patterns. However, even when `lambda` equals 1, our CV RMSE has no improvement over our previous XGBoost model.

Due to the low learning rate (`eta`), this cartesian grid search takes a long time. We stopped the search after 2 hours and only 98 of the 245 models had completed.

```
# hyperparameter grid
hyper_grid <- expand.grid(
  eta = 0.01,
  max_depth = 3,
  min_child_weight = 3,
  subsample = 0.5,
  colsample_bytree = 0.5,
  gamma = c(0, 1, 10, 100, 1000),
  lambda = c(0, 1e-2, 0.1, 1, 100, 1000, 10000),
  alpha = c(0, 1e-2, 0.1, 1, 100, 1000, 10000),
  rmse = 0,          # a place to dump RMSE results
  trees = 0          # a place to dump required number of trees
)
```

```
# grid search
for(i in seq_len(nrow(hyper_grid))) {
  set.seed(123)
  m <- xgb.cv(
    data = X,
    label = Y,
    nrounds = 4000,
    objective = "reg:linear",
    early_stopping_rounds = 50,
    nfold = 10,
    verbose = 0,
    params = list(
      eta = hyper_grid$eta[i],
      max_depth = hyper_grid$max_depth[i],
      min_child_weight = hyper_grid$min_child_weight[i],
      subsample = hyper_grid$subsample[i],
      colsample_bytree = hyper_grid$colsample_bytree[i],
      gamma = hyper_grid$gamma[i],
      lambda = hyper_grid$lambda[i],
      alpha = hyper_grid$alpha[i]
    )
  )
  hyper_grid$rmse[i] <- min(m$evaluation_log$test_rmse_mean)
  hyper_grid$trees[i] <- m$best_iteration
}

# results
hyper_grid %>%
  filter(rmse > 0) %>%
  arrange(rmse) %>%
  glimpse()
## Observations: 98
## Variables: 10
## $ eta              <dbl> 0.01, 0.01, 0.01, 0.01, 0.01, 0.0…
## $ max_depth        <dbl> 3, 3, 3, 3, 3, 3, 3, 3, 3, 3, 3, …
## $ min_child_weight <dbl> 3, 3, 3, 3, 3, 3, 3, 3, 3, 3, 3, …
## $ subsample        <dbl> 0.5, 0.5, 0.5, 0.5, 0.5, 0.5, 0.5…
## $ colsample_bytree <dbl> 0.5, 0.5, 0.5, 0.5, 0.5, 0.5, 0.5…
## $ gamma            <dbl> 0, 1, 10, 100, 1000, 0, 1, 10, 10…
## $ lambda           <dbl> 1, 1, 1, 1, 1, 1, 1, 1, 1, 1, 1, …
## $ alpha            <dbl> 0.00, 0.00, 0.00, 0.00, 0.00, 0.1…
## $ rmse             <dbl> 20488, 20488, 20488, 20488, 20488…
## $ trees            <dbl> 3944, 3944, 3944, 3944, 3944, 381…
```

Once you've found the optimal hyperparameters, fit the final model with `xgb.train` or `xgboost`. Be sure to use the optimal number of trees found during cross validation. In our example, adding regularization provides no improvement so we exclude them in our final model.

```
# optimal parameter list
params <- list(
  eta = 0.01,
  max_depth = 3,
  min_child_weight = 3,
  subsample = 0.5,
  colsample_bytree = 0.5
)

# train final model
xgb.fit.final <- xgboost(
  params = params,
  data = X,
  label = Y,
  nrounds = 3944,
  objective = "reg:linear",
  verbose = 0
)
```

12.6 Feature interpretation

Measuring GBM feature importance and effects follows the same construct as random forests. Similar to random forests, the **gbm** and **h2o** packages offer an impurity-based feature importance. **xgboost** actually provides three built-in measures for feature importance:

1. **Gain**: This is equivalent to the impurity measure in random forests (reference Section 11.6) and is the most common model-centric metric to use.
2. **Coverage**: The Coverage metric quantifies the relative number of observations influenced by this feature. For example, if you have 100 observations, 4 features and 3 trees, and suppose x_1 is used to decide the leaf node for 10, 5, and 2 observations in $tree_1$, $tree_2$ and $tree_3$ respectively; then the metric will count cover for this feature

as $10 + 5 + 2 = 17$ observations. This will be calculated for all the 4 features and expressed as a percentage.

3. **Frequency**: The percentage representing the relative number of times a particular feature occurs in the trees of the model. In the above example, if x_1 was used for 2 splits, 1 split and 3 splits in each of $tree_1$, $tree_2$ and $tree_3$ respectively; then the weightage for x_1 will be $2 + 1 + 3 = 6$. The frequency for x_1 is calculated as its percentage weight over weights of all x_p features.

If we examine the top 10 influential features in our final model using the impurity (gain) metric, we see very similar results as we saw with our random forest model (Section 11.6). The primary difference is we no longer see `Neighborhood` as a top influential feature, which is likely a result of how we label encoded the categorical features.

By default, `vip::vip()` uses the gain method for feature importance but you can assess the other types using the `type` argument. You can also use `xgboost::xgb.ggplot.importance()` to plot the various feature importance measures but you need to first run `xgb.importance()` on the final model.

```
# variable importance plot
vip::vip(xgb.fit.final)
```

12.7 Final thoughts

GBMs are one of the most powerful ensemble algorithms that are often first-in-class with predictive accuracy. Although they are less intuitive and more computationally demanding than many other machine learning algorithms, they are essential to have in your toolbox.

Although we discussed the most popular GBM algorithms, realize there are alternative algorithms not covered here. For example LightGBM (Ke et al., 2017) is a gradient boosting framework that focuses on *leaf-wise* tree growth versus the traditional level-wise tree growth. This means as a tree is grown deeper, it focuses on extending a single branch versus growing multiple branches (reference Figure 9.2. CatBoost (Dorogush et al., 2018) is another gradient boosting framework that focuses on using efficient methods for encoding categorical features during the gradient boosting process. Both frameworks are available in R.

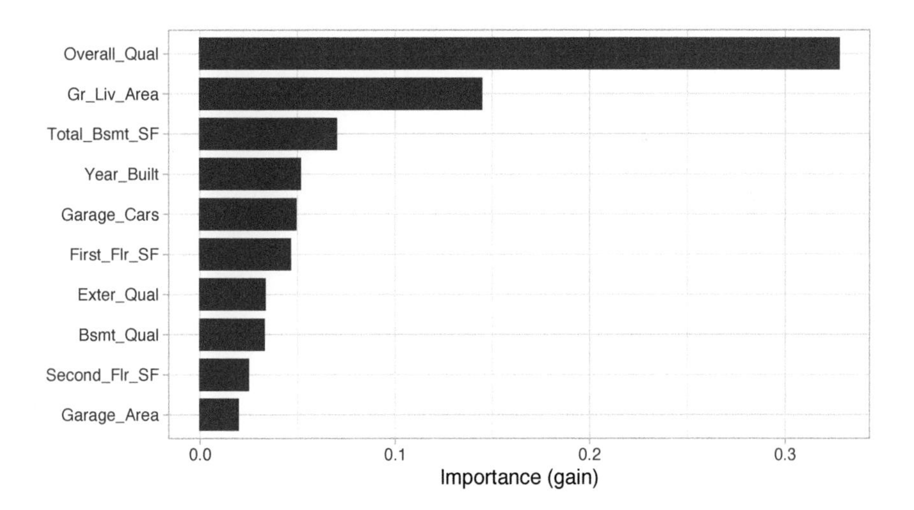

FIGURE 12.8: Top 10 most important variables based on the impurity (gain) metric.

13

Deep Learning

Machine learning algorithms typically search for the optimal representation of data using a feedback signal in the form of an objective function. However, most machine learning algorithms only have the ability to use one or two layers of data transformation to learn the output representation. We call these *shallow* models[1] since they only use 1–2 representations of the feature space. As data sets continue to grow in the dimensions of the feature space, finding the optimal output representation with a shallow model is not always possible. Deep learning provides a multi-layer approach to learn data representations, typically performed with a *multi-layer neural network*. Like other machine learning algorithms, *deep neural networks* (DNN) perform learning by mapping features to targets through a process of simple data transformations and feedback signals; however, DNNs place an emphasis on learning successive layers of meaningful representations. Although an intimidating subject, the overarching concept is rather simple and has proven highly successful across a wide range of problems (e.g., image classification, speech recognition, autonomous driving). This chapter will teach you the fundamentals of building a simple *feedforward* DNN, which is the foundation for the more advanced deep learning models.

Our online resources will provide content covering additional deep learning models such as convolutional, recurrent, and long short-term memory neural networks. Moreover, Chollet and Allaire (2018) is an excellent, in-depth text on applying deep learning methods with R.

13.1 Prerequisites

This tutorial will use a few supporting packages but the main emphasis will be on the **keras** package (Allaire and Chollet, 2019). Additional content provided

[1]Not to be confused with shallow decision trees.

online illustrates how to execute the same procedures we cover here with the **h2o** package. For more information on installing both CPU and GPU-based Keras and TensorFlow software, visit https://keras.rstudio.com[2].

```
# Helper packages
library(dplyr)         # for basic data wrangling

# Modeling packages
library(keras)         # for fitting DNNs
library(tfruns)        # addtl grid search & model training f(x)

# Modeling helper package - not necessary for reproducibility
library(tfestimators)  # addtl grid search & model run interface
```

We'll use the MNIST data to illustrate various DNN concepts. With DNNs, it is important to note a few items:

1. Feedforward DNNs require all feature inputs to be numeric. Consequently, if your data contains categorical features they will need to be numerically encoded (e.g., one-hot encoded, integer label encoded, etc.).
2. Due to the data transformation process that DNNs perform, they are highly sensitive to the individual scale of the feature values. Consequently, we should standardize our features first. Although the MNIST features are measured on the same scale (0–255), they are not standardized (i.e., have mean zero and unit variance); the code chunk below standardizes the MNIST data to resolve this. [3]
3. Since we are working with a multinomial response (0–9), **keras** requires our response to be a one-hot encoded matrix, which can be accomplished with the **keras** function `to_categorical()`.

```
# Import MNIST training data
mnist <- dslabs::read_mnist()
mnist_x <- mnist$train$images
mnist_y <- mnist$train$labels
```

[2] https://keras.rstudio.com/

[3] Standardization is not always necessary with neural networks. It largely depends on the type of network being trained. Feedforward networks, strictly speaking, do not require standardization; however, there are a variety of practical reasons why standardizing the inputs can make training faster and reduce the chances of getting stuck in local optima. Also, weight decay and Bayesian estimation can be applied more conveniently with standardized inputs (Sarle, Warren S., nd).

```
# Rename columns and standardize feature values
colnames(mnist_x) <- paste0("V", 1:ncol(mnist_x))
mnist_x <- mnist_x / 255

# One-hot encode response
mnist_y <- to_categorical(mnist_y, 10)

# Get number of features, which we'll use in our model
p <- ncol(mnist_x)
```

13.2 Why deep learning

Neural networks originated in the computer science field to answer questions that normal statistical approaches were not designed to answer at the time. The MNIST data is one of the most common examples you will find, where the goal is to to analyze hand-written digits and predict the numbers written. This problem was originally presented to AT&T Bell Lab's to help build automatic mail-sorting machines for the USPS (LeCun et al., 1990).

FIGURE 13.1: Sample images from MNIST test dataset (Wikipedia contributors, ndb).

This problem is quite unique because many different features of the data can be represented. As humans, we look at these numbers and consider features such as angles, edges, thickness, completeness of circles, etc. We interpret these different representations of the features and combine them to recognize the digit. In essence, neural networks perform the same task albeit in a far simpler manner than our brains. At their most basic levels, neural networks have three layers: an *input layer*, a *hidden layer*, and an *output layer*. The input layer consists of all of the original input features. The majority of the *learning* takes place in the hidden layer, and the output layer outputs the final predictions.

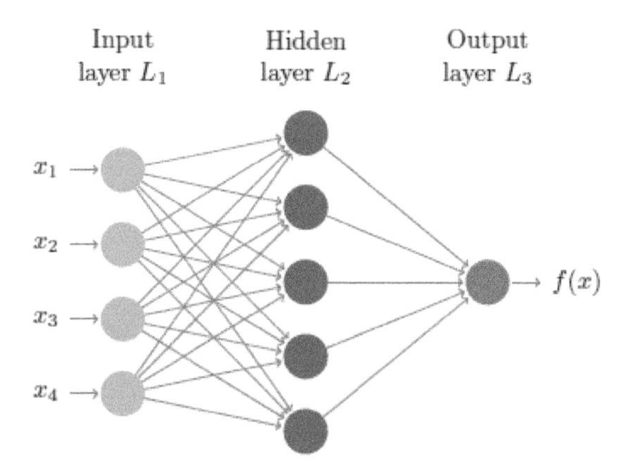

FIGURE 13.2: Representation of a simple feedforward neural network.

Although simple on the surface, the computations being performed inside a network require lots of data to learn and are computationally intense rendering them impractical to use in the earlier days. However, over the past several decades, advancements in computer hardware (off the shelf CPUs became faster and GPUs were created) made the computations more practical, the growth in data collection made them more relevant, and advancements in the underlying algorithms made the *depth* (number of hidden layers) of neural nets less of a constraint. These advancements have resulted in the ability to run very deep and highly parameterized neural networks (i.e., DNNs).

Such DNNs allow for very complex representations of data to be modeled, which has opened the door to analyzing high-dimensional data (e.g., images, videos, and sound bytes). In some machine learning approaches, features of the data need to be defined prior to modeling (e.g., ordinary linear regression). One can only imagine trying to create the features for the digit recognition problem above. However, with DNNs, the hidden layers provide the means to auto-identify useful features. A simple way to think of this is to go back to our digit recognition problem. The first hidden layer may learn about the angles of the line, the next hidden layer may learn about the thickness of the lines, the next may learn the location and completeness of the circles, etc. Aggregating these different attributes together by linking the layers allows the model to accurately predict what digit each image represents.

This is the reason that DNNs are so popular for very complex problems where feature engineering is important, but rather difficult to do by hand (e.g., facial recognition). However, at their core, DNNs perform successive non-linear

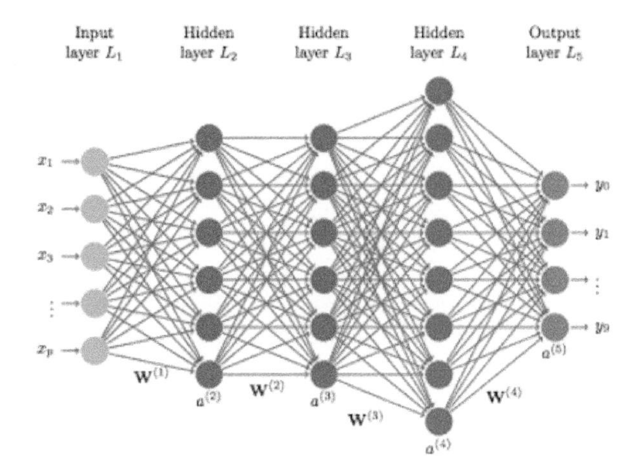

FIGURE 13.3: Representation of a deep feedforward neural network.

transformations across each layer, allowing DNNs to model very complex and non-linear relationships. This can make DNNs suitable machine learning approaches for traditional regression and classification problems as well. But it is important to keep in mind that deep learning thrives when dimensions of your data are sufficiently large (e.g., very large training sets). As the number of observations (n) and feature inputs (p) decrease, shallow machine learning approaches tend to perform just as well, if not better, and are more efficient.

13.3 Feedforward DNNs

Multiple DNN architectures exist and, as interest and research in this area increases, the field will continue to flourish. For example, convolutional neural networks (CNNs or ConvNets) have widespread applications in image and video recognition, recurrent neural networks (RNNs) are often used with speech recognition, and long short-term memory neural networks (LSTMs) are advancing automated robotics and machine translation. However, fundamental to all these methods is the feedforward DNN (aka multilayer perceptron). Feedforward DNNs are densely connected layers where inputs influence each successive layer which then influences the final output layer.

To build a feedforward DNN we need four key components:

1. Input data
2. A pre-defined network architecture;

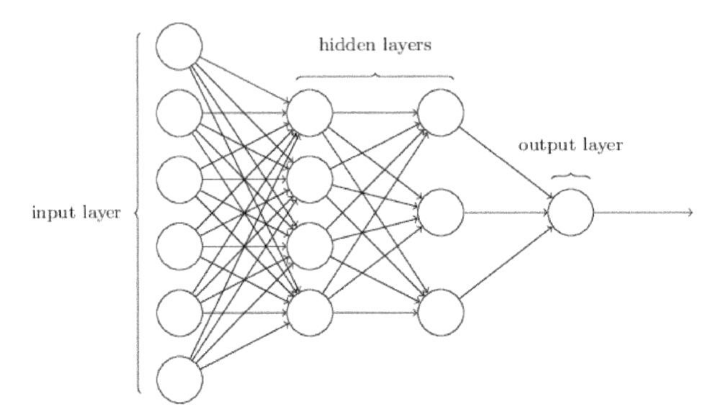

FIGURE 13.4: Feedforward neural network.

3. A feedback mechanism to help the network learn;
4. A model training approach.

The next few sections will walk you through steps 2)–4) to build a feedforward DNN to the MNIST data.

13.4 Network architecture

When developing the network architecture for a feedforward DNN, you really only need to worry about two features: (1) layers and nodes, and (2) activation.

13.4.1 Layers and nodes

The layers and nodes are the building blocks of our DNN and they decide how complex the network will be. Layers are considered *dense* (fully connected) when all the nodes in each successive layer are connected. Consequently, the more layers and nodes you add the more opportunities for new features to be learned (commonly referred to as the model's *capacity*).[4] Beyond the *input layer*, which is just our original predictor variables, there are two main types of layers to consider: *hidden layers* and an *output layer*.

[4]Often, the number of nodes in a layer is referred to as the network's *width* while the number of layers in a model is referred to as its *depth*.

13.4.1.1 Hidden layers

There is no well-defined approach for selecting the number of hidden layers and nodes; rather, these are the first of many hyperparameters to tune. With regular tabular data, 2–5 hidden layers are often sufficient but your best bet is to err on the side of more layers rather than fewer. The number of nodes you incorporate in these hidden layers is largely determined by the number of features in your data. Often, the number of nodes in each layer is equal to or less than the number of features but this is not a hard requirement. It is important to note that the number of hidden layers and nodes in your network can affect its computational complexity (e.g., training time). When dealing with many features and, therefore, many nodes, training deep models with many hidden layers can be computationally more efficient than training a single layer network with the same number of high volume nodes (Goodfellow et al., 2016). Consequently, the goal is to find the simplest model with optimal performance.

13.4.1.2 Output layers

The choice of output layer is driven by the modeling task. For regression problems, your output layer will contain one node that outputs the final predicted value. Classification problems are different. If you are predicting a binary output (e.g., True/False, Win/Loss), your output layer will still contain only one node and that node will predict the probability of success (however you define success). However, if you are predicting a multinomial output, the output layer will contain the same number of nodes as the number of classes being predicted. For example, in our MNIST data, we are predicting 10 classes (0–9); therefore, the output layer will have 10 nodes and the output would provide the probability of each class.

13.4.1.3 Implementation

The **keras** package allows us to develop our network with a layering approach. First, we initiate our sequential feedforward DNN architecture with `keras_model_sequential()` and then add some dense layers. This example creates two hidden layers, the first with 128 nodes and the second with 64, followed by an output layer with 10 nodes. One thing to point out is that the first layer needs the `input_shape` argument to equal the number of features in your data; however, the successive layers are able to dynamically interpret the number of expected inputs based on the previous layer.

```
model <- keras_model_sequential() %>%
  layer_dense(units = 128, input_shape = p) %>%
```

```
layer_dense(units = 64) %>%
layer_dense(units = 10)
```

13.4.2 Activation

A key component with neural networks is what's called *activation*. In the human brain, the biologic neuron receives inputs from many adjacent neurons. When these inputs accumulate beyond a certain threshold the neuron is *activated* suggesting there is a signal. DNNs work in a similar fashion.

13.4.2.1 Activation functions

As stated previously, each node is connected to all the nodes in the previous layer. Each connection gets a weight and then that node adds all the incoming inputs multiplied by its corresponding connection weight plus an extra *bias* parameter (w_0). The summed total of these inputs become an input to an *activation function*; see 13.5.

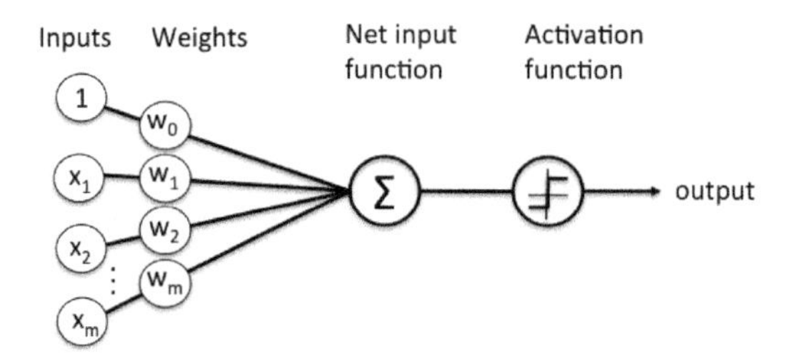

FIGURE 13.5: Flow of information in an artificial neuron.

The activation function is simply a mathematical function that determines whether or not there is enough informative input at a node to fire a signal to the next layer. There are multiple activation functions[5] to choose from but the most common ones include:

$$\texttt{Linear (identity):}\ f(x) = x \tag{13.1}$$

$$\texttt{Rectified linear unit (ReLU):}\ f(x) = \begin{cases} 0, & \text{for } x < 0. \\ x, & \text{for } x \geq 0. \end{cases} \tag{13.2}$$

$$\texttt{Sigmoid:}\ f(x) = \frac{1}{1 + e^{-x}} \tag{13.3}$$

[5] https://en.wikipedia.org/wiki/Activation_function

$$\texttt{Softmax:}\ f(x) = \frac{e^{x_i}}{\sum_j e^{x_j}} \tag{13.4}$$

When using rectangular data, the most common approach is to use ReLU activation functions in the hidden layers. The ReLU activation function is simply taking the summed weighted inputs and transforming them to a 0 (not fire) or > 0 (fire) if there is enough signal. For the output layers we use the linear activation function for regression problems, the sigmoid activation function for binary classification problems, and softmax for multinomial classification problems.

13.4.2.2 Implementation

To control the activation functions used in our layers we specify the `activation` argument. For the two hidden layers we add the ReLU activation function and for the output layer we specify `activation = softmax` (since MNIST is a multinomial classification problem).

```
model <- keras_model_sequential() %>%
  layer_dense(units = 128, activation = "relu", input_shape = p) %>%
  layer_dense(units = 64, activation = "relu") %>%
  layer_dense(units = 10, activation = "softmax")
```

Next, we need to incorporate a feedback mechanism to help our model learn.

13.5 Backpropagation

On the first run (or *forward pass*), the DNN will select a batch of observations, randomly assign weights across all the node connections, and predict the output. The engine of neural networks is how it assesses its own accuracy and automatically adjusts the weights across all the node connections to improve that accuracy. This process is called *backpropagation*. To perform backpropagation we need two things:

1. An objective function;
2. An optimizer.

First, you need to establish an objective (loss) function to measure performance. For regression problems this might be mean squared error (MSE) and for classification problems it is commonly binary and multi-categorical cross entropy (reference Section 2.6). DNNs can have multiple loss functions but we'll just focus on using one.

On each forward pass the DNN will measure its performance based on the loss function chosen. The DNN will then work backwards through the layers, compute the gradient[6] of the loss with regards to the network weights, adjust the weights a little in the opposite direction of the gradient, grab another batch of observations to run through the model, ...rinse and repeat until the loss function is minimized. This process is known as *mini-batch stochastic gradient descent*[7] (mini-batch SGD). There are several variants of mini-batch SGD algorithms; they primarily differ in how fast they descend the gradient (controlled by the *learning rate* as discussed in Section 12.2.2). These different variations make up the different *optimizers* that can be used.

Understanding the technical differences among the variants of gradient descent is beyond the intent of this book. An excellent source to learn more about these differences and appropriate scenarios to adjust this parameter is provided by Ruder (2016). For now, realize that sticking with the default optimizer (RMSProp) is often sufficient for most normal regression and classification problems; however, this is a tunable hyperparameter.

To incorporate the backpropagation piece of our DNN we include `compile()` in our code sequence. In addition to the optimizer and loss function arguments, we can also identify one or more metrics in addition to our loss function to track and report.

```
model <- keras_model_sequential() %>%

  # Network architecture
  layer_dense(units = 128, activation = "relu", input_shape = p) %>%
  layer_dense(units = 64, activation = "relu") %>%
  layer_dense(units = 10, activation = "softmax") %>%

  # Backpropagation
  compile(
    loss = 'categorical_crossentropy',
    optimizer = optimizer_rmsprop(),
```

[6]A gradient is the generalization of the concept of derivatives applied to functions of multidimensional inputs.

[7]It's considered stochastic because a random subset (batch) of observations is drawn for each forward pass.

```
    metrics = c('accuracy')
  )
```

13.6 Model training

We've created a base model, now we just need to train it with some data. To do so we feed our model into a `fit()` function along with our training data. We also provide a few other arguments that are worth mentioning:

- `batch_size`: As we mentioned in the last section, the DNN will take a batch of data to run through the mini-batch SGD process. Batch sizes can be between one and several hundred. Small values will be more computationally burdensome while large values provide less feedback signal. Values are typically provided as a power of two that fit nicely into the memory requirements of the GPU or CPU hardware like 32, 64, 128, 256, and so on.
- `epochs`: An *epoch* describes the number of times the algorithm sees the entire data set. So, each time the algorithm has seen all samples in the data set, an epoch has completed. In our training set, we have 60,000 observations so running batches of 128 will require 469 passes for one epoch. The more complex the features and relationships in your data, the more epochs you'll require for your model to learn, adjust the weights, and minimize the loss function.
- `validation_split`: The model will hold out XX% of the data so that we can compute a more accurate estimate of an out-of-sample error rate.
- `verbose`: We set this to `FALSE` for brevity; however, when `TRUE` you will see a live update of the loss function in your RStudio IDE.

Plotting the output shows how our loss function (and specified metrics) improve for each epoch. We see that our model's performance is optimized at 5–10 epochs and then proceeds to overfit, which results in a flatlined accuracy rate.

The training and validation below took ~30 seconds.

```
# Train the model
fit1 <- model %>%
  fit(
    x = mnist_x,
    y = mnist_y,
    epochs = 25,
    batch_size = 128,
    validation_split = 0.2,
    verbose = FALSE
  )

# Display output
fit1
## Trained on 48,000 samples (batch_size=128, epochs=25)
## Final epoch (plot to see history):
##      loss: 0.002156
##       acc: 0.9992
## val_loss: 0.1577
##   val_acc: 0.9777
plot(fit1)
```

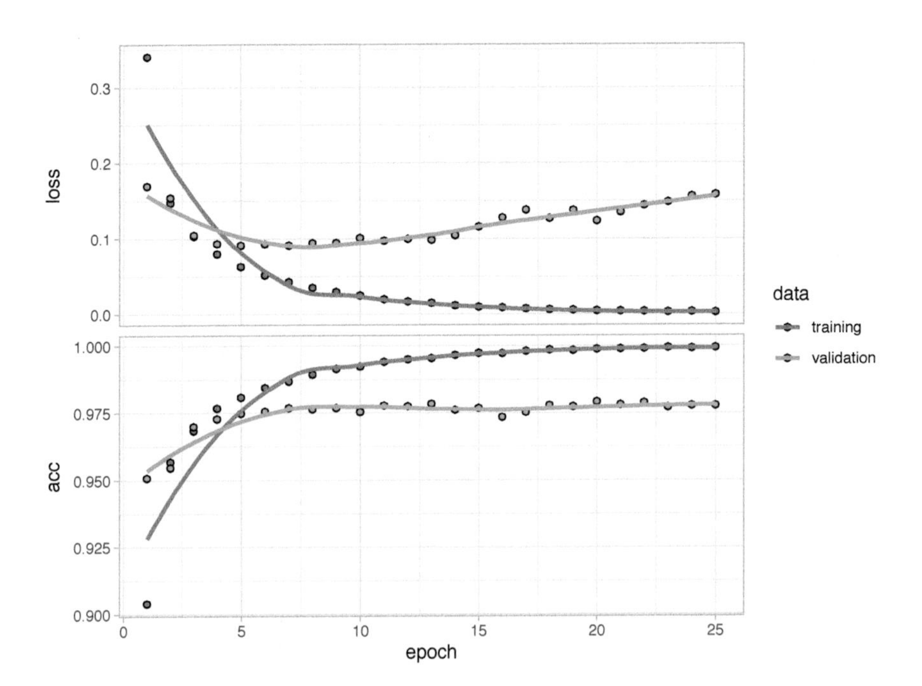

FIGURE 13.6: Training and validation performance over 25 epochs.

TABLE 13.1: Model capacities assessed represented as number of layers and nodes per layer.

	Hidden Layers		
Size	1	2	3
small	16	16, 8	16, 8, 4
medium	64	64, 32	64, 32, 16
large	256	256, 128	256, 128, 64

13.7 Model tuning

Now that we have an understanding of producing and running a DNN model, the next task is to find an optimal one by tuning different hyperparameters. There are many ways to tune a DNN. Typically, the tuning process follows these general steps; however, there is often a lot of iteration among these:

1. Adjust model capacity (layers & nodes);
2. Add batch normalization;
3. Add regularization;
4. Adjust learning rate.

13.7.1 Model capacity

Typically, we start by maximizing predictive performance based on model capacity. Higher model capacity (i.e., more layers and nodes) results in more *memorization capacity* for the model. On one hand, this can be good as it allows the model to learn more features and patterns in the data. On the other hand, a model with too much capacity will overfit to the training data. Typically, we look to maximize validation error performance while minimizing model capacity. As an example, we assessed nine different model capacity settings that include the following number of layers and nodes while maintaining all other parameters the same as the models in the previous sections (i.e.. our medium sized 2-hidden layer network contains 64 nodes in the first layer and 32 in the second.).

The models that performed best had 2–3 hidden layers with a medium to large number of nodes. All the "small" models underfit and would require more epochs to identify their minimum validation error. The large 3-layer model overfits extremely fast. Preferably, we want a model that overfits more slowly such as the 1- and 2-layer medium and large models (Chollet and Allaire, 2018).

If none of your models reach a flatlined validation error such as all the "small" models in Figure 13.7, increase the number of epochs trained. Alternatively, if your epochs flatline early then there is no reason to run so many epochs as you are just wasting computational energy with no gain. We can add a `callback()` function inside of `fit()` to help with this. There are multiple callbacks to help automate certain tasks. One such callback is early stopping, which will stop training if the loss function does not improve for a specified number of epochs.

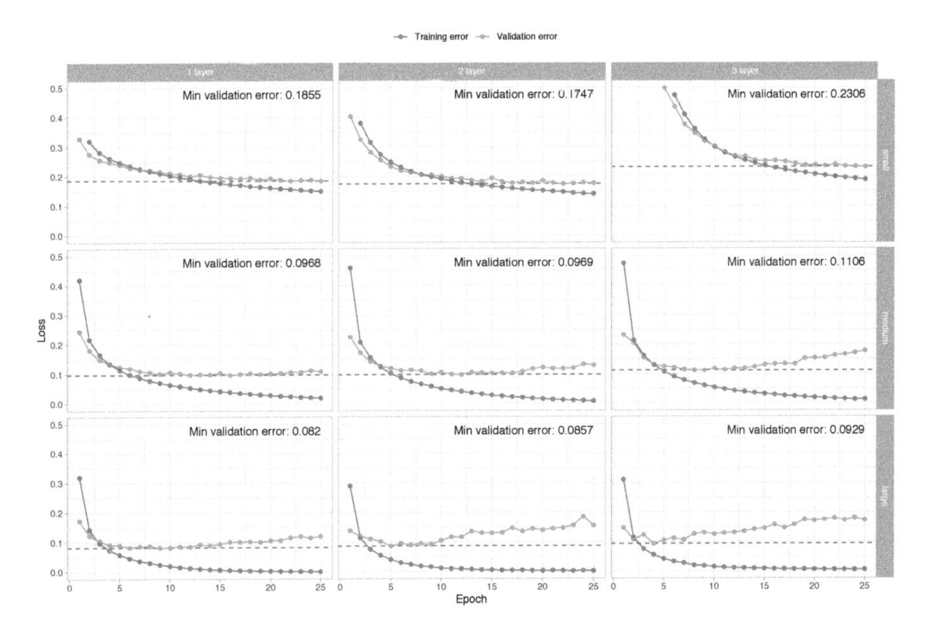

FIGURE 13.7: Training and validation performance for various model capacities.

13.7.2 Batch normalization

We've normalized the data before feeding it into our model, but data normalization should be a concern after every transformation performed by the network. Batch normalization (Ioffe and Szegedy, 2015) is a recent advancement that adaptively normalizes data even as the mean and variance change over time during training. The main effect of batch normalization is that it helps with gradient propagation, which allows for deeper networks. Consequently, as the depth of your networks increase, batch normalization becomes more important and can improve performance.

We can add batch normalization by including `layer_batch_normalization()` after each middle layer within the network architecture section of our code:

```
model_w_norm <- keras_model_sequential() %>%

  # Network architecture with batch normalization
  layer_dense(units = 256, activation = "relu", input_shape = p) %>%
  layer_batch_normalization() %>%
  layer_dense(units = 128, activation = "relu") %>%
  layer_batch_normalization() %>%
  layer_dense(units = 64, activation = "relu") %>%
  layer_batch_normalization() %>%
  layer_dense(units = 10, activation = "softmax") %>%

  # Backpropagation
  compile(
    loss = "categorical_crossentropy",
    optimizer = optimizer_rmsprop(),
    metrics = c("accuracy")
  )
```

If we add batch normalization to each of the previously assessed models, we see a couple patterns emerge. One, batch normalization often helps to minimize the validation loss sooner, which increases efficiency of model training. Two, we see that for the larger, more complex models (3-layer medium and 2- and 3-layer large), batch normalization helps to reduce the overall amount of overfitting. In fact, with batch normalization, our large 3-layer network now has the best validation error.

13.7.3 Regularization

As we've discussed in Chapters 6 and 12, placing constraints on a model's complexity with regularization is a common way to mitigate overfitting. DNNs models are no different and there are two common approaches to regularizing neural networks. We can use an L_1 or L_2 penalty to add a cost to the size of the node weights, although the most common penalizer is the L_2 *norm*, which is called *weight decay* in the context of neural networks.[8] Regularizing the weights will force small signals (noise) to have weights nearly equal to zero and only allow consistently strong signals to have relatively larger weights.

[8] Similar to the previous regularization discussions, the L_1 penalty is based on the absolute value of the weight coefficients, whereas the L_2 penalty is based on the square of the value of the weight coefficients.

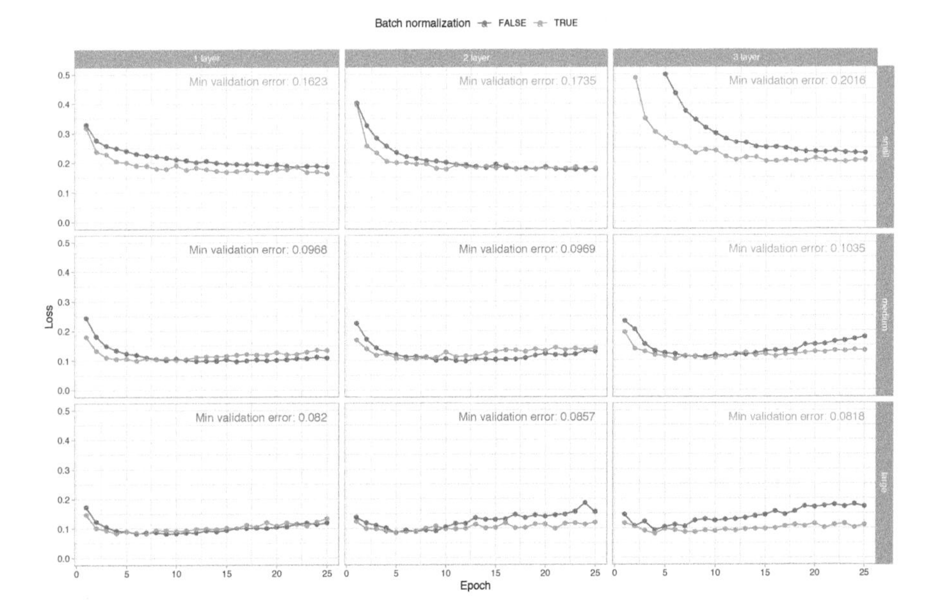

FIGURE 13.8: The effect of batch normalization on validation loss for various model capacities.

As you add more layers and nodes, regularization with L_1 or L_2 penalties tend to have a larger impact on performance. Since having too many hidden units runs the risk of overparameterization, L_1 or L_2 penalties can shrink the extra weights toward zero to reduce the risk of overfitting.

We can add an L_1, L_2, or a combination of the two by adding `regularizer_XX()` within each layer.

```
model_w_reg <- keras_model_sequential() %>%

  # Network architecture with L1 regularization and batch normalization
  layer_dense(units = 256, activation = "relu", input_shape = p,
              kernel_regularizer = regularizer_l2(0.001)) %>%
  layer_batch_normalization() %>%
  layer_dense(units = 128, activation = "relu",
              kernel_regularizer = regularizer_l2(0.001)) %>%
  layer_batch_normalization() %>%
  layer_dense(units = 64, activation = "relu",
              kernel_regularizer = regularizer_l2(0.001)) %>%
```

```
  layer_batch_normalization() %>%
  layer_dense(units = 10, activation = "softmax") %>%

  # Backpropagation
  compile(
    loss = "categorical_crossentropy",
    optimizer = optimizer_rmsprop(),
    metrics = c("accuracy")
  )
```

Dropout (Srivastava et al., 2014b; Hinton et al., 2012) is an additional regularization method that has become one of the most common and effectively used approaches to minimize overfitting in neural networks. Dropout in the context of neural networks randomly drops out (setting to zero) a number of output features in a layer during training. By randomly removing different nodes, we help prevent the model from latching onto happenstance patterns (noise) that are not significant. Typically, dropout rates range from 0.2–0.5 but can differ depending on the data (i.e., this is another tuning parameter). Similar to batch normalization, we can apply dropout by adding `layer_dropout()` in between the layers.

```
model_w_drop <- keras_model_sequential() %>%

  # Network architecture with 20% dropout
  layer_dense(units = 256, activation = "relu", input_shape = p) %>%
  layer_dropout(rate = 0.2) %>%
  layer_dense(units = 128, activation = "relu") %>%
  layer_dropout(rate = 0.2) %>%
  layer_dense(units = 64, activation = "relu") %>%
  layer_dropout(rate = 0.2) %>%
  layer_dense(units = 10, activation = "softmax") %>%

  # Backpropagation
  compile(
    loss = "categorical_crossentropy",
    optimizer = optimizer_rmsprop(),
    metrics = c("accuracy")
  )
```

For our MNIST data, we find that adding an L_1 or L_2 cost does not improve our loss function. However, adding dropout does improve performance. For example, our large 3-layer model with 256, 128, and 64 nodes per respective layer so far has the best performance with a cross-entropy loss of 0.0818. However, as illustrated in Figure 13.8, this network still suffers from overfitting. Figure 13.9 illustrates the same 3-layer model with 256, 128, and 64 nodes per respective layers, batch normalization, and dropout rates of 0.4, 0.3, and 0.2 between each respective layer. We see a significant improvement in overfitting, which results in an improved loss score.

Note that as you constrain overfitting, often you need to increase the number of epochs to allow the network enough iterations to find the global minimal loss.

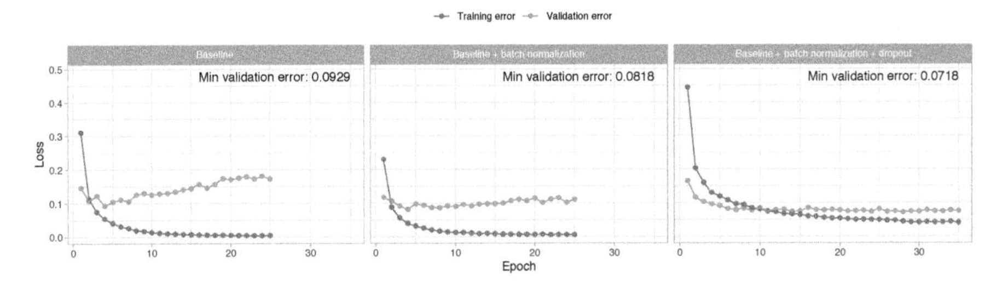

FIGURE 13.9: The effect of regularization with dropout on validation loss.

13.7.4 Adjust learning rate

Another issue to be concerned with is whether or not we are finding a global minimum versus a local minimum with our loss value. The mini-batch SGD optimizer we use will take incremental steps down our loss gradient until it no longer experiences improvement. The size of the incremental steps (i.e., the learning rate) will determine whether or not we get stuck in a local minimum instead of making our way to the global minimum.

There are two ways to circumvent this problem:

1. The different optimizers (e.g., RMSProp, Adam, Adagrad) have different algorithmic approaches for deciding the learning rate. We can adjust the learning rate of a given optimizer or we can adjust the optimizer used.
2. We can automatically adjust the learning rate by a factor of 2–10 once the validation loss has stopped improving.

The following builds onto our optimal model by changing the optimizer to

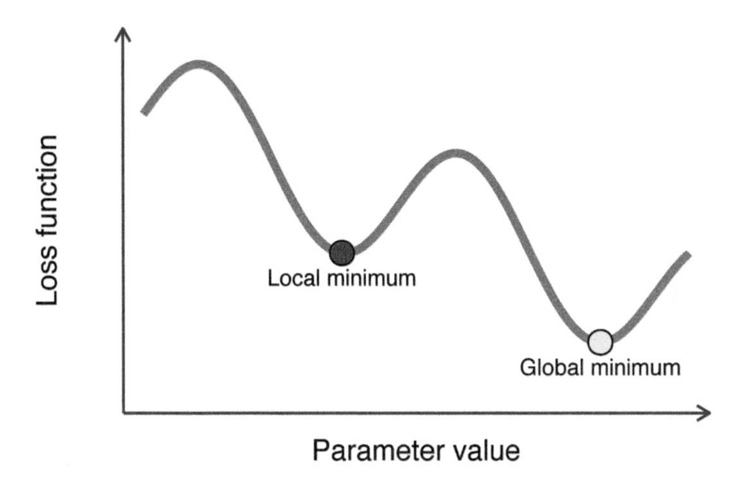

FIGURE 13.10: A local minimum and a global minimum.

Adam (Kingma and Ba, 2014) and reducing the learning rate by a factor of 0.05 as our loss improvement begins to stall. We also add an early stopping argument to reduce unnecessary runtime. We see a slight improvement in performance and our loss curve in Figure 13.11 illustrates how we stop model training just as we begin to overfit.

```
model_w_adj_lrn <- keras_model_sequential() %>%
  layer_dense(units = 256, activation = "relu", input_shape = p) %>%
  layer_batch_normalization() %>%
  layer_dropout(rate = 0.4) %>%
  layer_dense(units = 128, activation = "relu") %>%
  layer_batch_normalization() %>%
  layer_dropout(rate = 0.3) %>%
  layer_dense(units = 64, activation = "relu") %>%
  layer_batch_normalization() %>%
  layer_dropout(rate = 0.2) %>%
  layer_dense(units = 10, activation = "softmax") %>%
  compile(
    loss = 'categorical_crossentropy',
    optimizer = optimizer_adam(),
    metrics = c('accuracy')
  ) %>%
  fit(
    x = mnist_x,
    y = mnist_y,
```

```
    epochs = 35,
    batch_size = 128,
    validation_split = 0.2,
    callbacks = list(
      callback_early_stopping(patience = 5),
      callback_reduce_lr_on_plateau(factor = 0.05)
      ),
    verbose = FALSE
  )

model_w_adj_lrn
## Trained on 48,000 samples (batch_size=128, epochs=27)
## Final epoch (plot to see history):
##      loss: 0.04639
##       acc: 0.9851
##  val_loss: 0.07389
##   val_acc: 0.981
##        lr: 0.001

# Optimal
min(model_w_adj_lrn$metrics$val_loss)
## [1] 0.0696
max(model_w_adj_lrn$metrics$val_acc)
## [1] 0.981

# Learning rate
plot(model_w_adj_lrn)
```

13.8 Grid search

Hyperparameter tuning for DNNs tends to be a bit more involved than other ML models due to the number of hyperparameters that can/should be assessed and the dependencies between these parameters. For most implementations you need to predetermine the number of layers you want and then establish your search grid. If using **h2o**'s `h2o.deeplearning()` function, then creating and executing the search grid follows the same approach illustrated in Sections 11.5 and 12.4.2.

However, for **keras**, we use *flags* in a similar manner but their implementation provides added flexibility for tracking, visualizing, and managing training runs

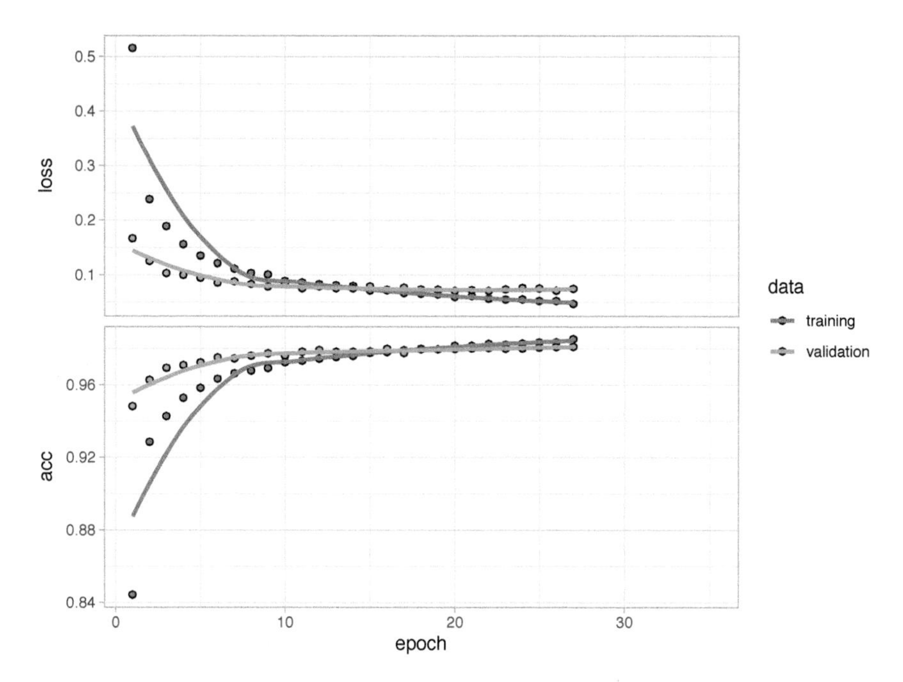

FIGURE 13.11: Training and validation performance on our 3-layer large network with dropout, adjustable learning rate, and using an Adam mini-batch SGD optimizer.

with the **tfruns** package (Allaire, 2018). For a full discussion regarding flags see the `https://tensorflow.rstudio.com/tools/` online resource. In this example we provide a training script mnist-grid-search.R[9] that will be sourced for the grid search.

To create and perform a grid search, we first establish flags for the different hyperparameters of interest. These are considered the default flag values:

```
FLAGS <- flags(
  # Nodes
  flag_numeric("nodes1", 256),
  flag_numeric("nodes2", 128),
  flag_numeric("nodes3", 64),
  # Dropout
  flag_numeric("dropout1", 0.4),
  flag_numeric("dropout2", 0.3),
  flag_numeric("dropout3", 0.2),
```

[9] `http://bit.ly/mnist-grid-search`

```
  # Learning paramaters
  flag_string("optimizer", "rmsprop"),
  flag_numeric("lr_annealing", 0.1)
)
```

Next, we incorporate the flag parameters within our model:

```
model <- keras_model_sequential() %>%
  layer_dense(units = FLAGS$nodes1, activation = "relu", input_shape =
                        p) %>%
  layer_batch_normalization() %>%
  layer_dropout(rate = FLAGS$dropout1) %>%
  layer_dense(units = FLAGS$nodes2, activation = "relu") %>%
  layer_batch_normalization() %>%
  layer_dropout(rate = FLAGS$dropout2) %>%
  layer_dense(units = FLAGS$nodes3, activation = "relu") %>%
  layer_batch_normalization() %>%
  layer_dropout(rate = FLAGS$dropout3) %>%
  layer_dense(units = 10, activation = "softmax") %>%
  compile(
    loss = 'categorical_crossentropy',
    metrics = c('accuracy'),
    optimizer = FLAGS$optimizer
  ) %>%
  fit(
    x = mnist_x,
    y = mnist_y,
    epochs = 35,
    batch_size = 128,
    validation_split = 0.2,
    callbacks = list(
      callback_early_stopping(patience = 5),
      callback_reduce_lr_on_plateau(factor = FLAGS$lr_annealing)
    ),
    verbose = FALSE
  )
```

To execute the grid search we use `tfruns::tuning_run()`. Since our grid search assesses 2,916 combinations, we perform a random grid search and assess only 5% of the total models (`sample = 0.05` which equates to 145 models). It becomes quite obvious that the hyperparameter search space explodes quickly with DNNs since there are so many model attributes that can be adjusted. Consequently, often a full Cartesian grid search is not possible due to time and computational constraints.

The optimal model has a validation loss of 0.0686 and validation accuracy rate of 0.9806 and the below code chunk shows the hyperparameter settings for this optimal model.

The following grid search took us over 1.5 hours to run!

```
# Run various combinations of dropout1 and dropout2
runs <- tuning_run("scripts/mnist-grid-search.R",
  flags = list(
    nodes1 = c(64, 128, 256),
    nodes2 = c(64, 128, 256),
    nodes3 = c(64, 128, 256),
    dropout1 = c(0.2, 0.3, 0.4),
    dropout2 = c(0.2, 0.3, 0.4),
    dropout3 = c(0.2, 0.3, 0.4),
    optimizer = c("rmsprop", "adam"),
    lr_annealing = c(0.1, 0.05)
  ),
  sample = 0.05
)

runs %>%
  filter(metric_val_loss == min(metric_val_loss)) %>%
  glimpse()
## Observations: 1
## Variables: 31
## $ run_dir            <chr> "runs/2019-04-27T14-44-38Z"
## $ metric_loss        <dbl> 0.0598
## $ metric_acc         <dbl> 0.9806
## $ metric_val_loss    <dbl> 0.0686
## $ metric_val_acc     <dbl> 0.9806
## $ flag_nodes1        <int> 256
## $ flag_nodes2        <int> 128
## $ flag_nodes3        <int> 256
## $ flag_dropout1      <dbl> 0.4
## $ flag_dropout2      <dbl> 0.2
## $ flag_dropout3      <dbl> 0.3
## $ flag_optimizer     <chr> "adam"
## $ flag_lr_annealing  <dbl> 0.05
## $ samples            <int> 48000
## $ validation_samples <int> 12000
```

```
## $ batch_size          <int> 128
## $ epochs              <int> 35
## $ epochs_completed    <int> 17
## $ metrics             <chr> "runs/2019-04-27T14-44-38Z/tfrun...
## $ model               <chr> "Model\n________________________...
## $ loss_function       <chr> "categorical_crossentropy"
## $ optimizer           <chr> "<tensorflow.python.keras.optimi...
## $ learning_rate       <dbl> 0.001
## $ script              <chr> "mnist-grid-search.R"
## $ start               <dttm> 2019-04-27 14:44:38
## $ end                 <dttm> 2019-04-27 14:45:39
## $ completed           <lgl> TRUE
## $ output              <chr> "\n> #' Trains a feedforward DL ...
## $ source_code         <chr> "runs/2019-04-27T14-44-38Z/tfrun...
## $ context             <chr> "local"
## $ type                <chr> "training"
```

13.9 Final thoughts

Training DNNs often requires more time and attention than other ML algorithms. With many other algorithms, the search space for finding an optimal model is small enough that Cartesian grid searches can be executed rather quickly. With DNNs, more thought, time, and experimentation is often required up front to establish a basic network architecture to build a grid search around. However, even with prior experimentation to reduce the scope of a grid search, the large number of hyperparameters still results in an exploding search space that can usually only be efficiently searched at random.

Historically, training neural networks was quite slow since runtime requires $O(NpML)$ operations where $N = \#$ observations, $p = \#$ features, $M = \#$ hidden nodes, and $L = \#$ epchos. Fortunately, software has advanced tremendously over the past decade to make execution fast and efficient. With open source software such as TensorFlow and Keras available via R APIs, performing state of the art deep learning methods is much more efficient, plus you get all the added benefits these open source tools provide (e.g., distributed computations across CPUs and GPUs, more advanced DNN architectures such as convolutional and recurrent neural nets, autoencoders, reinforcement learning, and more!).

14

Support Vector Machines

Support vector machines (SVMs) offer a direct approach to binary classification: try to find a *hyperplane* in some feature space that "best" separates the two classes. In practice, however, it is difficult (if not impossible) to find a hyperplane to perfectly separate the classes using just the original features. SVMs overcome this by extending the idea of finding a separating hyperplane in two ways: (1) loosen what we mean by "perfectly separates", and (2) use the so-called *kernel trick* to enlarge the feature space to the point that perfect separation of classes is (more) likely.

14.1 Prerequisites

Although there are a number of great packages that implement SVMs (e.g., **e1071** (Meyer et al., 2019) and **svmpath** (Hastie, 2016)), we'll focus on the most flexible implementation of SVMs in R: **kernlab** (Karatzoglou et al., 2004). We'll also use **caret** for tuning SVMs and pre-processing. In this chapter, we'll explicitly load the following packages:

```
# Helper packages
library(dplyr)     # for data wrangling
library(ggplot2)   # for awesome graphics
library(rsample)   # for data splitting

# Modeling packages
library(caret)     # for classification and regression training
library(kernlab)   # for fitting SVMs

# Model interpretability packages
library(pdp)       # for partial dependence plots, etc.
library(vip)       # for variable importance plots
```

To illustrate the basic concepts of fitting SVMs we'll use a mix of simulated data sets as well as the employee attrition data. The code for generating the simulated data sets and figures in this chapter are available on the book website. In the employee attrition example our intent is to predict on `Attrition` (coded as `"Yes"`/`"No"`). As in previous chapters, we'll set aside 30% of the data for assessing generalizability.

```
# Load attrition data
df <- attrition %>%
  mutate_if(is.ordered, factor, ordered = FALSE)

# Create training (70%) and test (30%) sets
set.seed(123)  # for reproducibility
churn_split <- initial_split(df, prop = 0.7, strata = "Attrition")
churn_train <- training(churn_split)
churn_test  <- testing(churn_split)
```

14.2 Optimal separating hyperplanes

Rather than diving right into SVMs we'll build up to them using concepts from basic geometry, starting with hyperplanes. A hyperplane in p-dimensional feature space is defined by the (linear) equation

$$f(X) = \beta_0 + \beta_1 X_1 + \cdots + \beta_p X_p = 0$$

When $p = 2$, this defines a line in 2-D space, and when $p = 3$, it defines a plane in 3-D space (see Figure 14.1). By definition, for points on one side of the hyperplane, $f(X) > 0$, and for points on the other side, $f(X) < 0$. For (mathematical) convenience, we'll re-encode the binary outcome Y_i using {-1, 1} so that $Y_i \times f(X_i) > 0$ for points on the correct side of the hyperplane. In this context the hyperplane represents a *decision boundary* that partitions the feature space into two sets, one for each class. The SVM will classify all the points on one side of the decision boundary as belonging to one class and all those on the other side as belonging to the other class.

While SVMs may seem mathematically frightening at first, the fundamental ideas behind them are incredibly intuitive and easy to understand. We'll illustrate these simple ideas using simulated binary classification data with two features. In this hypothetical example, we have two classes: households that own a riding lawn mower ($Y = +1$) and (2) households that do not ($Y = -1$). We also have two features, household income (X_1) and lot size (X_2), that

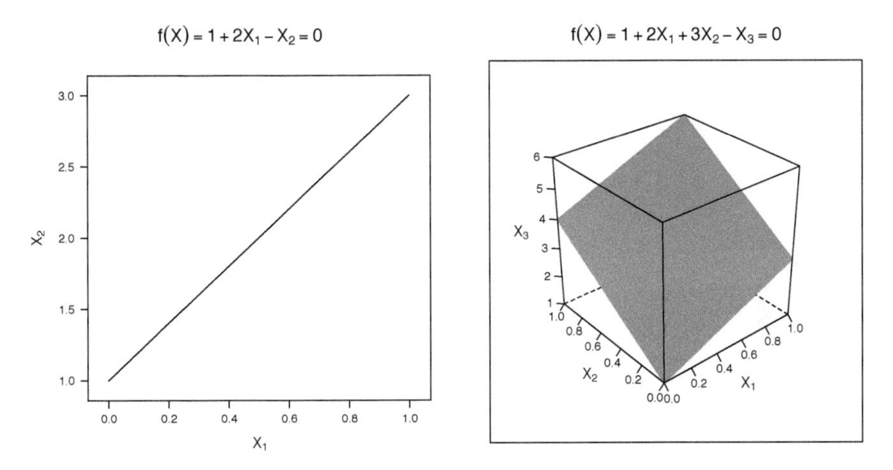

FIGURE 14.1: Examples of hyperplanes in 2-D and 3-D feature space.

have been standardized (i.e., centered around zero with a standard deviation of one). Intuitively, we might expect households with a larger lot and a higher income to be more likely to own a riding mower. In fact, the two classes in the left side of Figure 14.2 are perfectly separable by a straight line (i.e., a hyperplane in 2-D space).

14.2.1 The hard margin classifier

As you might imagine, for two separable classes, there are an infinite number of separating hyperplanes! This is illustrated in the right side of Figure 14.2 where we show the hyperplanes (i.e., decision boundaries) that result from a simple logistic regression model (GLM), a *linear discriminant analysis* (LDA; another popular classification tool), and an example of a *hard margin classifier* (HMC)—which we'll define in a moment. So which decision boundary is "best"? Well, it depends on how we define "best". If you were asked to draw a decision boundary with good generalization performance on the left side of Figure 14.2, how would it look to you? Naturally, you would probably draw a boundary that provides the maximum separation between the two classes, and that's exactly what the HMC is doing!

Although we can draw an unlimited number of separating hyperplanes, what we want is a separating hyperplane with good generalization performance! The HMC is one such "optimal" separating hyperplane and the simplest type of SVM. The HMC is optimal in the sense that it separates the two classes while maximizing the distance to the closest points from either class; see Figure 14.3 below. The decision boundary (i.e., hyperplane) from the HMC separates the two classes by maximizing the distance between them. This maximized distance

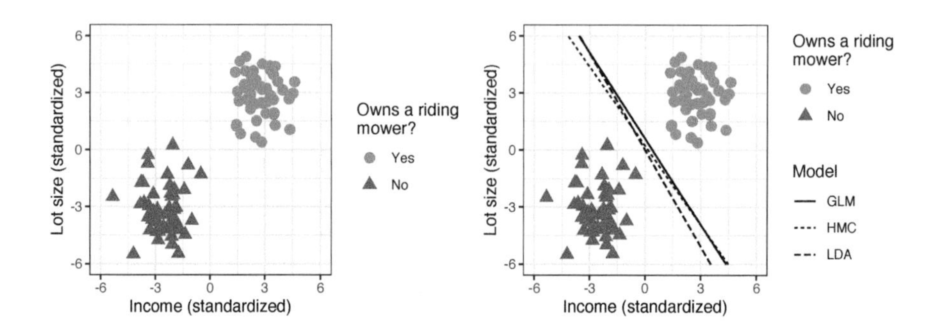

FIGURE 14.2: Simulated binary classification data with two separable classes. Left: Raw data. Right: Raw data with example decision boundaries (in this case, separating hyperplanes) from various machine learning algorithms.

is referred to as the margin M (the shaded areas in Figure 14.3). Finding this decision boundary can also be done with simple geometry. Geometrically, finding the HMC for two separable classes amounts to the following:

1. Draw the *convex hull*[1] around each class (these are the polygons surrounding each class in Figure 14.3).
2. Draw the shortest line segment that connects the two convex hulls (this is the dotted line segment in Figure 14.3).
3. The perpendicular bisector of this line segment is the HMC!
4. The margin boundaries are formed by drawing lines that pass through the support vectors and are parallel to the separating hyperplane (these are the dashed line segments in Figure 14.3).

This can also be formulated as an optimization problem. Mathematically speaking, the HMC estimates the coefficients of the hyperplane by solving a quadratic programming problem with linear inequality constraints, in particular:

[1]The convex hull of a set of points in 2-D space can be thought of as the shape formed by a rubber band stretched around the data. This of course can be generalized to higher dimensions (e.g., a rubber membrane stretched around a cloud of points in 3-D space).

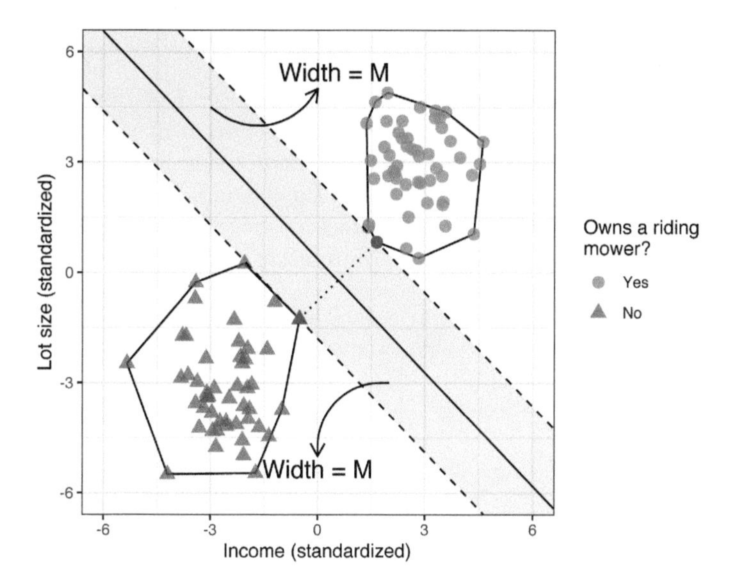

FIGURE 14.3: HMC for the simulated riding mower data. The solid black line forms the decision boundary (in this case, a separating hyperplane), while the dashed lines form the boundaries of the margins (shaded regions) on each side of the hyperplane. The shortest distance between the two classes (i.e., the dotted line connecting the two convex hulls) has length $2M$. Two of the training observations (solid red points) fall on the margin boundaries; in the context of SVMs (which we discuss later), these two points form the support vectors.

$$\underset{\beta_0, \beta_1, \dots, \beta_p}{\text{maximize}} \quad M \tag{14.1}$$

$$\text{subject to} \quad \begin{cases} \sum_{j=1}^{p} \beta_j^2 = 1, \\ y_i\left(\beta_0 + \beta_1 x_{i1} + \cdots + \beta_p x_{ip}\right) \geq M, \quad i = 1, 2, \dots, n \end{cases} \tag{14.2}$$

Put differently, the HMC finds the separating hyperplane that provides the largest margin/gap between the two classes. The width of both margin boundaries is M. With the constraint $\sum_{j=1}^{p} \beta_j^2 = 1$, the quantity $y_i\left(\beta_0 + \beta_1 x_{i1} + \cdots + \beta_p x_{ip}\right)$ represents the distance from the i-th data point to the decision boundary. Note that the solution to the optimization problem above does not allow any points to be on the wrong side of the margin; hence the term hard margin classifier.

14.2.2 The soft margin classifier

Sometimes perfect separation is achievable, but not desirable! Take, for example, the data in Figure 14.4. Here we added a single outlier at the point $(0.5, 1)$. While the data are still perfectly separable, the decision boundaries obtained using logistic regression and the HMC will not generalize well to new data and accuracy will suffer (i.e., these models are not robust to outliers in the feature space). The LDA model seems to produce a more reasonable decision boundary.

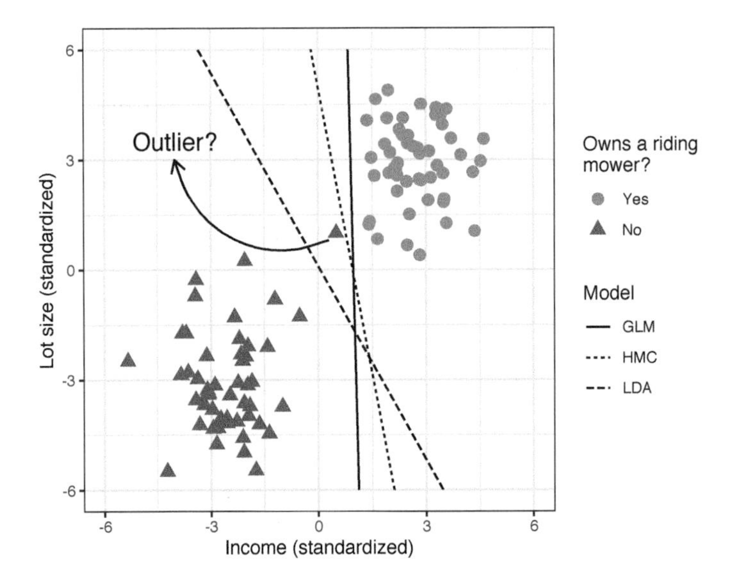

FIGURE 14.4: Simulated binary classification data with an outlier at the point $(0.5, 1)$.

In this situation, we can loosen the constraints (or *soften the margin*) by allowing some points to be on the wrong side of the margin; this is referred to as the the *soft margin classifier* (SMC). The SMC, similar to the HMC, estimates the coefficients of the hyperplane by solving the slightly modified optimization problem:

$$\underset{\beta_0, \beta_1, \dots, \beta_p}{\text{maximize}} \quad M \tag{14.3}$$

$$\text{subject to} \quad \begin{cases} \sum_{j=1}^{p} \beta_j^2 = 1, \\ y_i \left(\beta_0 + \beta_1 x_{i1} + \dots + \beta_p x_{ip}\right) \ge M\left(1 - \xi_i\right), \quad i = 1, 2, \dots, n \\ \xi_i \ge 0, \\ \sum_{i=1}^{n} \xi_i \le C \end{cases} \tag{14.4}$$

Similar to before, the SMC finds the separating hyperplane that provides the largest margin/gap between the two classes, but allows for some of the points to cross over the margin boundaries. Here C is the allowable budget for the total amount of overlap and is our first tunable hyperparameter for the SVM.

By varying C, we allow points to violate the margin which helps make the SVM robust to outliers. For example, in Figure 14.5, we fit the SMC at both extremes: $C = 0$ (the HMC) and $C = \infty$ (maximum overlap). Ideally, the hyperplane giving the decision boundary with the best generalization performance lies somewhere in between these two extremes and can be determined using, for example, k-fold CV.

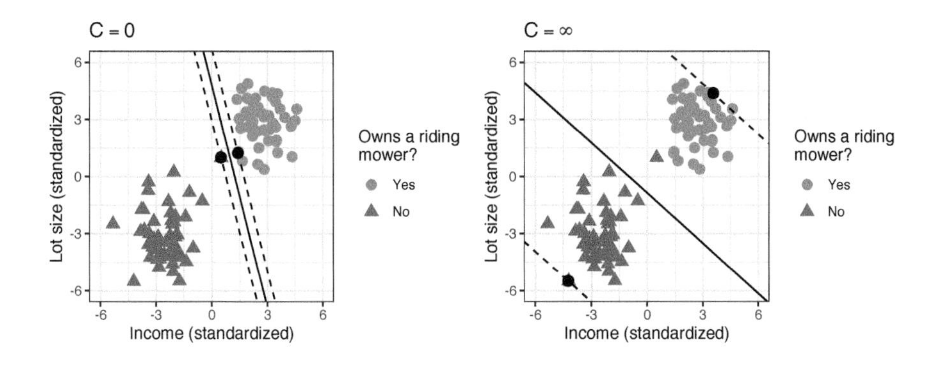

FIGURE 14.5: Soft margin classifier. Left: Zero budget for overlap (i.e., the HMC). Right: Maximumn allowable overlap. The solid black points represent the support vectors that define the margin boundaries.

14.3 The support vector machine

So far, we've only used linear decision boundaries. Such a classifier is likely too restrictive to be useful in practice, especially when compared to other algorithms that can adapt to nonlinear relationships. Fortunately, we can use a simple trick, called the *kernel trick*, to overcome this. A deep understanding of the kernel trick requires an understanding of *kernel functions* and *reproducing kernel Hilbert spaces* . Fortunately, we can use a couple illustrations in 2-D/3-D feature space to drive home the key idea.

Consider, for example, the circle data on the left side of Figure 14.6. This is another binary classification problem. The first class forms a circle in the middle of a square, the remaining points form the second class. Although these two classes do not overlap (although they appear to overlap slightly due to the size of the plotted points), they are not perfectly separable by a hyperplane (i.e., a straight line). However, we can enlarge the feature space by adding a third feature, say $X_3 = X_1^2 + X_2^2$—this is akin to using the polynomial kernel function discussed below with $d = 2$. The data are plotted in the enlarged feature space in the middle of Figure 14.6. In this new three dimensional feature space, the two classes are perfectly separable by a hyperplane (i.e., a flat plane); though it is hard to see (see the middle of Figure 14.6), the green points form the tip of the hyperboloid in 3-D feature space (i.e., X_3 is smaller for all the green points leaving a small gap between the two classes). The resulting decision boundary is then projected back onto the original feature space resulting in a non-linear decision boundary which perfectly separates the original data (see the right side of Figure 14.6)!

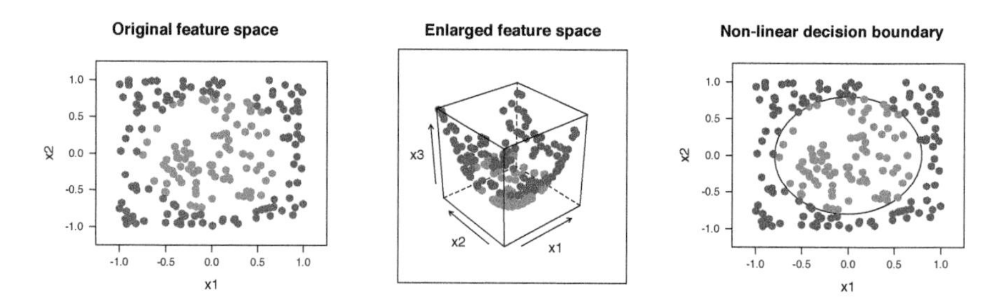

FIGURE 14.6: Simulated nested circle data. Left: The two classes in the original (2-D) feature space. Middle: The two classes in the enlarged (3-D) feature space. Right: The decision boundary from the HMC in the enlarged feature space projected back into the original feature space.

In essence, SVMs use the kernel trick to enlarge the feature space using basis functions (e.g., like in MARS or polynomial regression). In this enlarged (kernel-induced) feature space, a hyperplane can often separate the two classes. The resulting decision boundary, which is linear in the enlarged feature space, will be nonlinear when transformed back onto the original feature space.

Popular kernel functions used by SVMs include:

- d-th degree polynomial: $K(x, x') = \gamma (1 + \langle x, x' \rangle)^d$
- Radial basis function: $K(x, x') = \exp(\gamma \|x - x'\|^2)$
- Hyperbolic tangent: $K(x, x') = \tanh(k_1 \|x - x'\| + k_2)$

Here $\langle x, x' \rangle = \sum_{i=1}^{n} x_i x_i'$ is called an *inner product*. Notice how each of these kernel functions include hyperparameters that need to be tuned. For example, the polynomial kernel includes a degree term d and a scale parameter γ. Similarly, the radial basis kernel includes a γ parameter related to the inverse of the σ parameter of a normal distribution. In R, you can use **caret**'s `getModelInfo()` to extract the hyperparameters from various SVM implementations with different kernel functions, for example:

```
# Linear (i.e., soft margin classifier)
caret::getModelInfo("svmLinear")$svmLinear$parameters
##   parameter   class label
## 1         C numeric  Cost

# Polynomial kernel
caret::getModelInfo("svmPoly")$svmPoly$parameters
##   parameter   class             label
## 1    degree numeric Polynomial Degree
## 2     scale numeric             Scale
## 3         C numeric              Cost

# Radial basis kernel
caret::getModelInfo("svmRadial")$svmRadial$parameters
##   parameter   class label
## 1     sigma numeric Sigma
## 2         C numeric  Cost
```

Through the use of various kernel functions, SVMs are extremely flexible and capable of estimating complex nonlinear decision boundaries. For example, the right side of Figure 14.7 demonstrates the flexibility of an SVM using a radial basis kernel applied to the two spirals benchmark problem (see `?mlbench::mlbench.spirals` for details). As a reference, the left side of Figure 14.7 shows the decision boundary from a default random forest fit using the **ranger** package. The random forest decision boundary, while flexible, has trouble capturing smooth decision boundaries (like a spiral). The SVM with a radial basis kernel, on the other hand, does a great job (and in this case is more accurate).

The radial basis kernel is extremely flexible and as a rule of thumb, we generally start with this kernel when fitting SVMs in practice.

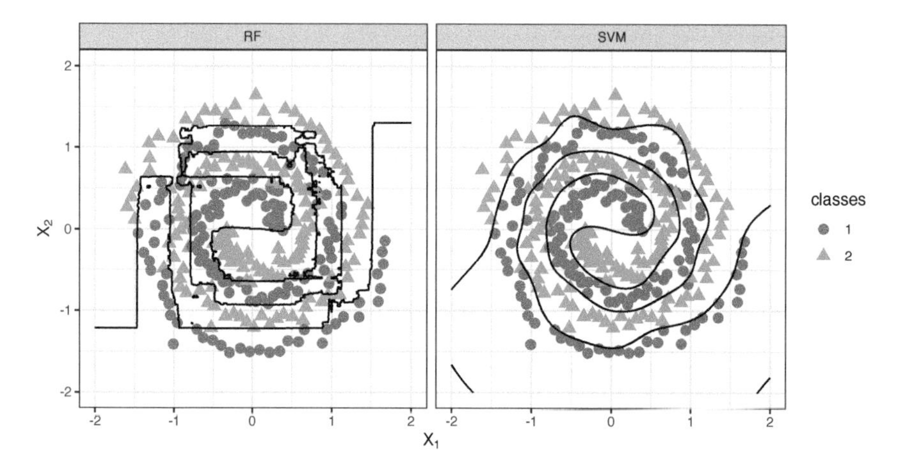

FIGURE 14.7: Two spirals benchmark problem. Left: Decision boundary from a random forest. Right: Decision boundary from an SVM with radial basis kernel.

14.3.1 More than two classes

The SVM, as introduced, is applicable to only two classes! What do we do when we have more than two classes? There are two general approaches: *one-versus-all* (OVA) and one-versus-one (OVO). In OVA, we fit an SVM for each class (one class versus the rest) and classify to the class for which the margin is the largest. In OVO, we fit all $\binom{\# \ classes}{2}$ pairwise SVMs and classify to the class that wins the most pairwise competitions. All the popular implementations of SVMs, including **kernlab**, provide such approaches to multinomial classification.

14.3.2 Support vector regression

SVMs can also be extended to regression problems (i.e., when the outcome is continuous). In essence, SVMs find a separating hyperplane in an enlarged feature space that generally results in a nonlinear decision boundary in the original feature space with good generalization performance. This enlarged feature spaced is constructed using special functions called kernel functions. The idea behind support vector regression (SVR) is very similar: find a good fitting hyperplane in a kernel-induced feature space that will have good generalization performance using the original features. Although there are many flavors of SVR, we'll introduce the most common: *ϵ-insensitive loss regression.*

Recall that the least squares (LS) approach to function estimation (Chapter 4) minimizes the sum of the squared residuals, where in general we define the residual as $r(x, y) = y - f(x)$. (In ordinary linear regression $f(x) = \beta_0 + \beta_1 x_1 + \cdots + \beta_p x_p$). The problem with LS is that it involves squaring the residuals which gives outliers undue influence on the fitted regression function. Although we could rely on the MAE metric (which looks at the absolute value as opposed to the squared residuals), another intuitive loss metric, called *ϵ-insensitive loss*, is more robust to outliers:

$$L_\epsilon = max\left(0, |r(x, y)| - \epsilon\right)$$

Here ϵ is a threshold set by the analyst. In essence, we're forming a margin around the regression curve of width ϵ (see Figure 14.8), and trying to contain as many data points within the margin as possible with a minimal number of violations. The data points that satisfy $r(x, y) \pm \epsilon$ form the support vectors that define the margin. The model is said to be *ϵ-insensitive* because the points within the margin have no influence on the fitted regression line! Similar to SVMs, we can use kernel functions to capture nonlinear relationships (in this case, the support vectors are those points who's residuals satisfy $r(x, y) \pm \epsilon$ in the kernel-induced feature space).

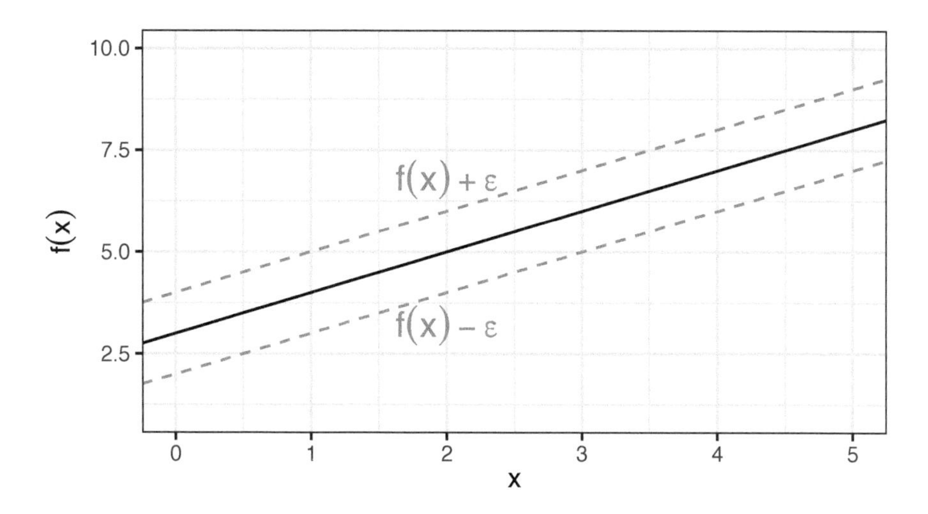

FIGURE 14.8: ϵ-insensitive regression band. The solid black line represents the estimated regression curve $f(x)$.

To illustrate, we simulated data from the sinc function $\sin(x)/x$ with added Gaussian nose (i.e., random errors from a normal distribution). The simulated data are shown in Figure 14.9. This is a highly nonlinear, but smooth function of x.

We fit three regression models to these data: a default MARS model (Chapter 7), a default RF (Chapter 11), and an SVR model using ϵ-insensitive loss and a radial basis kernel with default tuning parameters (technically, we set `kpar =`

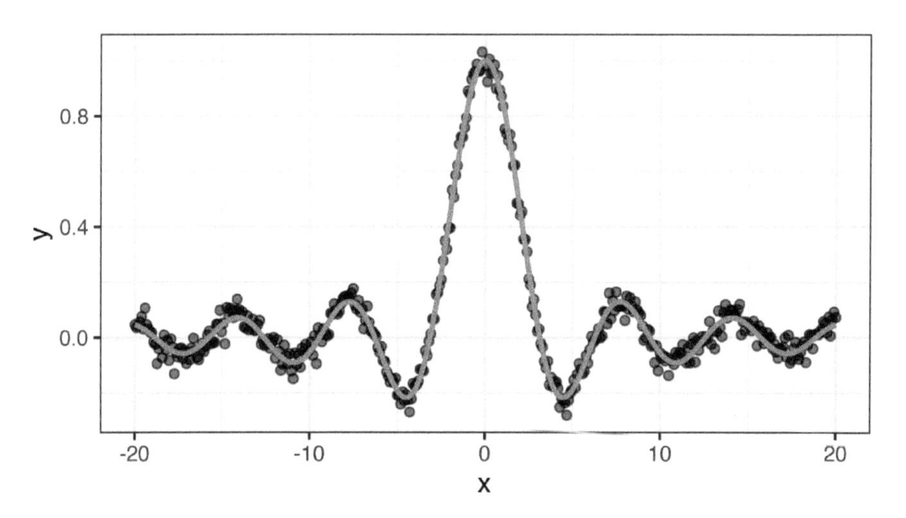

FIGURE 14.9: Simulated data from a sinc function with added noise.

`"automatic"` which tells `kernlab::ksvm()` to use `kernlab::sigest()` to find a reasonable estimate for the kernel's scale parameter). To use ϵ-insensitive loss regression, specify `type = "eps-svr"` in the call to `kernlab::ksvm()` (the default for ϵ is `epsilon = 0.1`. The results are displayed in Figure 14.10. Although this is a simple one-dimensional problem, the MARS and RF struggle to adapt to the smooth, but highly nonlinear function. The MARS model, while probably effective, is too rigid and fails to adequately capture the relationship towards the left-hand side. The RF is too wiggly and is indicative of slight overfitting (perhaps tuning the minimum observations per node can help here and is left as an exercise on the book website). The SVR model, on the other hand, works quite well and provides a smooth fit to the data!

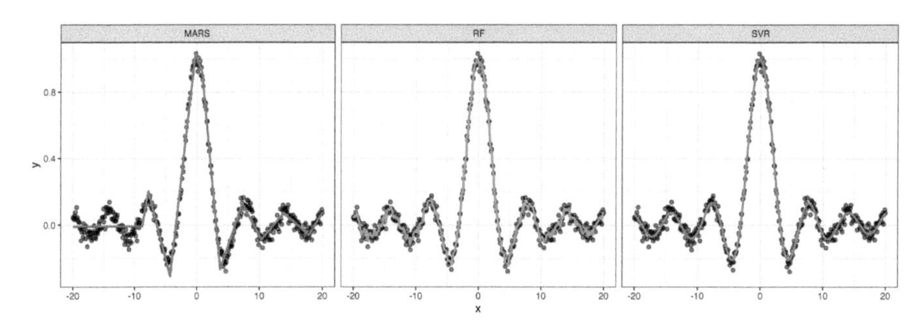

FIGURE 14.10: Simulated sine

Applying support vector regression to the Ames housing example is left as an exercise for the reader on the book's website.

14.4 Job attrition example

Returning to the employee attrition example, we tune and fit an SVM with a radial basis kernel (recall our earlier rule of thumb regarding kernel functions). Recall that the radial basis kernel has two hyperparameters: σ and C. While we can use k-fold CV to find good estimates of both parameters, hyperparameter tuning can be time consuming for SVMs[2]. Fortunately, it is possible to use the training data to find a good estimate of σ. This is provided by the `kernlab::sigest()` function. This function estimates the range of σ values which would return good results when used with a radial basis SVM. Ideally, any value within the range of estimates returned by this function should produce reasonable results. This is the approach taken by **caret**'s `train()` function when `method = "svmRadialSigma"`, which we use below. Also, note that a reasonable search grid for the cost parameter C is an exponentially growing series, for example $2^{-2}, 2^{-1}, 2^{0}, 2^{1}, 2^{2}$, etc. See `caret::getModelInfo("svmRadialSigma")` for details.

Next, we'll use **caret**'s `train()` function to tune and train an SVM using the radial basis kernel function with autotuning for the σ parameter (i.e., `"svmRadialSigma"`) and 10-fold CV.

```
# Tune an SVM with radial basis kernel
set.seed(1854)  # for reproducibility
churn_svm <- train(
  Attrition ~ .,
  data = churn_train,
  method = "svmRadial",
  preProcess = c("center", "scale"),
  trControl = trainControl(method = "cv", number = 10),
  tuneLength = 10
)
```

Plotting the results, we see that smaller values of the cost parameter ($C \approx$ 2–8) provide better cross-validated accuracy scores for these training data:

```
# Plot results
ggplot(churn_svm) + theme_light()
```

[2]SVMs typically have to estimate at least as many parameters as there are rows in the training data. Hence, SVMs are more commonly used in wide data situations.

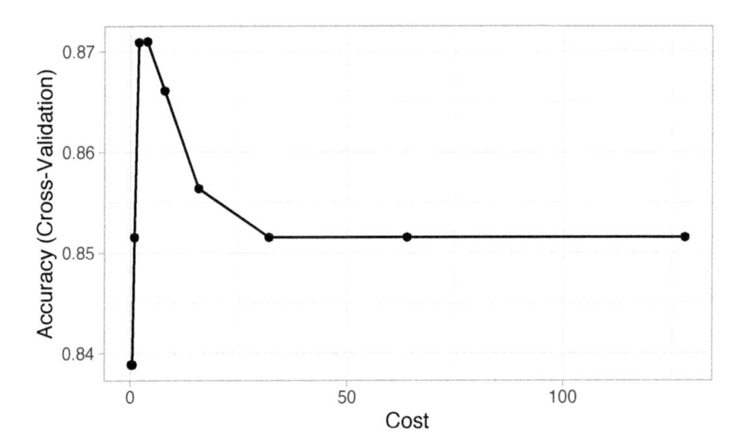


```
# Print results
churn_svm$results
##      sigma      C Accuracy Kappa AccuracySD KappaSD
## 1  0.00959   0.25    0.839 0.000    0.00409   0.000
## 2  0.00959   0.50    0.839 0.000    0.00409   0.000
## 3  0.00959   1.00    0.852 0.130    0.01443   0.101
## 4  0.00959   2.00    0.871 0.353    0.02375   0.145
## 5  0.00959   4.00    0.871 0.417    0.02664   0.130
## 6  0.00959   8.00    0.866 0.424    0.02627   0.121
## 7  0.00959  16.00    0.856 0.401    0.02687   0.130
## 8  0.00959  32.00    0.852 0.383    0.02872   0.134
## 9  0.00959  64.00    0.852 0.383    0.02872   0.134
## 10 0.00959 128.00    0.852 0.383    0.02872   0.134
```

14.4.1 Class weights

By default, most classification algorithms treat misclassification costs equally. This is not ideal in situations where one type of misclassification is more important than another or there is a severe class imbalance (which is usually the case). SVMs (as well as most tree-based methods) allow you to assign specific misclassification costs to the different outcomes. In **caret** and **kernlab**, this is accomplished via the `class.weights` argument, which is just a named vector of weights for the different classes. In the employee attrition example, for instance, we might specify

```
class.weights = c("No" = 1, "Yes" = 10)
```

in the call to `caret::train()` or `kernlab::ksvm()` to make false negatives (i.e., predicting "Yes" when the truth is "No") ten times more costly than false positives (i.e., predicting "No" when the truth is "Yes"). Cost-sensitive training with SVMs is left as an exercise on the book website.

14.4.2 Class probabilities

SVMs classify new observations by determining which side of the decision boundary they fall on; consequently, they do not automatically provide predicted class probabilities! In order to obtain predicted class probabilities from an SVM, additional parameters need to be estimated as described in Platt (1999). In practice, predicted class probabilities are often more useful than the predicted class labels. For instance, we would need the predicted class probabilities if we were using an optimization metric like AUC (Chapter 2), as opposed to classification accuracy. In that case, we can set `prob.model = TRUE` in the call to `kernlab::ksvm()` or `classProbs = TRUE` in the call to `caret::trainControl()` (for details, see `?kernlab::ksvm` and the references therein):

```
# Control params for SVM
ctrl <- trainControl(
  method = "cv",
  number = 10,
  classProbs = TRUE,
  summaryFunction = twoClassSummary  # also needed for AUC/ROC
)

# Tune an SVM
set.seed(5628)  # for reproducibility
churn_svm_auc <- train(
  Attrition ~ .,
  data = churn_train,
  method = "svmRadial",
  preProcess = c("center", "scale"),
  metric = "ROC",  # area under ROC curve (AUC)
  trControl = ctrl,
  tuneLength = 10
)
```

```
# Print results
churn_svm_auc$results
##        sigma      C   ROC  Sens  Spec  ROCSD SensSD SpecSD
## 1  0.00973   0.25 0.838 0.968 0.393 0.0670 0.0121 0.1147
## 2  0.00973   0.50 0.838 0.965 0.376 0.0669 0.0109 0.1478
## 3  0.00973   1.00 0.838 0.965 0.406 0.0673 0.0078 0.0987
## 4  0.00973   2.00 0.834 0.976 0.346 0.0680 0.0127 0.1432
## 5  0.00973   4.00 0.820 0.975 0.345 0.0719 0.0131 0.1208
## 6  0.00973   8.00 0.812 0.970 0.333 0.0758 0.0135 0.1182
## 7  0.00973  16.00 0.792 0.976 0.285 0.0779 0.0101 0.1070
## 8  0.00973  32.00 0.785 0.975 0.285 0.0775 0.0106 0.0892
## 9  0.00973  64.00 0.785 0.975 0.285 0.0774 0.0106 0.0985
## 10 0.00973 128.00 0.785 0.973 0.278 0.0774 0.0109 0.1091
```

Similar to before, we see that smaller values of the cost parameter C ($C \approx 2-4$) provide better cross-validated AUC scores on the training data. Also, notice how in addition to ROC we also get the corresponding sensitivity (true positive rate) and specificity (true negative rate). In this case, sensitivity (column `Sens`) refers to the proportion of `No`s correctly predicted as `No` and specificity (column `Spec`) refers to the proportion of `Yes`s correctly predicted as `Yes`. We can succinctly describe the different classification metrics using **caret**'s `confusionMatrix()` function (see `?caret::confusionMatrix` for details):

```
confusionMatrix(churn_svm_auc)
## Cross-Validated (10 fold) Confusion Matrix
##
## (entries are percentual average cell counts across resamples)
##
##           Reference
## Prediction   No  Yes
##        No  81.2  9.8
##        Yes  2.7  6.3
##
##  Accuracy (average) : 0.8748
```

In this case it is clear that we do a far better job at predicting the `No`s.

14.5 Feature interpretation

Like many other ML algorithms, SVMs do not emit any natural measures of feature importance; however, we can use the **vip** package to quantify the importance of each feature using the permutation approach described later on in Chapter 16 (the **iml** and **DALEX** packages could also be used).

Our metric function should reflect the fact that we trained the model using AUC. Any custom metric function provided to `vip()` should have the arguments `actual` and `predicted` (in that order). We illustrate this below where we wrap the `auc()` function from the **ModelMetrics** package (Hunt, 2018):

Since we are using AUC as our metric, our prediction wrapper function should return the predicted class probabilities for the reference class of interest. In this case, we'll use `"Yes"` as the reference class (to do this we'll specify `reference_class = "Yes"` in the call to `vip::vip()`). Our prediction function looks like:

```
prob_yes <- function(object, newdata) {
  predict(object, newdata = newdata, type = "prob")[, "Yes"]
}
```

To compute the variable importance scores we just call `vip()` with `method = "permute"` and pass our previously defined predictions wrapper to the `pred_wrapper` argument:

```
# Variable importance plot
set.seed(2827)  # for reproducibility
vip(churn_svm_auc, method = "permute", nsim = 5, train = churn_train,
    target = "Attrition", metric = "auc", reference_class = "Yes",
    pred_wrapper = prob_yes)
```

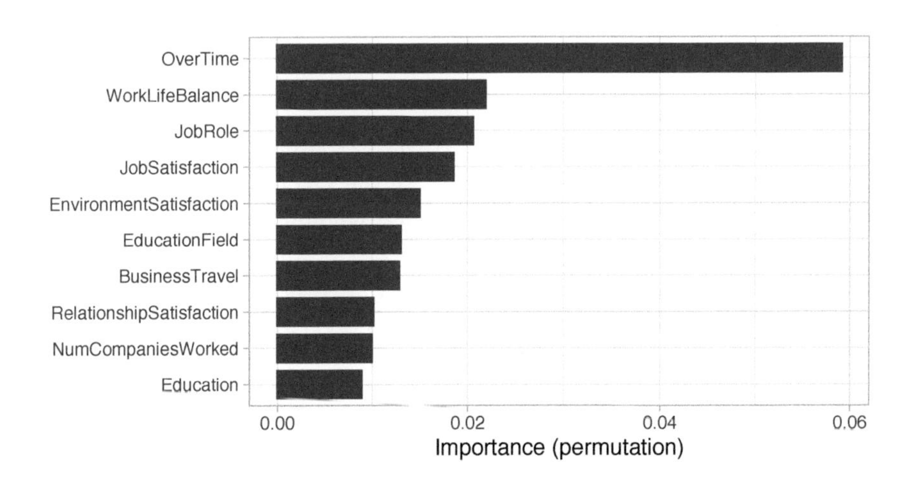

The results indicate that `OverTime` (Yes/No) is the most important feature in predicting attrition. Next, we use the **pdp** package to construct PDPs for the top four features according to the permutation-based variable importance scores (notice we set `prob = TRUE` in the call to `pdp::partial()` so that the feature effect plots are on the probability scale; see `?pdp::partial` for details). Additionally, since the predicted probabilities from our model come in two columns (`No` and `Yes`), we specify `which.class = 2` so that our interpretation is in reference to predicting `Yes`:

```
features <- c("OverTime", "WorkLifeBalance",
              "JobSatisfaction", "JobRole")
pdps <- lapply(features, function(x) {
  partial(churn_svm_auc, pred.var = x, which.class = 2,
          prob = TRUE, plot = TRUE, plot.engine = "ggplot2") +
    coord_flip()
})
grid.arrange(grobs = pdps,  ncol = 2)
```

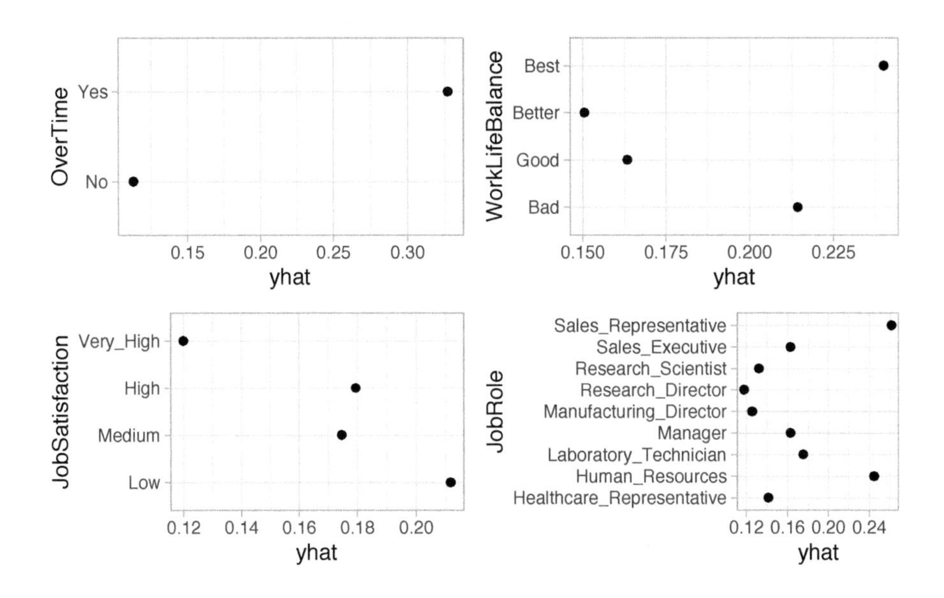

For instance, we see that employees with a low job satisfaction level have the highest probability of attriting, while those with a very high level of satisfaction tend to have the lowest probability.

14.6 Final thoughts

SVMs have a number of advantages compared to other ML algorithms described in this book. First off, they attempt to directly maximize generalizability (i.e., accuracy). Since SVMs are essentially just convex optimization problems, we're always guaranteed to find a global optimum (as opposed to potentially getting stuck in local optima as with DNNs). By softening the margin using a budget (or cost) parameter (C), SVMs are relatively robust to outliers. And finally, using kernel functions, SVMs are flexible enough to adapt to complex nonlinear decision boundaries (i.e., they can flexibly model nonlinear relationships). However, SVMs do carry a few disadvantages as well. For starters, they can be slow to train on tall data (i.e., $n \gg p$). This is because SVMs essentially have to estimate at least one parameter for each row in the training data! Secondly, SVMs only produce predicted class labels; obtaining predicted class probabilities requires additional adjustments and computations not covered in this chapter. Lastly, special procedures (e.g., OVA and OVO) have to be used to handle multinomial classification problems with SVMs.

15

Stacked Models

In the previous chapters, you've learned how to train individual learners, which in the context of this chapter will be referred to as *base learners*. ***Stacking*** (sometimes called "stacked generalization") involves training a new learning algorithm to combine the predictions of several base learners. First, the base learners are trained using the available training data, then a combiner or meta algorithm, called the *super learner*, is trained to make a final prediction based on the predictions of the base learners. Such stacked ensembles tend to outperform any of the individual base learners (e.g., a single RF or GBM) and have been shown to represent an asymptotically optimal system for learning (van der Laan et al., 2003).

15.1 Prerequisites

This chapter leverages the following packages, with the emphasis on **h2o**:

```
# Helper packages
library(rsample)    # for creating our train-test splits
library(recipes)    # for minor feature engineering tasks

# Modeling packages
library(h2o)        # for fitting stacked models
```

To illustrate key concepts we continue with the Ames housing example from previous chapters:

```
# Load and split the Ames housing data
ames <- AmesHousing::make_ames()
set.seed(123)  # for reproducibility
split <- initial_split(ames, strata = "Sale_Price")
```

```
ames_train <- training(split)
ames_test <- testing(split)

# Make sure we have consistent categorical levels
blueprint <- recipe(Sale_Price ~ ., data = ames_train) %>%
  step_other(all_nominal(), threshold = 0.005)

# Create training & test sets for h2o
train_h2o <- prep(blueprint, training = ames_train, retain = TRUE) %>%
  juice() %>%
  as.h2o()
test_h2o <- prep(blueprint, training = ames_train) %>%
  bake(new_data = ames_test) %>%
  as.h2o()

# Get response and feature names
Y <- "Sale_Price"
X <- setdiff(names(ames_train), Y)
```

15.2 The Idea

Leo Breiman, known for his work on classification and regression trees and random forests, formalized stacking in his 1996 paper on *Stacked Regressions* (Breiman, 1996b). Although the idea originated in (Wolpert, 1992) under the name "Stacked Generalizations", the modern form of stacking that uses internal k-fold CV was Breiman's contribution.

However, it wasn't until 2007 that the theoretical background for stacking was developed, and also when the algorithm took on the cooler name, ***Super Learner*** (Van der Laan et al., 2007). Moreover, the authors illustrated that super learners will learn an optimal combination of the base learner predictions and will typically perform as well as or better than any of the individual models that make up the stacked ensemble. Until this time, the mathematical reasons for why stacking worked were unknown and stacking was considered a black art.

15.2.1 Common ensemble methods

Ensemble machine learning methods use multiple learning algorithms to obtain better predictive performance than could be obtained from any of the constituent learning algorithms. The idea of combining multiple models rather than selecting the single best is well-known and has been around for a long time. In fact, many of the popular modern machine learning algorithms (including ones in previous chapters) are actually ensemble methods.

For example, bagging (Chapter 10) and random forests (Chapter 11) are ensemble approaches that average the predictions from many decision trees to reduce prediction variance and are robust to outliers and noisy data; ultimately leading to greater predictive accuracy. Boosted decision trees (Chapter 12) are another ensemble approach that slowly learns unique patterns in the data by sequentially combining individual, shallow trees.

Stacking, on the other hand, is designed to ensemble a *diverse group of strong learners*.

15.2.2 Super learner algorithm

The super learner algorithm consists of three phases:

1. Set up the ensemble
    - Specify a list of L base learners (with a specific set of model parameters).
    - Specify a meta learning algorithm. This can be any one of the algorithms discussed in the previous chapters but most often is some form of regularized regression.
2. Train the ensemble
    - Train each of the L base learners on the training set.
    - Perform k-fold CV on each of the base learners and collect the cross-validated predictions from each (the same k-folds must be used for each base learner). These predicted values represent $p_1, \dots, p_L$ in Eq. (15.1).
    - The N cross-validated predicted values from each of the L algorithms can be combined to form a new $N \times L$ feature matrix (represented by Z in Eq. (15.1). This matrix, along with the original response vector (y), are called the "level-one" data. ($N =$ number of rows in the training set.)

$$n\left\{ \begin{bmatrix} p_1 \end{bmatrix} \cdots \begin{bmatrix} p_L \end{bmatrix} \begin{bmatrix} y \end{bmatrix} \right. \rightarrow n\left\{ \overbrace{\begin{bmatrix} \quad Z \quad \end{bmatrix}}^{L} \begin{bmatrix} y \end{bmatrix} \right. \tag{15.1}$$

- Train the meta learning algorithm on the level-one data ($y = f(Z)$). The "ensemble model" consists of the L base learning models and the meta learning model, which can then be used to generate predictions on new data.

3. Predict on new data.
   - To generate ensemble predictions, first generate predictions from the base learners.
   - Feed those predictions into the meta learner to generate the ensemble prediction.

Stacking never does worse than selecting the single best base learner on the training data (but not necessarily the validation or test data). The biggest gains are usually produced when stacking base learners that have high variability, and uncorrelated, predicted values. The more similar the predicted values are between the base learners, the less advantage there is to combining them.

15.2.3 Available packages

There are a few package implementations for model stacking in the R ecosystem. **SuperLearner** (Polley et al., 2019) provides the original Super Learner and includes a clean interface to 30+ algorithms. Package **subsemble** (LeDell et al., 2014) also provides stacking via the super learner algorithm discussed above; however, it also offers improved parallelization over the **SuperLearner** package and implements the subsemble algorithm (Sapp et al., 2014).[1] Unfortunately, **subsemble** is currently only available via GitHub and is primarily maintained for backward compatibility rather than forward development. A third package, **caretEnsemble** (Deane-Mayer and Knowles, 2016), also provides an approach for stacking, but it implements a bootsrapped (rather than cross-validated) version of stacking. The bootstrapped version will train faster since bootsrapping (with a train/test set) requires a fraction of the work of k-fold CV; however, the the ensemble performance often suffers as a result of this shortcut.

This chapter focuses on the use of **h2o** for model stacking. **h2o** provides an efficient implementation of stacking and allows you to stack existing base learners, stack a grid search, and also implements an automated machine learning search with stacked results. All three approaches will be discussed.

[1]The subsemble algorithm is a general subset ensemble prediction method, which can be used for small, moderate, or large data sets. Subsemble partitions the full data set into subsets of observations, fits a specified underlying algorithm on each subset, and uses a unique form of k-fold CV to output a prediction function that combines the subset-specific fits.

15.3 Stacking existing models

The first approach to stacking is to train individual base learner models separately and then stack them together. For example, say we found the optimal hyperparameters that provided the best predictive accuracy for the following algorithms:

1. Regularized regression base learner.
2. Random forest base learner.
3. GBM base learner.
4. XGBoost base learner.

We can train each of these models individually (see the code chunk below). However, to stack them later we need to do a few specific things:

1. All models must be trained on the same training set.
2. All models must be trained with the same number of CV folds.
3. All models must use the same fold assignment to ensure the same observations are used (we can do this by using `fold_assignment = "Modulo"`).
4. The cross-validated predictions from all of the models must be preserved by setting `keep_cross_validation_predictions = TRUE`. This is the data which is used to train the meta learner algorithm in the ensemble.

```
# Train & cross-validate a GLM model
best_glm <- h2o.glm(
  x = X, y = Y, training_frame = train_h2o, alpha = 0.1,
  remove_collinear_columns = TRUE, nfolds = 10,
  fold_assignment = "Modulo",
  keep_cross_validation_predictions = TRUE, seed = 123
)

# Train & cross-validate a RF model
best_rf <- h2o.randomForest(
  x = X, y = Y, training_frame = train_h2o, ntrees = 1000,
  mtries = 20, max_depth = 30, min_rows = 1, sample_rate = 0.8,
  nfolds = 10, fold_assignment = "Modulo",
  keep_cross_validation_predictions = TRUE, seed = 123,
  stopping_rounds = 50, stopping_metric = "RMSE",
```

```
  stopping_tolerance = 0
)

# Train & cross-validate a GBM model
best_gbm <- h2o.gbm(
  x = X, y = Y, training_frame = train_h2o, ntrees = 5000,
  learn_rate = 0.01, max_depth = 7, min_rows = 5, sample_rate = 0.8,
  nfolds = 10, fold_assignment = "Modulo",
  keep_cross_validation_predictions = TRUE, seed = 123,
  stopping_rounds = 50, stopping_metric = "RMSE",
  stopping_tolerance = 0
)

# Train & cross-validate an XGBoost model
best_xgb <- h2o.xgboost(
  x = X, y = Y, training_frame = train_h2o, ntrees = 5000,
  learn_rate = 0.05, max_depth = 3, min_rows = 3, sample_rate = 0.8,
  categorical_encoding = "Enum", nfolds = 10,
  fold_assignment = "Modulo",
  keep_cross_validation_predictions = TRUE, seed = 123,
  stopping_rounds = 50, stopping_metric = "RMSE",
  stopping_tolerance = 0
)
```

We can now use `h2o.stackedEnsemble()` to stack these models. Note how we feed the base learner models into the `base_models = list()` argument. Here, we apply a random forest model as the metalearning algorithm. However, you could also apply regularized regression, GBM, or a neural network as the metalearner (see `?h2o.stackedEnsemble` for details).

```
# Train a stacked tree ensemble
ensemble_tree <- h2o.stackedEnsemble(x = X, y = Y,
  training_frame = train_h2o, model_id = "my_tree_ensemble",
  base_models = list(best_glm, best_rf, best_gbm, best_xgb),
  metalearner_algorithm = "drf"
)
```

Since our ensemble is built on the CV results of the base learners, but has no cross-validation results of its own, we'll use the test data to compare our results. If we assess the performance of our base learners on the test data we see that the stochastic GBM base learner has the lowest RMSE of 20859.92. The stacked model achieves a small 1% performance gain with an RMSE of 20664.56.

```r
# Get results from base learners
get_rmse <- function(model) {
  results <- h2o.performance(model, newdata = test_h2o)
  results@metrics$RMSE
}
list(best_glm, best_rf, best_gbm, best_xgb) %>%
  purrr::map_dbl(get_rmse)
## [1] 30024.67 23075.24 20859.92 21391.20

# Stacked results
h2o.performance(ensemble_tree, newdata = test_h2o)@metrics$RMSE
## [1] 20664.56
```

We previously stated that the biggest gains are usually produced when we are stacking base learners that have high variability, and uncorrelated, predicted values. If we assess the correlation of the CV predictions we can see strong correlation across the base learners, especially with three tree-based learners. Consequentley, stacking provides less advantage in this situation since the base learners have highly correlated predictions; however, a 1% performance improvement can still be considerable improvement depending on the business context.

```r
glm_id <- best_glm@model$cross_validation_holdout_predictions_frame_id
rf_id  <- best_rf@model$cross_validation_holdout_predictions_frame_id
gbm_id <- best_gbm@model$cross_validation_holdout_predictions_frame_id
xgb_id <- best_xgb@model$cross_validation_holdout_predictions_frame_id

data.frame(
  GLM_pred = as.vector(h2o.getFrame(glm_id$name)),
  RF_pred  = as.vector(h2o.getFrame(rf_id$name)),
  GBM_pred = as.vector(h2o.getFrame(gbm_id$name)),
  XGB_pred = as.vector(h2o.getFrame(xgb_id$name))
) %>% cor()
##           GLM_pred   RF_pred  GBM_pred  XGB_pred
## GLM_pred 1.0000000 0.9390229 0.9291982 0.9345048
## RF_pred  0.9390229 1.0000000 0.9920349 0.9821944
## GBM_pred 0.9291982 0.9920349 1.0000000 0.9854160
## XGB_pred 0.9345048 0.9821944 0.9854160 1.0000000
```

15.4 Stacking a grid search

An alternative ensemble approach focuses on stacking multiple models generated from the same base learner. In each of the previous chapters, you learned how to perform grid searches to automate the tuning process. Often we simply select the best performing model in the grid search but we can also apply the concept of stacking to this process.

Many times, certain tuning parameters allow us to find unique patterns within the data. By stacking the results of a grid search, we can capitalize on the benefits of each of the models in our grid search to create a meta model. For example, the following performs a random grid search across a wide range of GBM hyperparameter settings. We set the search to stop after 25 models have run.

```
# Define GBM hyperparameter grid
hyper_grid <- list(
  max_depth = c(1, 3, 5),
  min_rows = c(1, 5, 10),
  learn_rate = c(0.01, 0.05, 0.1),
  learn_rate_annealing = c(0.99, 1),
  sample_rate = c(0.5, 0.75, 1),
  col_sample_rate = c(0.8, 0.9, 1)
)

# Define random grid search criteria
search_criteria <- list(
  strategy = "RandomDiscrete",
  max_models = 25
)

# Build random grid search
random_grid <- h2o.grid(
  algorithm = "gbm", grid_id = "gbm_grid", x = X, y = Y,
  training_frame = train_h2o, hyper_params = hyper_grid,
  search_criteria = search_criteria, ntrees = 5000,
  stopping_metric = "RMSE", stopping_rounds = 10,
  stopping_tolerance = 0, nfolds = 10, fold_assignment = "Modulo",
  keep_cross_validation_predictions = TRUE, seed = 123
)
```

If we look at the grid search models we see that the cross-validated RMSE ranges from 20756–57826

```
# Sort results by RMSE
h2o.getGrid(
  grid_id = "gbm_grid",
  sort_by = "rmse"
)
## H2O Grid Details
## ================
##
## Grid ID: gbm_grid
## Used hyper parameters:
##   -  col_sample_rate
##   -  learn_rate
##   -  learn_rate_annealing
##   -  max_depth
##   -  min_rows
##   -  sample_rate
## Number of models: 25
## Number of failed models: 0
##
## Hyper-Parameter Search Summary: ordered by increasing rmse
##   col_sample_rate learn_rate learn_rate_annealing ...       rmse
## 1             0.9       0.01                  1.0 ... 20756.167
## 2             0.9       0.01                  1.0 ... 21188.696
## 3             0.9        0.1                  1.0 ... 21203.754
## 4             0.8       0.01                  1.0 ... 21704.258
## 5             1.0        0.1                 0.99 ... 21710.276
##
## ---
##    col_sample_rate learn_rate learn_rate_annealing ...       rmse
## 20             1.0       0.01                  1.0 ... 26164.879
## 21             0.8       0.01                 0.99 ... 44805.638
## 22             1.0       0.01                 0.99 ... 44854.611
## 23             0.8       0.01                 0.99 ... 57797.874
## 24             0.9       0.01                 0.99 ... 57809.603
## 25             0.8       0.01                 0.99 ... 57826.304
```

If we apply the best performing model to our test set, we achieve an RMSE of 21599.8.

```
# Grab the model_id for the top model, chosen by validation error
best_model_id <- random_grid_perf@model_ids[[1]]
best_model <- h2o.getModel(best_model_id)
h2o.performance(best_model, newdata = test_h2o)
## H2ORegressionMetrics: gbm
##
## MSE:  466551295
## RMSE:  21599.8
## MAE:  13697.78
## RMSLE:  0.1090604
## Mean Residual Deviance :  466551295
```

Rather than use the single best model, we can combine all the models in our grid search using a super learner. In this example, our super learner does not provide any performance gains because the hyperparameter settings of the leading models have low variance which results in predictions that are highly correlated. However, in cases where you see high variability across hyperparameter settings for your leading models, stacking the grid search or even the leaders in the grid search can provide significant performance gains.

Stacking a grid search provides the greatest benefit when leading models from the base learner have high variance in their hyperparameter settings.

```
# Train a stacked ensemble using the GBM grid
ensemble <- h2o.stackedEnsemble(x = X, y = Y,
  training_frame = train_h2o, model_id = "ensemble_gbm_grid",
  base_models = random_grid@model_ids, metalearner_algorithm = "gbm"
)

# Eval ensemble performance on a test set
h2o.performance(ensemble, newdata = test_h2o)
## H2ORegressionMetrics: stackedensemble
##
## MSE:  469579433
## RMSE:  21669.78
## MAE:  13499.93
## RMSLE:  0.1061244
## Mean Residual Deviance :  469579433
```

15.5 Automated machine learning

Our final topic to discuss involves performing an automated search across multiple base learners and then stack the resulting models (this is sometimes referred to as *automated machine learning* or AutoML). This is very much like the grid searches that we have been performing for base learners and discussed in Chapters 4-14; however, rather than search across a variety of parameters for a *single base learner*, we want to perform a search across a variety of hyperparameter settings for many *different base learners*.

There are several competitors that provide licensed software that help automate the end-to-end machine learning process to include feature engineering, model validation procedures, model selection, hyperparameter optimization, and more. Open source applications are more limited and tend to focus on automating the model building, hyperparameter configurations, and comparison of model performance.

Although AutoML has made it easy for non-experts to experiment with machine learning, there is still a significant amount of knowledge and background in data science that is required to produce high-performing machine learning models. AutoML is more about freeing up your time (which is quite valuable). The machine learning process is often long, iterative, and repetitive and AutoML can also be a helpful tool for the advanced user, by simplifying the process of performing a large number of modeling-related tasks that would typically require hours/days writing many lines of code. This can free up the user's time to focus on other tasks in the data science pipeline such as data-preprocessing, feature engineering, model interpretability, and model deployment.

h2o provides an open source implementation of AutoML with the `h2o.automl()` function. The current version of `h2o.automl()` trains and cross-validates a random forest, an *extremely-randomized forest*, a random grid of GBMs, a random grid of DNNs, and then trains a stacked ensemble using all of the models; see `?h2o::h2o.automl` for details.

By default, `h2o.automl()` will search for 1 hour but you can control how long it searches by adjusting a variety of stopping arguments (e.g., `max_runtime_secs`, `max_models`, and `stopping_tolerance`).

The following performs an automated search for two hours, which ended up assessing 80 models. `h2o.automl()` will automatically use the same folds for stacking so you do not need to specify `fold_assignment = "Modulo"`. This allows for consistent model comparison across the same CV sets. We see that most of the leading models are GBM variants and achieve an RMSE in the 22000–23000 range. As you probably noticed, this was not as good as some of our best models we found using our own GBM grid searches (reference Chapter 12). However, we could start this AutoML procedure and then spend our two hours performing other tasks while **h2o** automatically assesses these 80 models. The AutoML procedure then provides us direction for further analysis. In this case, we could start by further assessing the hyperparameter settings in the top five GBM models to see if there were common attributes that could point us to additional grid searches worth exploring.

```
# Use AutoML to find a list of candidate models (i.e., leaderboard)
auto_ml <- h2o.automl(x = X, y = Y,
  training_frame = train_h2o, nfolds = 5,
  max_runtime_secs = 60 * 120, max_models = 50,
  keep_cross_validation_predictions = TRUE, sort_metric = "RMSE",
  stopping_rounds = 50, stopping_metric = "RMSE",
  stopping_tolerance = 0, seed = 123
)

# Assess the leader board; the following truncates the results to
# show the top and bottom 15 models. You can get the top model
# with auto_ml@leader
auto_ml@leaderboard %>%
  as.data.frame() %>%
  dplyr::select(model_id, rmse) %>%
  dplyr::slice(1:25)
##                                             model_id     rmse
## 1                    XGBoost_1_AutoML_20190220_084553 22229.97
## 2           GBM_grid_1_AutoML_20190220_084553_model_1 22437.26
## 3           GBM_grid_1_AutoML_20190220_084553_model_3 22777.57
## 4                        GBM_2_AutoML_20190220_084553 22785.60
## 5                        GBM_3_AutoML_20190220_084553 23133.59
## 6                        GBM_4_AutoML_20190220_084553 23185.45
## 7                    XGBoost_2_AutoML_20190220_084553 23199.68
## 8                    XGBoost_1_AutoML_20190220_075753 23231.28
## 9                        GBM_1_AutoML_20190220_084553 23326.57
## 10          GBM_grid_1_AutoML_20190220_075753_model_2 23330.42
## 11                   XGBoost_3_AutoML_20190220_084553 23475.23
## 12      XGBoost_grid_1_AutoML_20190220_084553_model_3 23550.04
## 13     XGBoost_grid_1_AutoML_20190220_075753_model_15 23640.95
## 14      XGBoost_grid_1_AutoML_20190220_084553_model_8 23646.66
```

```
## 15        XGBoost_grid_1_AutoML_20190220_084553_model_6   23682.37
## ...                                                 ...        ...
## 65            GBM_grid_1_AutoML_20190220_084553_model_5   33971.32
## 66            GBM_grid_1_AutoML_20190220_075753_model_8   34489.39
## 67  DeepLearning_grid_1_AutoML_20190220_084553_model_3   36591.73
## 68            GBM_grid_1_AutoML_20190220_075753_model_6   36667.56
## 69       XGBoost_grid_1_AutoML_20190220_084553_model_13   40416.32
## 70            GBM_grid_1_AutoML_20190220_075753_model_9   47744.43
## 71     StackedEnsemble_AllModels_AutoML_20190220_084553   49856.66
## 72     StackedEnsemble_AllModels_AutoML_20190220_075753   59127.09
## 73 StackedEnsemble_BestOfFamily_AutoML_20190220_084553   76714.90
## 74 StackedEnsemble_BestOfFamily_AutoML_20190220_075753   76748.40
## 75            GBM_grid_1_AutoML_20190220_075753_model_5   78465.26
## 76            GBM_grid_1_AutoML_20190220_075753_model_3   78535.34
## 77            GLM_grid_1_AutoML_20190220_075753_model_1   80284.34
## 78            GLM_grid_1_AutoML_20190220_084553_model_1   80284.34
## 79        XGBoost_grid_1_AutoML_20190220_075753_model_4   92559.44
## 80       XGBoost_grid_1_AutoML_20190220_075753_model_10  125384.88
```

16

Interpretable Machine Learning

In the previous chapters you learned how to train several different forms of advanced ML models. Often, these models are considered "black boxes" due to their complex inner-workings. However, because of their complexity, they are typically more accurate for predicting nonlinear, faint, or rare phenomena. Unfortunately, more accuracy often comes at the expense of interpretability, and interpretability is crucial for business adoption, model documentation, regulatory oversight, and human acceptance and trust. Luckily, several advancements have been made to aid in interpreting ML models over the years and this chapter demonstrates how you can use them to extract important insights. Interpreting ML models is an emerging field that has become known as *interpretable machine learning* (IML).

16.1 Prerequisites

There are multiple packages that provide robust machine learning interpretation capabilities. Unfortunately there is not one single package that is optimal for all IML applications; rather, when performing IML you will likely use a combination of packages. The following packages are used in this chapter.

```
# Helper packages
library(dplyr)    # for data wrangling
library(ggplot2)  # for awesome graphics

# Modeling packages
library(h2o)       # for interfacing with H2O
library(recipes)   # for ML recipes
library(rsample)   # for data splitting
library(xgboost)   # for fitting GBMs

# Model interpretability packages
```

```r
library(pdp)      # for partial dependence plots (and ICE curves)
library(vip)      # for variable importance plots
library(iml)      # for general IML-related functions
library(DALEX)    # for general IML-related functions
library(lime)     # for local interpretable model-agnostic explanations
```

To illustrate various concepts we'll continue working with the **h2o** version of the Ames housing example from Section 15.1. We'll also use the stacked ensemble model (`ensemble_tree`) created in Section 15.3.

16.2 The idea

It is not enough to identify a machine learning model that optimizes predictive performance; understanding and trusting model results is a hallmark of good science and necessary for our model to be adopted. As we apply and embed ever-more complex predictive modeling and machine learning algorithms, both we (the analysts) and the business stakeholders need methods to interpret and understand the modeling results so we can have trust in its application for business decisions (Doshi-Velez and Kim, 2017).

Advancements in interpretability now allow us to extract key insights and actionable information from the most advanced ML models. These advancements allow us to answer questions such as:

- What are the most important customer attributes driving behavior?
- How are these attributes related to the behavior output?
- Do multiple attributes interact to drive different behavior among customers?
- Why do we expect a customer to make a particular decision?
- Are the decisions we are making based on predicted results fair and reliable?

Approaches to model interpretability to answer the exemplar questions above can be broadly categorized as providing *global* or *local* explanations. It is important to understand the entire model that you've trained on a global scale, and also to zoom in on local regions of your data or your predictions and derive explanations. Being able to answer such questions and provide both levels of explanation is key to any ML project becoming accepted, adopted, embedded, and properly utilized.

16.2.1 Global interpretation

Global interpretability is about understanding how the model makes predictions, based on a holistic view of its features and how they influence the underlying model structure. It answers questions regarding which features are relatively influential, how these features influence the response variable, and what kinds of potential interactions are occurring. Global model interpretability helps to understand the relationship between the response variable and the individual features (or subsets thereof). Arguably, comprehensive global model interpretability is very hard to achieve in practice. Any model that exceeds a handful of features will be hard to fully grasp as we will not be able to comprehend the whole model structure at once.

While global model interpretability is usually out of reach, there is a better chance to understand at least some models on a modular level. This typically revolves around gaining understanding of which features are the most influential (via *feature importance*) and then focusing on how the most influential variables drive the model output (via *feature effects*). Although you may not be able to fully grasp a model with a hundred features, typically only a dozen or so of these variables are really influential in driving the model's performance. And it is possible to have a firm grasp of how a dozen variables are influencing a model.

16.2.2 Local interpretation

Global interpretability methods help us understand the inputs and their overall relationship with the response variable, but they can be highly deceptive in some cases (e.g., when strong interactions are occurring). Although a given feature may influence the predictive accuracy of our model as a whole, it does not mean that that feature has the largest influence on a predicted value for a given observation (e.g., a customer, house, or employee) or even a group of observations. Local interpretations help us understand what features are influencing the predicted response for a given observation (or small group of observations). These techniques help us to not only answer what we expect a customer to do, but also why our model is making a specific prediction for a given observation.

There are three primary approaches to local interpretation:

- Local interpretable model-agnostic explanations (LIME)
- Shapley values
- Localized step-wise procedures

These techniques have the same objective: to explain which variables are most influential in predicting the target for a set of observations. To illustrate, we'll focus on two observations. The first is the observation that our ensemble produced the highest predicted `Sale_Price` for (i.e., observation 1825 which has a predicted `Sale_Price` of $663,136), and the second is the observation with the lowest predicted `Sale_Price` (i.e., observation 139 which has a predicted `Sale_Price` of $47,245.45). Our goal with local interpretation is to explain what features are driving these two predictions.

```
# Compute predictions
predictions <- predict(ensemble_tree, train_h2o) %>% as.vector()

# Print the highest and lowest predicted sales price
paste("Observation", which.max(predictions),
      "has a predicted sale price of", scales::dollar(max(predictions)))
## [1] "Observation 1825 has a predicted sale price of $663,136"

paste("Observation", which.min(predictions),
      "has a predicted sale price of", scales::dollar(min(predictions)))
## [1] "Observation 139 has a predicted sale price of $47,245.45"

# Grab feature values for observations with min/max predicted
# sales price
high_ob <- as.data.frame(train_h2o)[which.max(predictions), ] %>%
  select(-Sale_Price)
low_ob  <- as.data.frame(train_h2o)[which.min(predictions), ] %>%
  select(-Sale_Price)
```

16.2.3 Model-specific vs. model-agnostic

It's also important to understand that there are *model-specific* and *model-agnostic* approaches for interpreting your model. Many of the approaches you've seen in the previous chapters for understanding feature importance are model-specific. For example, in linear models we can use the absolute value of the t–statistic as a measure of feature importance (though this becomes complicated when your linear model involves interaction terms and transformations). Random forests, on the other hand, can record the prediction accuracy on the OOB portion of the data, then the same is done after permuting each predictor variable, and the difference between the two accuracies are then averaged over all trees, and normalized by the standard error. These model-specific interpretation tools are limited to their respective model classes. There can be advantages to using model-specific approaches as they are more closely tied to the model performance and they may be able to more accurately incorporate the correlation structure between the predictors (Kuhn and Johnson, 2013).

However, there are also some disadvantages. For example, many ML algorithms (e.g., stacked ensembles) have no natural way of measuring feature importance:

```
vip(ensemble_tree, method = "model")
## Error in vi_model.default(ensemble_tree, method = "model") :
##   model-specific variable importance scores are currently not
##   available for objects of class "h2o.stackedEnsemble.summary".
```

Furthermore, comparing model-specific feature importance across model classes is difficult since you are comparing different measurements (e.g., the magnitude of t-statistics in linear models vs. degradation of prediction accuracy in random forests). In model-agnostic approaches, the model is treated as a "black box". The separation of interpretability from the specific model allows us to easily compare feature importance across different models.

Ultimately, there is no one best approach for model interpretability. Rather, only by applying multiple approaches (to include comparing model specific and model agnostic results) can we really gain full trust in the interpretations we extract.

An important item to note is that when using model agnostic procedures, additional code preparation is often required. For example, the **iml** (Molnar, 2019), **DALEX** (Biecek, 2019), and **LIME** (Pedersen and Benesty, 2018) packages use purely model agnostic procedures. Consequently, we need to create a model agnostic object that contains three components:

1) A data frame with just the features (must be of class `"data.frame"`, cannot be an `"H2OFrame"` or other object).
2) A vector with the actual responses (must be numeric—0/1 for binary classification problems).
3) A custom function that will take the features from 1), apply the ML algorithm, and return the predicted values as a vector.

The following code extracts these items for the **h2o** example:

```
# 1) create a data frame with just the features
features <- as.data.frame(train_h2o) %>% select(-Sale_Price)

# 2) Create a vector with the actual responses
```

```
response <- as.data.frame(train_h2o) %>% pull(Sale_Price)

# 3) Create custom predict function that returns the predicted
#    values as a vector
pred <- function(object, newdata)  {
  results <- as.vector(h2o.predict(object, as.h2o(newdata)))
  return(results)
}

# Example of prediction output
pred(ensemble_tree, features) %>% head()
## [1] 207144 108958 164248 241984 190001 202796
```

Once we have these three components we can create our model agnostic objects for the **iml**[1] and **DALEX** packages, which will just pass these downstream components (along with the ML model) to other functions.

```
# iml model agnostic object
components_iml <- Predictor$new(
  model = ensemble_tree,
  data = features,
  y = response,
  predict.fun = pred
)

# DALEX model agnostic object
components_dalex <- DALEX::explain(
  model = ensemble_tree,
  data = features,
  y = response,
  predict_function = pred
)
```

[1]Note that the **iml** package uses the R6 class, which is less common than the normal S3 and S4 classes. For more information on R6 see Wickham (2014), Chapter 14.

16.3 Permutation-based feature importance

In previous chapters we illustrated a few model-specific approaches for measuring feature importance (e.g., for linear models we used the absolute value of the t-statistic). For SVMs, on the other hand, we had to rely on a model-agnostic approach which was based on the permutation feature importance measurement introduced for random forests by Breiman (2001) (see Section 11.6) and expanded on by Fisher et al. (2018).

16.3.1 Concept

The permutation approach measures a feature's importance by calculating the increase of the model's prediction error after permuting the feature. The idea is that if we randomly permute the values of an important feature in the training data, the training performance would degrade (since permuting the values of a feature effectively destroys any relationship between that feature and the target variable). The permutation approach uses the difference (or ratio) between some baseline performance measure (e.g., RMSE) and the same performance measure obtained after permuting the values of a particular feature in the training data. From an algorithmic perspective, the approach follows these steps:

```
For any given loss function do the following:
1. Compute loss function for original model
2. For variable i in {1,...,p} do
     | randomize values
     | apply given ML model
     | estimate loss function
     | compute feature importance (some difference/ratio measure
       between permuted loss & original loss)
   End
3. Sort variables by descending feature importance
```

Algorithm 2 A simple algorithm for computing permutation-based variable importance for the feature set X.

A feature is "important" if permuting its values increases the model error relative to the other features, because the model relied on the feature for the prediction. A feature is "unimportant" if permuting its values keeps the model error relatively unchanged, because the model ignored the feature for the prediction.

This type of variable importance is tied to the model's performance. Therefore, it is assumed that the model has been properly tuned (e.g., using cross-validation) and is not over fitting.

16.3.2 Implementation

Permutation-based feature importance is available with the **DALEX**, **iml**, and **vip** packages; each providing unique benefits.

The **iml** package provides the `FeatureImp()` function which computes feature importance for general prediction models using the permutation approach. It is written in R6 and allows the user to specify a generic loss function or select from a pre-defined list (e.g., `loss = "mse"` for mean squared error). It also allows the user to specify whether importance is measures as the difference or as the ratio of the original model error and the model error after permutation. The user can also specify the number of repetitions used when permuting each feature to help stabilize the variability in the procedure.

The **DALEX** package also provides permutation-based variable importance scores through the `variable_importance()` function. Similar to `iml::FeatureImp()`, this function allows the user to specify a loss function and how the importance scores are computed (e.g., using the difference or ratio). It also provides an option to sample the training data before shuffling the data to compute importance (the default is to use `n_sample = 1000`. This can help speed up computation.

The **vip** package specifically focuses on variable importance plots (VIPs) and provides both model-specific and a number of model-agnostic approaches to computing variable importance, including the permutation approach. With **vip** you can use customized loss functions (or select from a pre-defined list), perform a Monte Carlo simulation to stabilize the procedure, sample observations prior to permuting features, perform the computations in parallel which can speed up runtime on large data sets, and more.

The following executes a permutation-based feature importance via **vip**. To speed up execution we sample 50% of the training data but repeat the simulations 5 times to increase stability of our estimates (whenever `nsim >`, you also get an estimated standard deviation for each importance score). We see that many of the same features that have been influential in model-specific approaches illustrated in previous chapters (e.g., `Gr_Liv_Area`, `Overall_Qual`, `Total_Bsmt_SF`, and `Neighborhood`) are also considered influential in our stacked model using the permutation approach.

Permutation-based approaches can become slow as the number of predictors grows. This implementation took 9 minutes. You can speed up execution by parallelizing, reducing the sample size, or reducing the number of simulations. However, note that the last two options also increases the variability of the feature importance estimates.

```
vip(
  ensemble_tree,
  train = as.data.frame(train_h2o),
  method = "permute",
  target = "Sale_Price",
  metric = "RMSE",
  nsim = 5,
  sample_frac = 0.5,
  pred_wrapper = pred
)
```

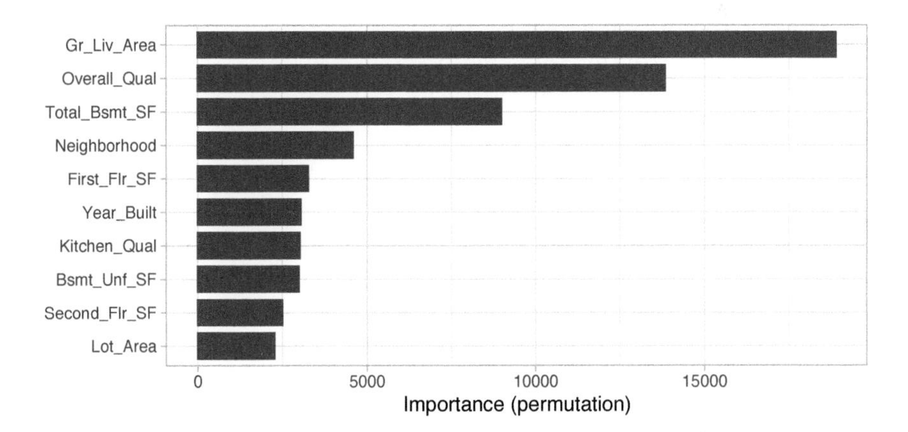

FIGURE 16.1: Top 10 most influential variables for the stacked H2O model using permutation-based feature importance.

16.4 Partial dependence

Partial dependence helps to understand the marginal effect of a feature (or subset thereof) on the predicted outcome. In essence, it allows us to understand how the response variable changes as we change the value of a feature while taking into account the average effect of all the other features in the model.

16.4.1 Concept

The procedure follows the traditional methodology documented in Friedman (2001). The algorithm (illustrated below) will split the feature of interest into j equally spaced values. For example, the `Gr_Liv_Area` feature ranges from 334–5095 square feet. Say the user selects $j = 20$. The algorithm will first create an evenly spaced grid consisting of 20 values across the distribution of `Gr_Liv_area` (e.g., $334.00, 584.58, \dots, 5095.00$). Then the algorithm will make 20 copies of the original training data (one copy for each value in the grid). The algorithms will then set `Gr_Liv_Area` for all observations in the first copy to 334, 585 in the second copy, 835 in the third copy, ..., and finally to 5095 in the 20-th copy (all other features remain unchanged). The algorithm then predicts the outcome for each observation in each of the 20 copies, and then averages the predicted values for each set. These averaged predicted values are known as partial dependence values and are plotted against the 20 evenly spaced values for `Gr_Liv_Area`.

```
For a selected predictor (x)
1. Construct a grid of j evenly spaced values across the distribution
   of x: {x1, x2, ..., xj}
2. For i in {1,...,j} do
     | Copy the training data and replace the original values of x
       with the constant xi
     | Apply given ML model (i.e., obtain vector of predictions)
     | Average predictions together
   End
3. Plot the averaged predictions against x1, x2, ..., xj
```

Algorithm 2 A simple algorithm for constructing the partial dependence of the response on a single predictor x.

Algorithm 1 can be quite computationally intensive since it involves j passes over the training records (and therefore j calls to the prediction function). Fortunately, the algorithm can be parallelized quite easily (see (Greenwell, 2018) for an example). It can also be easily extended to larger subsets of two or more features as well (i.e., to visualize interaction effects).

If we plot the partial dependence values against the grid values we get what's known as a *partial dependence plot* (PDP) (Figure 16.2) where the line represents the average predicted value across all observations at each of the j values of x.

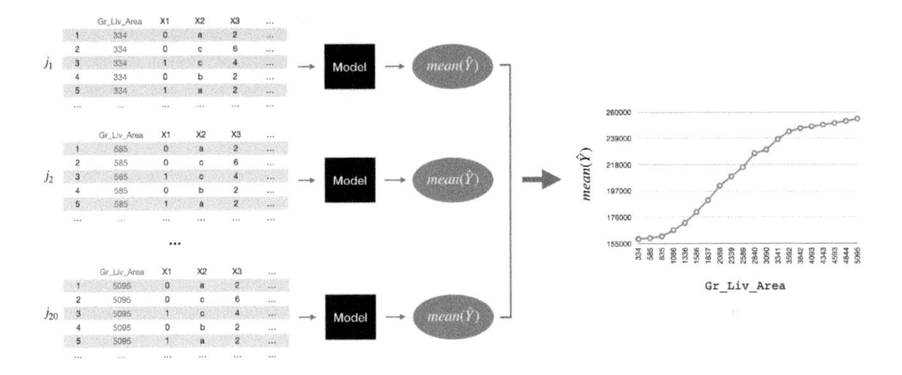

FIGURE 16.2: Illustration of the partial dependence process.

16.4.2 Implementation

The **pdp** package (Greenwell, 2018) is a widely used, mature, and flexible package for constructing PDPs. The **iml** and **DALEX** packages also provide PDP capabilities.[2] **pdp** has built-in support for many packages but for models that are not supported (such as **h2o** stacked models) we need to create a custom prediction function wrapper, as illustrated below. First, we create a custom prediction function similar to that which we created in Section 16.2.3; however, here we return the mean of the predicted values. We then use `pdp::partial()` to compute the partial dependence values.

We can use `autoplot()` to view PDPs using **ggplot2**. The `rug` argument provides markers for the decile distribution of `Gr_Liv_Area` and when you include `rug = TRUE` you must also include the training data.

```
# Custom prediction function wrapper
pdp_pred <- function(object, newdata)  {
  results <- mean(as.vector(h2o.predict(object, as.h2o(newdata))))
  return(results)
}

# Compute partial dependence values
pd_values <- partial(
  ensemble_tree,
  train = as.data.frame(train_h2o),
```

[2]In fact, previous versions of the **DALEX** package relied on **pdp** under the hood.

```
  pred.var = "Gr_Liv_Area",
  pred.fun = pdp_pred,
  grid.resolution = 20
)
head(pd_values)  # take a peak
##   Gr_Liv_Area   yhat
## 1         334 158858
## 2         584 159567
## 3         835 160878
## 4        1085 165897
## 5        1336 171666
## 6        1586 180505

# Partial dependence plot
autoplot(pd_values, rug = TRUE, train = as.data.frame(train_h2o))
```

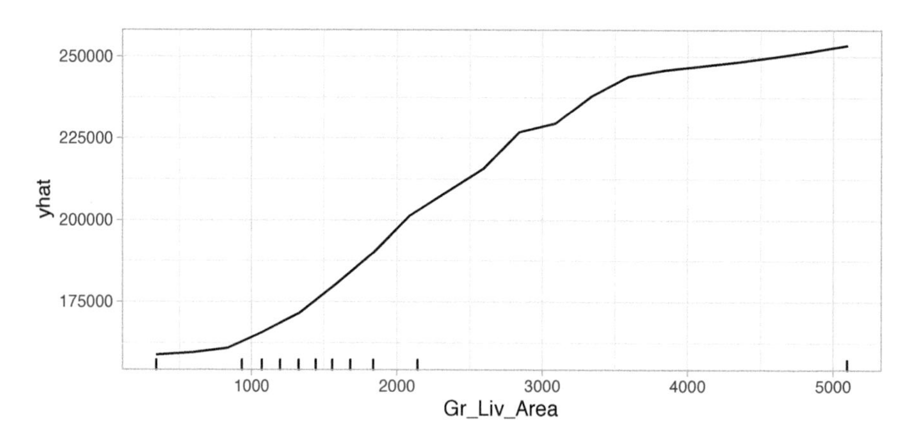

FIGURE 16.3: Partial dependence plot for the main living area feature illustrating the average increase in predicted sale price as main living area increases.

16.4.3 Alternative uses

PDPs have primarily been used to illustrate the marginal effect a feature has on the predicted response value. However, Greenwell et al. (2018b) illustrate an approach that uses a measure of the relative "flatness" of the partial dependence function as a measure of variable importance. The idea is that those features with larger marginal effects on the response have greater importance. You can implement a PDP-based measure of feature importance by using the **vip** package and setting `method = "pdp"`. The resulting variable importance scores also retain the computed partial dependence values (so you can easily view plots of both feature importance and feature effects).

16.5 Individual conditional expectation

Individual conditional expectation (ICE) curves (Goldstein et al., 2015) are very similar to PDPs; however, rather than averaging the predicted values across all observations we observe and plot the individual observation-level predictions.

16.5.1 Concept

An ICE plot visualizes the dependence of the predicted response on a feature for *each* instance separately, resulting in multiple lines, one for each observation, compared to one line in partial dependence plots. A PDP is the average of the lines of an ICE plot. Note that the following algorithm is the same as the PDP algorithms except for the last line where PDPs averaged the predicted values.

```
For a selected predictor (x)
1. Construct a grid of j evenly spaced values across the distribution
   of x: {x1, x2, ..., xj}
2. For i in {1,...,j} do
      | Copy the training data and replace the original values of x
        with the constant xi
      | Apply given ML model (i.e., obtain vector of predictions)
   End
3. Plot the predictions against x1, x2, ..., xj with lines connecting
   oberservations that correspond to the same row number in the
   original training data
```

Algorithm 3 A simple algorithm for constructing the individual conditional expectation of the response on a single predictor x.

So, what do you gain by looking at individual expectations, instead of partial dependencies? PDPs can obfuscate heterogeneous relationships that result from strong interaction effects. PDPs can show you what the average relationship between feature x_s and the predicted value ($\hat{y}$) looks like. This works only well in cases where the interactions between features are weak but in cases where interactions exist, ICE curves will help to highlight this. One issue to be aware of, often differences in ICE curves can only be identified by centering the feature. For example, ~ref(fig:ice-illustration) below displays ICE curves for the `Gr_Liv_Area` feature. The left plot makes it appear that all observations have very similar effects across `Gr_Liv_Area` values. However, the right plot shows centered ICE (c-ICE) curves which helps to highlight heterogeneity more clearly and also draws more attention to those observations that deviate from the general pattern.

You will typically see ICE curves centered at the minimum value of the feature. This allows you to see how effects change as the feature value increases.

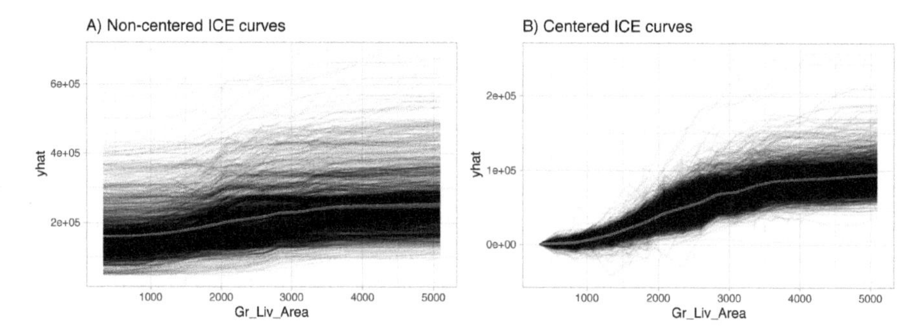

FIGURE 16.4: Non-centered (A) and centered (B) ICE curves for main living area illustrating the observation-level effects (black lines) in predicted sale price as the living area increases. The plot also illustrates the PDP line (red), representing the average values across all observations.

16.5.2 Implementation

Similar to PDPs, the premier package to use for ICE curves is the **pdp** package; however, the **iml** package also provides ICE curves. To create ICE curves with the **pdp** package we follow the same procedure as with PDPs; however, we exclude the averaging component (applying `mean()`) in the custom prediction function. By default, `autoplot()` will plot all observations; we also include `center = TRUE` to center the curves at the first value.

Note that we use `pred.fun = pred`. This is using the same custom prediction function created in Section 16.2.3.

```r
# Construct c-ICE curves
partial(
  ensemble_tree,
  train = as.data.frame(train_h2o),
  pred.var = "Gr_Liv_Area",
  pred.fun = pred,
  grid.resolution = 20,
  plot = TRUE,
  center = TRUE,
  plot.engine = "ggplot2"
)
```

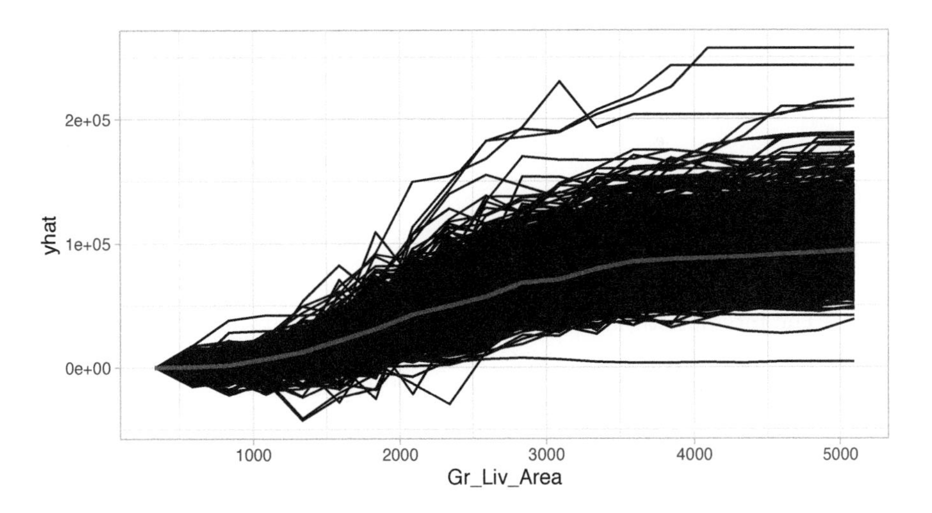

FIGURE 16.5: Centered ICE curve for the main living area feature illustrating the observation-level effects in predicted sale price as main living area increases.

PDPs for classification models are typically plotted on a logit-type scale, rather than on the probability scale (see Greenwell (2018) for details). This is more important for ICE curves and c-ICE curves, which can be more difficult to interpret. For example, c-ICE curves can result in negative probabilities. The ICE curves will also be more clumped together and harder to interpret when the predicted probabilities are close to zero or one.

16.6 Feature interactions

When features in a prediction model interact with each other, the influence of the features on the prediction surface is not additive but more complex. In real life, most relationships between features and some response variable are complex and include interactions. This is largely why more complex algorithms (especially tree-based algorithms) tend to perform very well—the nature of their complexity often allows them to naturally capture complex interactions. However, identifying and understanding the nature of these interactions is difficult.

One way to estimate the interaction strength is to measure how much of the variation of the predicted outcome depends on the interaction of the features. This measurement is called the *H*-statistic and was introduced by Friedman et al. (2008).

16.6.1 Concept

There are two main approaches to assessing interactions with the *H*-statistic:

1. The interaction between two features, which tells us how strongly two specific features interact with each other in the model;
2. The interaction between a feature and all other features, which tells us how strongly (in total) the specific feature interacts in the model with all the other features.

To measure both types of interactions, we leverage partial dependence values for the features of interest. For the first approach, which measures how a feature (x_i) interacts with all other features. The algorithm performs the following steps:

```
1. For variable i in {1,...,p} do
     | f(x) = estimate predicted values with original model
     | pd(x) = partial dependence of variable i
     | pd(!x) = partial dependence of all features excluding i
     | upper = sum(f(x) - pd(x) - pd(!x))
     | lower = variance(f(x))
     | rho = upper / lower
   End
2. Sort variables by descending rho (interaction strength)
```

Algorithm 4 A simple algorithm for measuring the interaction strength between x_i and all other features.

For the second approach, which measures the two-way interaction strength of feature x_i and x_j, the algorithm performs the following steps:

```
1. i = a selected variable of interest
2. For remaining variables j in {1,...,p} do
      | pd(ij) = interaction partial dependence of variables i and j
      | pd(i) = partial dependence of variable i
      | pd(j) = partial dependence of variable j
      | upper = sum(pd(ij) - pd(i) - pd(j))
      | lower = variance(pd(ij))
      | rho = upper / lower
   End
3. Sort interaction relationship by descending rho (interaction
   strength)
```

Algorithm 5 A simple algorithm for measuring the interaction strength between x_i and x_j.

In essence, the H-statistic measures how much of the variation of the predicted outcome depends on the interaction of the features. In both cases, $\rho =$ rho represents the interaction strength, which will be between 0 (when there is no interaction at all) and 1 (if all of variation of the predicted outcome depends on a given interaction).

16.6.2 Implementation

Currently, the **iml** package provides the only viable implementation of the H-statistic as a model-agnostic application. We use `Interaction$new()` to compute the one-way interaction to assess if and how strongly two specific features interact with each other in the model. We find that `First_Flr_SF` has the strongest interaction (although it is a weak interaction since $\rho < 0.139$).

Unfortunately, due to the algorithm complexity, the H-statistic is very computationally demanding as it requires $2n^2$ runs. This example of computing the one-way interaction H-statistic took two hours to complete! However, **iml** does allow you to speed up computation by reducing the `grid.size` or by parallelizing computation with `parallel = TRUE`. See `vignette("parallel", package = "iml")` for more info.

```
interact <- Interaction$new(components_iml)

interact$results %>%
  arrange(desc(.interaction)) %>%
  head()
##          .feature .interaction
## 1   First_Flr_SF   0.13917718
## 2   Overall_Qual   0.11077722
## 3   Kitchen_Qual   0.10531653
## 4 Second_Flr_SF    0.10461824
## 5       Lot_Area   0.10389242
## 6    Gr_Liv_Area   0.09833997

plot(interact)
```

Once we've identified the variable(s) with the strongest interaction signal (`First_Flr_SF` in our case), we can then compute the h-statistic to identify which features it mostly interacts with. This second iteration took over two hours and identified `Overall_Qual` as having the strongest interaction effect with `First_Flr_SF` (again, a weak interaction effect given $\rho = 0.144$).

```
# feature of interest
feat <- "First_Flr_SF"

interact_2way <- Interaction$new(components_iml, feature = feat)
interact_2way$results %>%
  arrange(desc(.interaction)) %>%
  top_n(10)
##                       .feature .interaction
## 1    Overall_Qual:First_Flr_SF   0.14385963
## 2      Year_Built:First_Flr_SF   0.09314573
## 3    Kitchen_Qual:First_Flr_SF   0.06567883
## 4       Bsmt_Qual:First_Flr_SF   0.06228321
## 5   Bsmt_Exposure:First_Flr_SF   0.05900530
## 6   Second_Flr_SF:First_Flr_SF   0.05747438
## 7   Kitchen_AbvGr:First_Flr_SF   0.05675684
## 8     Bsmt_Unf_SF:First_Flr_SF   0.05476509
## 9      Fireplaces:First_Flr_SF   0.05470992
## 10   Mas_Vnr_Area:First_Flr_SF   0.05439255
```

Identifying these interactions can help point us in the direction of assessing how the interactions relate to the response variable. We can use PDPs or ICE

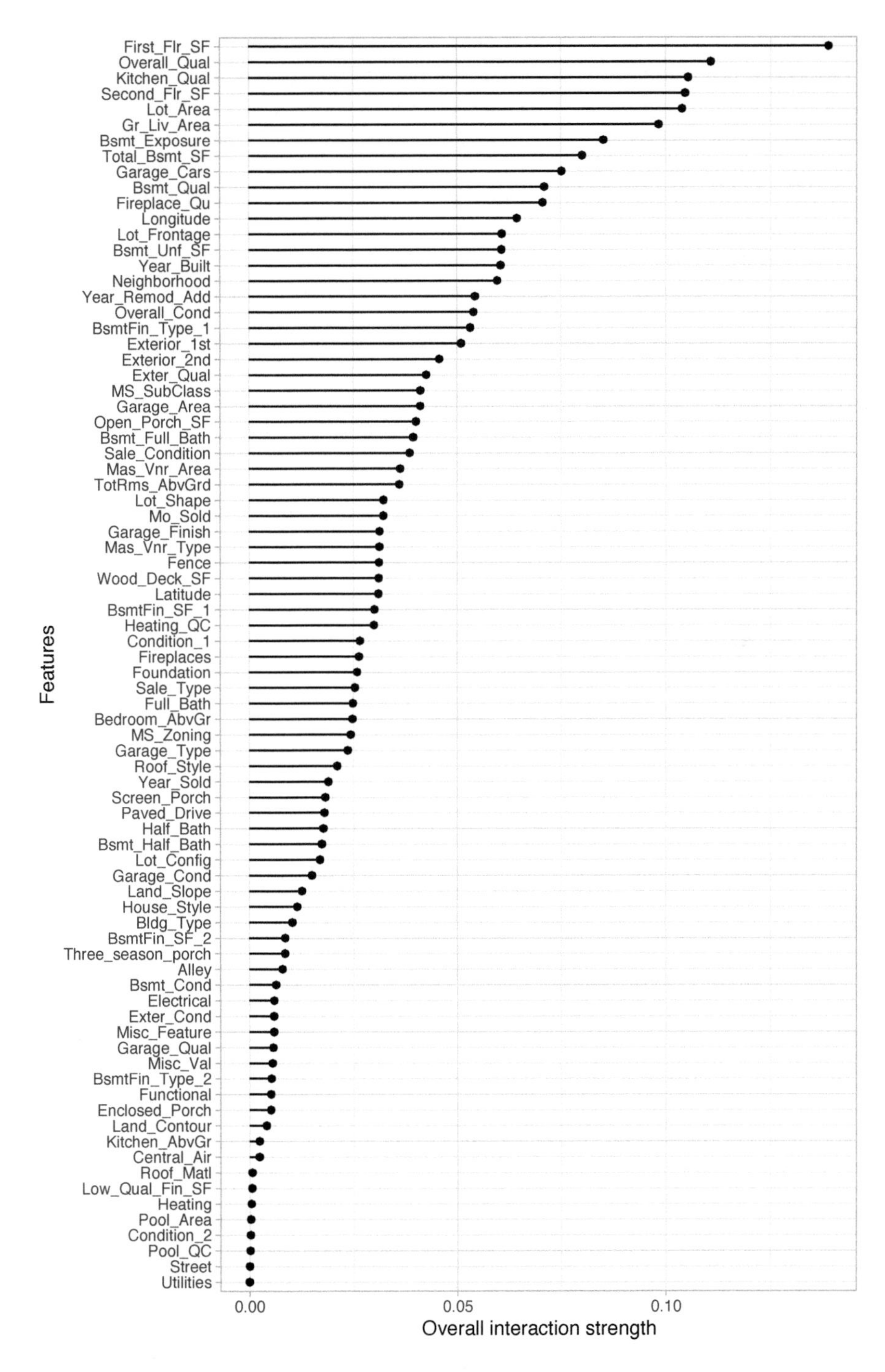

FIGURE 16.6: H-statistics for the 80 predictors in the Ames Housing data based on the H2O ensemble model.

curves with interactions to see their effect on the predicted response. Since the above process pointed out that `First_Flr_SF` and `Overall_Qual` had the highest interaction effect, the code plots this interaction relationship with predicted `Sale_Price`. We see that properties with "good" or lower `Overall_Qual` values tend have their `Sale_Prices` level off as `First_Flr_SF` increases moreso than properties with really strong `Overall_Qual` values. Also, you can see that properties with "very good" `Overall_Qual` tend to have a much larger increase in `Sale_Price` as `First_Flr_SF` increases from 1500–2000 than most other properties. (Although **pdp** allows more than one predictor, we take this opportunity to illustrate PDPs with the **iml** package.)

```
# Two-way PDP using iml
interaction_pdp <- Partial$new(
  components_iml,
  c("First_Flr_SF", "Overall_Qual"),
  ice = FALSE,
  grid.size = 20
)
plot(interaction_pdp)
```

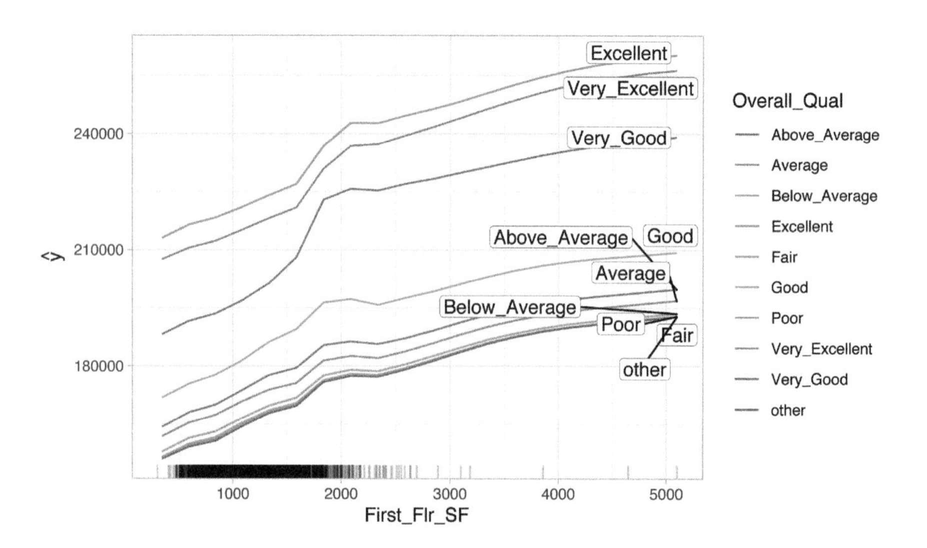

FIGURE 16.7: Interaction PDP illustrating the joint effect of the first floor square footage and overall quality features on predicted sales price.

16.6.3 Alternatives

Obviously computational time constraints are a major issue in identifying potential interaction effects. Although the H-statistic is the most statistically sound approach to detecting interactions, there are alternatives. The PDP-based variable importance measure discussed in Greenwell et al. (2018b) can also be used to quantify the strength of potential interaction effects. A thorough discussion of this approach is provided by Greenwell, Brandon M. and Boehmke, Bradley C. (2019) and can be implemented with `vip::vint()`. Also, Kuhn and Johnson (2019) provide a fairly comprehensive chapter discussing alternative approaches for identifying interactions.

16.7 Local interpretable model-agnostic explanations

Local Interpretable Model-agnostic Explanations (LIME) is an algorithm that helps explain individual predictions and was introduced by Ribeiro et al. (2016). Behind the workings of LIME lies the assumption that every complex model is linear on a local scale (i.e. in a small neighborhood around an observation of interest) and asserting that it is possible to fit a simple surrogate model around a single observation that will mimic how the global model behaves at that locality.

16.7.1 Concept

To do so, LIME samples the training data multiple times to identify observations that are similar to the individual record of interest. It then trains an interpretable model (often a LASSO model) weighted by the proximity of the sampled observations to the instance of interest. The resulting model can then be used to explain the predictions of the more complex model at the locality of the observation of interest.

The general algorithm LIME applies is:

1. ***Permute*** your training data to create replicated feature data with slight value modifications.
2. Compute ***proximity measure*** (e.g., 1 - distance) between the observation of interest and each of the permuted observations.
3. Apply selected machine learning model to ***predict outcomes*** of permuted data.
4. ***Select m number of features*** to best describe predicted outcomes.

5. ***Fit a simple model*** to the permuted data, explaining the complex model outcome with m features from the permuted data weighted by its similarity to the original observation.
6. Use the resulting ***feature weights to explain local behavior***.

Algorithm 6 The generalized LIME algorithm.

Each of these steps will be discussed in further detail as we proceed.

Although the **iml** package implements the LIME algorithm, the **lime** package provides the most comprehensive implementation.

16.7.2 Implementation

The implementation of **Algorithm 6** via the **lime** package is split into two operations: `lime::lime()` and `lime::explain()`. The `lime::lime()` function creates an `"explainer"` object, which is just a list that contains the fitted machine learning model and the feature distributions for the training data. The feature distributions that it contains includes distribution statistics for each categorical variable level and each continuous variable split into n bins (the current default is four bins). These feature attributes will be used to permute data.

```
# Create explainer object
components_lime <- lime(
  x = features,
  model = ensemble_tree,
  n_bins = 10
)

class(components_lime)
## [1] "data_frame_explainer" "explainer"
## [3] "list"
summary(components_lime)
##                  Length Class
## model             1     H2ORegressionModel
## preprocess        1     -none-
## bin_continuous    1     -none-
## n_bins            1     -none-
## quantile_bins     1     -none-
## use_density       1     -none-
## feature_type     80     -none-
```

```
## bin_cuts             80      -none-
## feature_distribution 80      -none-
##                      Mode
## model                S4
## preprocess           function
## bin_continuous       logical
## n_bins               numeric
## quantile_bins        logical
## use_density          logical
## feature_type         character
## bin_cuts             list
## feature_distribution list
```

Once we've created our lime object (i.e., `components_lime`), we can now perform the LIME algorithm using the `lime::explain()` function on the observation(s) of interest.

Recall that for local interpretation we are focusing on the two observations identified in Section 16.2.2 that contain the highest and lowest predicted sales prices.

This function has several options, each providing flexibility in how we perform **Algorithm 6**:

- `x`: Contains the observation(s) you want to create local explanations for. (See step 1 in **Algorithm 6**.)
- `explainer`: Takes the explainer object created by `lime::lime()`, which will be used to create permuted data. Permutations are sampled from the variable distributions created by the `lime::lime()` explainer object. (See step 1 in **Algorithm 6**.)
- `n_permutations`: The number of permutations to create for each observation in `x` (default is 5,000 for tabular data). (See step 1 in **Algorithm 6**.)
- `dist_fun`: The distance function to use. The default is Gower's distance but can also use Euclidean, Manhattan, or any other distance function allowed by the `dist()` function (see `?dist()` for details). To compute similarities, categorical features will be recoded based on whether or not they are equal to the actual observation. If continuous features are binned (the default) these features will be recoded based on whether they are in the same bin as the observation to be explained. Using the recoded data the distance to the original observation is then calculated based on a user-chosen distance measure. (See step 2 in **Algorithm 6**.)

- `kernel_width`: To convert the distance measure to a similarity score, an exponential kernel of a user defined width (defaults to 0.75 times the square root of the number of features) is used. Smaller values restrict the size of the local region. (See step 2 in **Algorithm 6**.)
- `n_features`: The number of features to best describe the predicted outcomes. (See step 4 in **Algorithm 6**.)
- `feature_select`: `lime::lime()` can use forward selection, ridge regression, lasso, or a decision tree to select the "best" `n_features` features. In the next example we apply a ridge regression model and select the m features with highest absolute weights. (See step 4 in **Algorithm 6**.)

For classification models we need to specify a couple of additional arguments:

- `labels`: The specific labels (classes) to explain (e.g., 0/1, "Yes"/"No")?
- `n_labels`: The number of labels to explain (e.g., Do you want to explain both success and failure or just the reason for success?)

```
# Use LIME to explain previously defined instances: high_ob & low_ob
lime_explanation <- lime::explain(
  x = rbind(high_ob, low_ob),
  explainer = components_lime,
  n_permutations = 5000,
  dist_fun = "gower",
  kernel_width = 0.25,
  n_features = 10,
  feature_select = "highest_weights"
)
```

If the original ML model is a regressor, the local model will predict the output of the complex model directly. If it is a classifier, the local model will predict the probability of the chosen class(es).

The output from `lime::explain()` is a data frame containing various information on the local model's predictions. Most importantly, for each observation supplied it contains the fitted explainer model (`model_r2`) and the weighted importance (`feature_weight`) for each important feature (`feature_desc`) that best describes the local relationship.

```
glimpse(lime_explanation)
## Observations: 20
## Variables: 11
```

```
## $ model_type       <chr> "regression", "regression"...
## $ case             <chr> "1825", "1825", "1825", "1...
## $ model_r2         <dbl> 0.4353, 0.4353, 0.4353, 0....
## $ model_intercept  <dbl> 170824, 170824, 170824, 17...
## $ model_prediction <dbl> 422391, 422391, 422391, 42...
## $ feature          <chr> "Gr_Liv_Area", "Overall_Qu...
## $ feature_value    <int> 3627, 8, 1930, 1831, 1796,...
## $ feature_weight   <dbl> 58305, 50071, 41102, 21368...
## $ feature_desc     <chr> "2141 < Gr_Liv_Area", "Ove...
## $ data             <list> [[Two_Story_1946_and_Newe...
## $ prediction       <dbl> 663136, 663136, 663136, 66...
```

Visualizing the results in Figure (16.8) we see that size and quality of the home appears to be driving the predictions for both `high_ob` (high `Sale_Price` observation) and `low_ob` (low `Sale_Price` observation). However, it's important to note the low R^2 ("Explanation Fit") of the models. The local model appears to have a fairly poor fit and, therefore, we shouldn't put too much faith in these explanations.

```
plot_features(lime_explanation, ncol = 1)
```

16.7.3 Tuning

Considering there are several knobs we can adjust when performing LIME, we can treat these as tuning parameters to try to tune the local model. This helps to maximize the amount of trust we can have in the local region explanation. As an example, the following code block changes the distance function to be Euclidean, increases the kernel width to create a larger local region, and changes the feature selection approach to a LARS-based LASSO model.

The result is a fairly substantial increase in our explanation fits, giving us much more confidence in their explanations.

```
# Tune the LIME algorithm a bit
lime_explanation2 <- explain(
  x = rbind(high_ob, low_ob),
  explainer = components_lime,
  n_permutations = 5000,
  dist_fun = "euclidean",
```

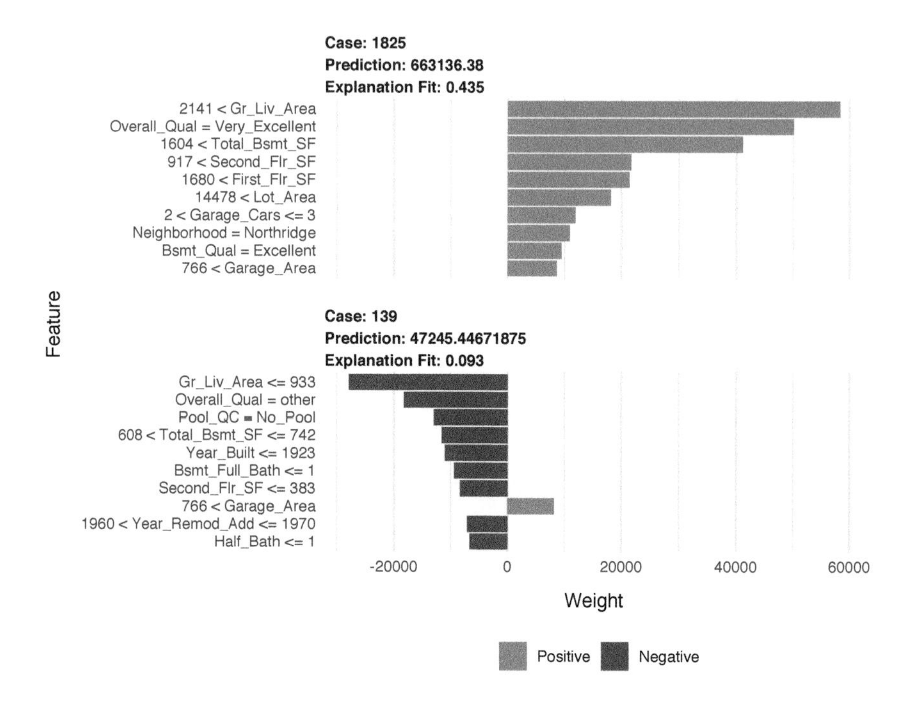

FIGURE 16.8: Local explanation for observations 1825 (high probability observation) and 139 (low probability observation) using LIME.

```
  kernel_width = 0.75,
  n_features = 10,
  feature_select = "lasso_path"
)

# Plot the results
plot_features(lime_explanation2, ncol = 1)
```

16.7.4 Alternative uses

The discussion above revolves around using LIME for tabular data sets. However, LIME can also be applied to non-traditional data sets such as text and images. For text, LIME creates a new *document term matrix* with perturbed text (e.g., it generates new phrases and sentences based on existing text). It then follows a similar procedure of weighting the similarity of the generated text to the original. The localized model then helps to identify which words in

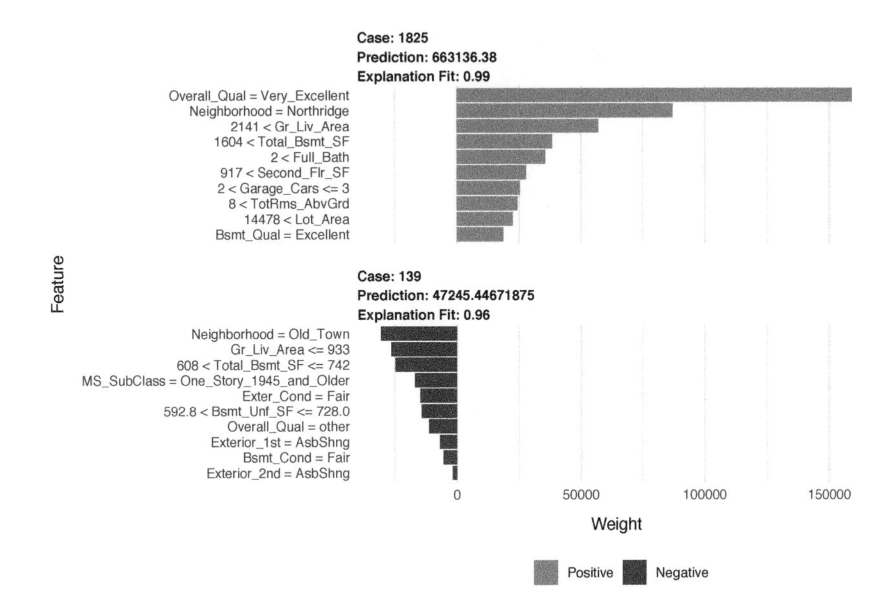

FIGURE 16.9: Local explanation for observations 1825 (case 1) and 139 (case 2) after tuning the LIME algorithm.

the perturbed text are producing the strongest signal. For images, variations of the images are created by replacing certain groupings of pixels with a constant color (e.g., gray). LIME then assesses the predicted labels for the given group of pixels not perturbed. For more details on such use cases see Molnar et al. (2018).

16.8 Shapley values

Another method for explaining individual predictions borrows ideas from coalitional (or cooperative) game theory to produce whats called Shapley values (Lundberg and Lee, 2016, 2017). By now you should realize that when a model gives a prediction for an observation, all features do not play the same role: some of them may have a lot of influence on the model's prediction, while others may be irrelevant. Consequently, one may think that the effect of each feature can be measured by checking what the prediction would have been if that feature was absent; the bigger the change in the model's output, the more important that feature is. This is exactly what happens with permutation-based variable importance (since LIME most often uses a ridge or lasso model, it also uses a similar approach to identify localized feature importance).

However, observing only single feature effects at a time implies that dependencies between features are not taken into account, which could produce inaccurate and misleading explanations of the model's internal logic. Therefore, to avoid missing any interaction between features, we should observe how the prediction changes for each possible subset of features and then combine these changes to form a unique contribution for each feature value.

16.8.1 Concept

The concept of Shapley values is based on the idea that the feature values of an individual observation work together to cause a change in the model's prediction with respect to the model's expected output, and it divides this total change in prediction among the features in a way that is "fair" to their contributions across all possible subsets of features.

To do so, Shapley values assess every combination of predictors to determine each predictors impact. Focusing on feature x_j, the approach will test the accuracy of every combination of features not including x_j and then test how adding x_j to each combination improves the accuracy. Unfortunately, computing Shapley values is very computationally expensive. Consequently, the **iml** package implements an approximate Shapley value.

To compute the approximate Shapley contribution of feature x_j on x we need to construct two new "Frankenstein" instances and take the difference between their corresponding predictions. This is outlined in the brief algorithm below. Note that this is often repeated several times (e.g., 10–100) for each feature/observation combination and the results are averaged together. See `http://bit.ly/fastshap` and Štrumbelj and Kononenko (2014) for details.

```
ob = single observation of interest
1. For variables j in {1,...,p} do
     | m = draw random sample(s) from data set
     | randomly shuffle the feature names, perm <- sample(names(x))
     | Create two new instances b1 and b2 as follows:
       | b1 = x, but all the features in perm that appear after
       |      feature xj get their values swapped with the
       |      corresponding values in z.
       | b2 = x, but feature xj, as well as all the features in perm
       |      that appear after xj, get their values swapped with the
       |      corresponding values in z.
     | f(b1) = compute predictions for b1
```

```
    | f(b2) = compute predictions for b2
    | shap_ind = f(b1) - f(b2)
    | phi = mean(shap_ind)
   End
2. Sort phi in decreasing order
```

Algorithm 7 A simple algorithm for computing approximate Shapley values.

The aggregated Shapley values ($\phi =$ `phi`) represent the contribution of each feature towards a predicted value compared to the average prediction for the data set. Figure 16.10, represents the first iteration of our algorithm where we focus on the impact of feature X_1. In step (A) we sample the training data. In step (B) we create two copies of an individually sampled row and randomize the order of the features. Then in one copy we include all values from the observation of interest for the values from the first column feature up to *and including* X_1. We then include the values from the sampled row for all the other features. In the second copy, we include all values from the observation of interest for the values from the first column feature up to *but not including* X_1. We use values from the sample row for X_1 and all the other features. Then in step (C), we apply our model to both copies of this row and in step (D) compute the difference between the predicted outputs.

We follow this procedure for all the sampled rows and the average difference across all sampled rows is the Shapley value. It should be obvious that the more observations we include in our sampling procedure the closer our approximate Shapley computation will be to the true Shapley value.

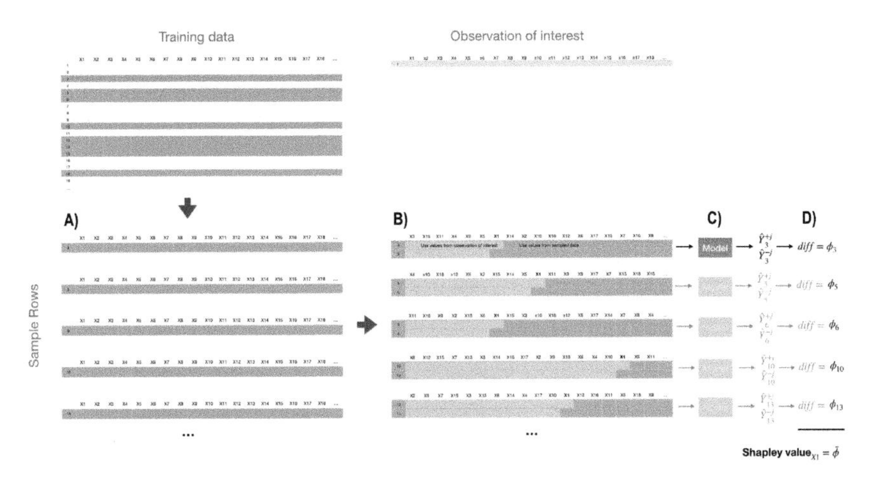

FIGURE 16.10: Generalized concept behind approximate Shapley value computation.

16.8.2 Implementation

The **iml** package provides one of the few Shapley value implementations in R. We use `Shapley$new()` to create a new Shapley object. The time to compute is largely driven by the number of predictors and the sample size drawn. By default, `Shapley$new()` will only use a sample size of 100 but you can control this to either reduce compute time or increase confidence in the estimated values.

In this example we increased the sample size to 1000 for greater confidence in the estimated values; it took roughly 3.5 minutes to compute.

Looking at the results we see that the predicted sale price of \$663,136.38 is \$481,797.42 larger than the average predicted sale price of \$181,338.96; Figure 16.11 displays the contribution each predictor played in this difference. We see that `Gr_Liv_Area`, `Overall_Qual`, and `Second_Flr_SF` are the top three features positively influencing the predicted sale price; all of which contributed close to, or over, \$75,000 towards the \$481.8K difference.

```
# Compute (approximate) Shapley values
(shapley <- Shapley$new(components_iml, x.interest = high_ob,
                        sample.size = 1000))
## Interpretation method:  Shapley
## Predicted value: 663136.380000, Average prediction: 181338.96359...
##
## Analysed predictor:
## Prediction task: unknown
##
##
## Analysed data:
## Sampling from data.frame with 2199 rows and 80 columns.
##
## Head of results:
##        feature     phi  phi.var
## 1   MS_SubClass  1878.4 4.14e+07
## 2     MS_Zoning    80.4 2.48e+06
## 3 Lot_Frontage  1457.3 8.26e+07
## 4      Lot_Area 15649.7 4.12e+08
## 5        Street     0.0 0.00e+00
## 6         Alley    53.6 4.21e+05
##                          feature.value
## 1 MS_SubClass=Two_Story_1946_and_Newer
```

```
## 2     MS_Zoning=Residential_Low_Density
## 3                     Lot_Frontage=118
## 4                       Lot_Area=35760
## 5                          Street=Pave
## 6               Alley=No_Alley_Access

# Plot results
plot(shapley)
```

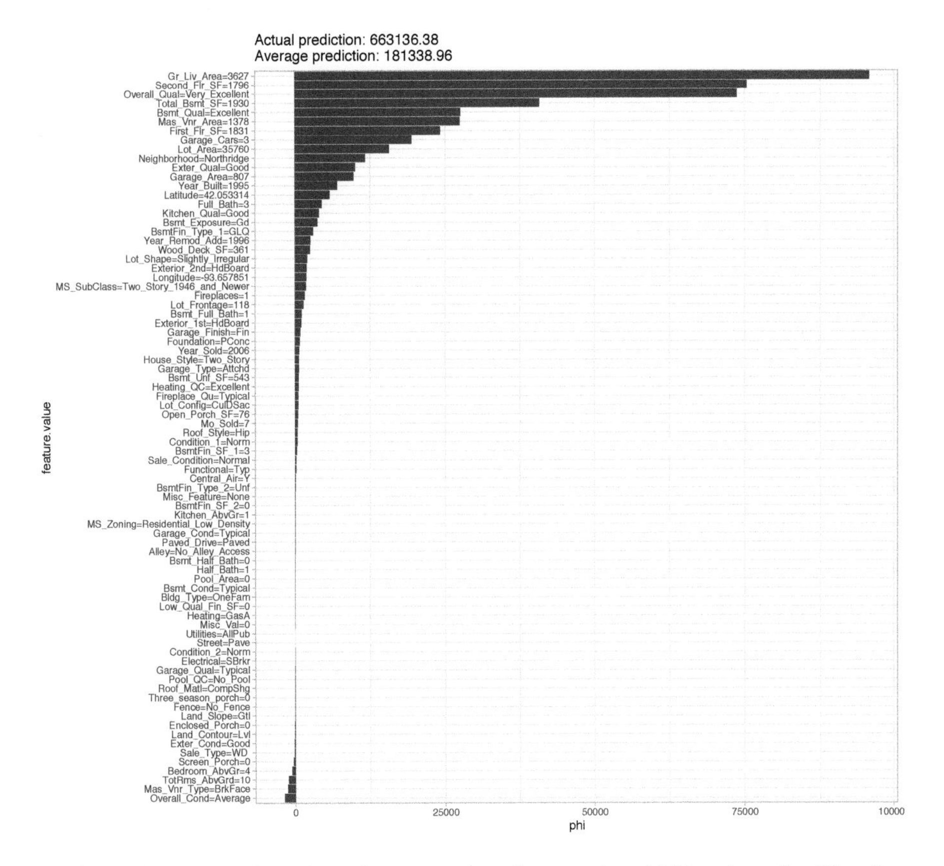

FIGURE 16.11: Local explanation for observation 1825 using the Shapley value algorithm.

Since **iml** uses R6, we can reuse the Shapley object to identify the influential predictors that help explain the low `Sale_Price` observation. In Figure 16.12 we see similar results to LIME in that `Overall_Qual` and `Gr_Liv_Area` are the most influential predictors driving down the price of this home.

```
# Reuse existing object
shapley$explain(x.interest = low_ob)

# Plot results
shapley$results %>%
  top_n(25, wt = abs(phi)) %>%
  ggplot(aes(phi, reorder(feature.value, phi), color = phi > 0)) +
  geom_point(show.legend = FALSE)
```

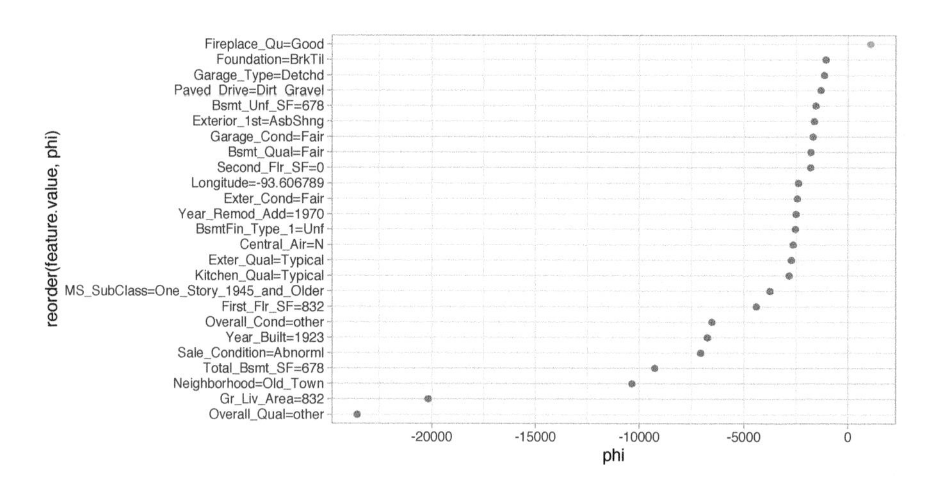

FIGURE 16.12: Local explanation for observation 139 using the Shapley value algorithm.

16.8.3 XGBoost and built-in Shapley values

True Shapley values are considered theoretically optimal (Lundberg and Lee, 2016); however, as previously discussed they are computationally challenging. The approximate Shapley values provided by **iml** are much more computationally feasible. Another common option is discussed by Lundberg and Lee (2017) and, although not purely model-agnostic, is applicable to tree-based models and is fully integrated in most XGBoost implementations (including the **xgboost** package). Similar to **iml**'s approximation procedure, this tree-based Shapley value procedure is also an approximation, but allows for polynomial runtime instead of exponential runtime.

To demonstrate, we'll use the features used and the final XGBoost model created in Section 12.5.2.

```
# Compute tree SHAP for a previously obtained XGBoost model
X <- readr::read_rds("data/xgb-features.rds")
xgb.fit.final <- readr::read_rds("data/xgb-fit-final.rds")
```

The benefit of this expedient approach is we can reasonably compute Shapley values for every observation and every feature in one fell swoop. This allows us to use Shapley values for more than just local interpretation. For example, the following computes and plots the Shapley values for every feature and observation in our Ames housing example; see Figure 16.13. The left plot displays the individual Shapley contributions. Each dot represents a feature's contribution to the predicted `Sale_Price` for an individual observation. This allows us to see the general magnitude and variation of each feature's contributions across all observations. We can use this information to compute the average absolute Shapley value across all observations for each features and use this as a global measure of feature importance (right plot).

There's a fair amount of general data wrangling going on here but the key line of code is `predict(newdata = X, predcontrib = TRUE)`. This line computes the prediction contribution for each feature and observation in the data supplied via `newdata`.

```
# Try to re-scale features (low to high)
feature_values <- X %>%
  as.data.frame() %>%
  mutate_all(scale) %>%
  gather(feature, feature_value) %>%
  pull(feature_value)

# Compute SHAP values, wrangle a bit, compute SHAP-based
# importance, etc.
shap_df <- xgb.fit.final %>%
  predict(newdata = X, predcontrib = TRUE) %>%
  as.data.frame() %>%
  select(-BIAS) %>%
  gather(feature, shap_value) %>%
  mutate(feature_value = feature_values) %>%
  group_by(feature) %>%
  mutate(shap_importance = mean(abs(shap_value)))

# SHAP contribution plot
```

```
p1 <- ggplot(shap_df,
             aes(x = shap_value,
                 y = reorder(feature, shap_importance))) +
  ggbeeswarm::geom_quasirandom(groupOnX = FALSE, varwidth = TRUE,
                               size = 0.4, alpha = 0.25) +
  xlab("SHAP value") +
  ylab(NULL)

# SHAP importance plot
p2 <- shap_df %>%
  select(feature, shap_importance) %>%
  filter(row_number() == 1) %>%
  ggplot(aes(x = reorder(feature, shap_importance),
             y = shap_importance)) +
    geom_col() +
    coord_flip() +
    xlab(NULL) +
    ylab("mean(|SHAP value|)")

# Combine plots
gridExtra::grid.arrange(p1, p2, nrow = 1)
```

We can also use this information to create an alternative to PDPs. Shapley-based dependence plots (Figure 16.14) show the Shapley values of a feature on the y-axis and the value of the feature for the x-axis. By plotting these values for all observations in the data set we can see how the feature's attributed importance changes as its value varies.

```
shap_df %>%
  filter(feature %in% c("Overall_Qual", "Gr_Liv_Area")) %>%
  ggplot(aes(x = feature_value, y = shap_value)) +
    geom_point(aes(color = shap_value)) +
    scale_colour_viridis_c(name = "Feature value\n(standardized)",
                           option = "C") +
    facet_wrap(~ feature, scales = "free") +
    scale_y_continuous('Shapley value', labels = scales::comma) +
    xlab('Normalized feature value')
```

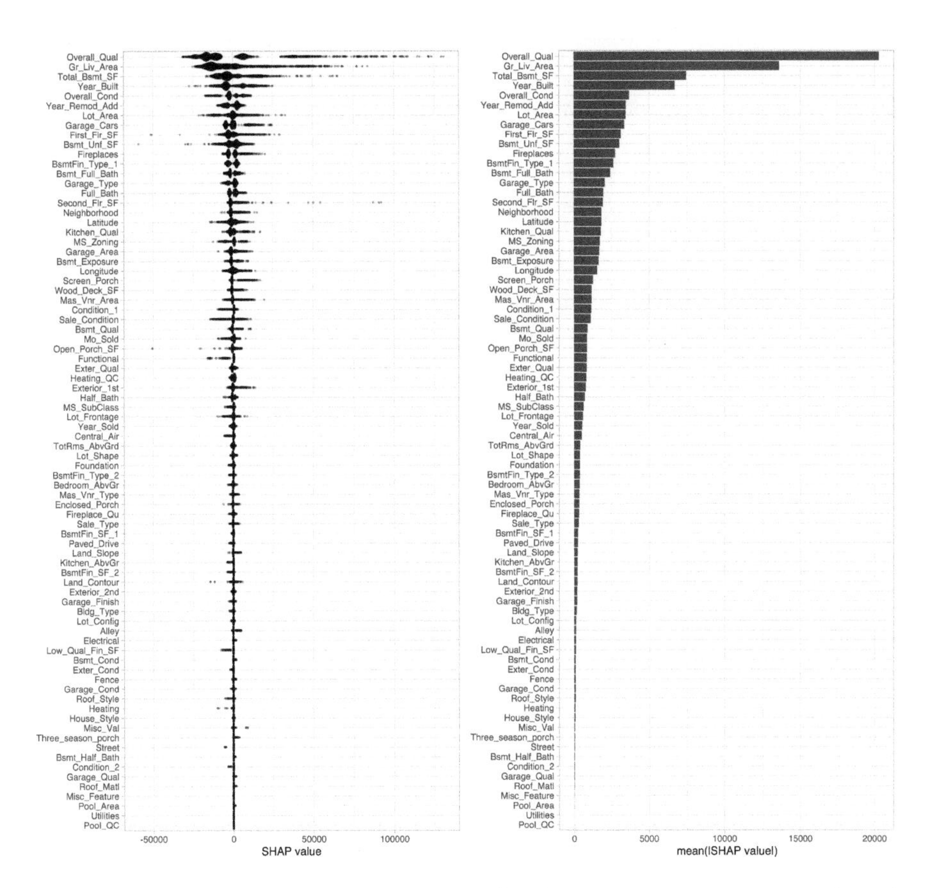

FIGURE 16.13: Shapley contribution (left) and global importance (right) plots.

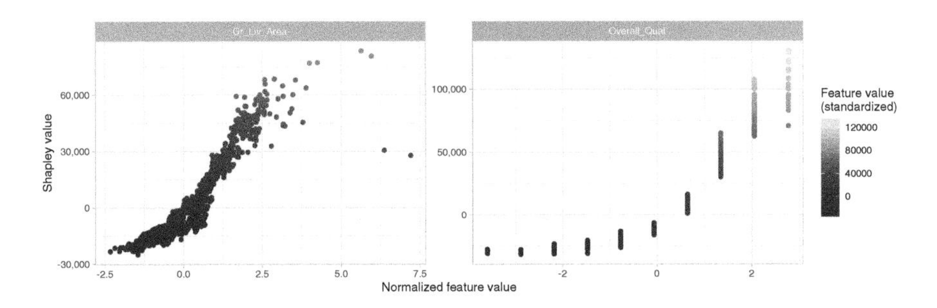

FIGURE 16.14: Shapley-based dependence plot illustrating the variability in contribution across the range of main living area and overall quality values.

16.9 Localized step-wise procedure

An additional approach for localized explanation is a procedure that is loosely related to the partial dependence algorithm with an added step-wise procedure. The procedure was introduced by Staniak and Biecek (2018) and is known as the *Break Down* method, which uses a greedy strategy to identify and remove features iteratively based on their influence on the overall average predicted response.

16.9.1 Concept

The Break Down method provides two sequential approaches; the default is called *step up*. This procedure, essentially, takes the value for a given feature in the single observation of interest, replaces all the observations in the training data set, and identifies how it effects the prediction error. It performs this process iteratively and independently for each feature, identifies the column with the largest difference score, and adds that variable to the list as the most important. This feature's signal is then removed (via randomization), and the procedure sweeps through the remaining predictors and applies the same process until all variables have been assessed.

```
existing_data = validation data set used in explainer
new_ob = single observation to perform local interpretation on
p = number of predictors
l = list of predictors
baseline = mean predicted response of existing_data

for variable i in {1,...,p} do
  for variable j in {1,...,l} do
    | exchange variable j in existing_data with variable j value
    | in new_ob predicted_j = mean predicted response of altered
    | existing_data diff_j = absolute difference between
    | baseline - predicted reset existing_data
    end
  | t = variable j with largest diff value
  | contribution for variable t = diff value for variable t
  | remove variable t from l
  end
```

Algorithm 8 A simple algorithm for computing Break Down values with the step up method.

An alternative approach is called the *step down* method which follows a similar algorithm but rather than remove the variable with the largest difference score on each sweep, it removes the variable with the smallest difference score. Both approaches are analogous to backward stepwise selection where *step up* removes variables with largest impact and *step down* removes variables with the smallest impact.

16.9.2 Implementation

To perform the Break Down algorithm on a single observation, use the `DALEX::prediction_breakdown()` function. The output is a data frame with class `"prediction_breakdown_explainer"` that lists the contribution for each variable. Similar to Shapley values, the results display the contribution that each feature value for the given observation has on the difference between the overall average response (`Sale_Price` in this example) and the response for the given observation of interest.

The default approach is ***step up*** but you can perform ***step down*** by specifying `direction = "down"`.

If you look at the contribution output, realize the feature ordering is in terms of importance. Consequently, `Gr_Liv_Area` was identified as most influential followed by `Second_Flr_SF` and `Total_Bsmt_SF`. However, if you look at the contribution value, you will notice that `Second_Flr_SF` appears to have a larger contribution to the above average price than `Gr_Liv_Area`. However, the `Second_Flr_SF` contribution is based on having already taken `Gr_Liv_Area`'s contribution into effect.

The break down algorithm is the most computationally intense of all methods discussed in this chapter. Since the number of required iterations increases by $p \times (p - 1)$ for every additional feature, wider data sets cause this algorithm to become burdensome. For example, this single application took over 6 hours to compute!

```
high_breakdown <- prediction_breakdown(components_dalex,
                                       observation = high_ob)

# class of prediction_breakdown output
```

```
class(high_breakdown)
## [1] "prediction_breakdown_explainer" "data.frame"

# check out the top 10 influential variables for this observation
high_breakdown[1:10, 1:5]
##                                                   variable contribution ...
## 1                                              (Intercept)    181338.96 ...
## Gr_Liv_Area                         + Gr_Liv_Area = 4316     46971.64 ...
## Second_Flr_SF                     + Second_Flr_SF = 1872     52997.40 ...
## Total_Bsmt_SF                     + Total_Bsmt_SF = 2444     41339.89 ...
## Overall_Qual    + Overall_Qual = Very_Excellent     47690.10 ...
## First_Flr_SF                       + First_Flr_SF = 2444     56780.92 ...
## Bsmt_Qual                        + Bsmt_Qual = Excellent     49341.73 ...
## Neighborhood          + Neighborhood = Northridge     54289.27 ...
## Garage_Cars                           + Garage_Cars = 3     41959.23 ...
## Kitchen_Qual                  + Kitchen_Qual = Excellent     59805.57 ...
```

We can plot the entire list of contributions for each variable using `plot(high_breakdown)`.

16.10 Final thoughts

Since this book focuses on hands-on applications, we have focused on only a small sliver of IML. IML is a rapidly expanding research space that covers many more topics including moral and ethical considerations such as fairness, accountability, and transparency along with many more analytic procedures to interpret model performance, sensitivity, bias identification, and more. Moreover, the above discussion only provides a high-level understanding of the methods. To gain deeper understanding around these methods and to learn more about the other areas of IML (like not discussed in this book) we highly recommend Molnar et al. (2018) and Hall, Patrick (2018).

Part III

Dimension Reduction

17

Principal Components Analysis

Principal components analysis (PCA) is a method for finding low-dimensional representations of a data set that retain as much of the original variation as possible. The idea is that each of the n observations lives in p-dimensional space, but not all of these dimensions are equally interesting. In PCA we look for a smaller number of dimensions that are as interesting as possible, where the concept of *interesting* is measured by the amount that the observations vary along each dimension. Each of the new dimensions found in PCA is a linear combination of the original p features. The hope is to use a small subset of these linear feature combinations in further analysis while retaining most of the information present in the original data.

17.1 Prerequisites

This chapter leverages the following packages.

```
library(dplyr)       # basic data manipulation and plotting
library(ggplot2)     # data visualization
library(h2o)         # performing dimension reduction
```

To illustrate dimension reduction techniques, we'll use the `my_basket` data set (Section 1.4). This data set identifies items and quantities purchased for 2,000 transactions from a grocery store. The objective is to identify common groupings of items purchased together.

```
url <- "https://koalaverse.github.io/homlr/data/my_basket.csv"
my_basket <- readr::read_csv(url)
dim(my_basket)
## [1] 2000   42
```

To perform dimension reduction techniques in R, generally, the data should be prepared as follows:

1. Data are in tidy format per Wickham et al. (2014);
2. Any missing values in the data must be removed or imputed;
3. Typically, the data must all be numeric values (e.g., one-hot, label, ordinal encoding categorical features);
4. Numeric data should be standardized (e.g., centered and scaled) to make features comparable.

The my_basket data already fullfills these requirements. However, some of the packages we'll use to perform dimension reduction tasks have built-in capabilities to impute missing data, numerically encode categorical features (typically one-hot encode), and standardize the features.

17.2 The idea

Dimension reduction methods, such as PCA, focus on reducing the feature space, allowing most of the information or variability in the data set to be explained using fewer features; in the case of PCA, these new features will also be uncorrelated. For example, among the 42 variables within the my_basket data set, 23 combinations of variables have moderate correlation (≥ 0.25) with each other. Looking at the table below, we see that some of these combinations may be represented with smaller dimension categories (e.g., soda, candy, breakfast, and italian food)

We often want to explain common attributes such as these in a lower dimensionality than the original data. For example, when we purchase soda we may often buy multiple types at the same time (e.g., Coke, Pepsi, and 7UP). We could reduce these variables to one *latent variable* (i.e., unobserved feature) called "soda". This can help in describing many features in our data set and it can also remove multicollinearity, which can often improve predictive accuracy in downstream supervised models.

So how do we identify variables that could be grouped into a lower dimension? One option includes examining pairwise scatterplots of each variable against every other variable and identifying co-variation. Unfortunately, this is tedious and becomes excessive quickly even with a small number of variables (given p variables there are $p(p-1)/2$ possible scatterplot combinations). For example, since the my_basket data has 42 numeric variables, we would need to examine

TABLE 17.1: Various items in our my basket data that are correlated.

Item 1	Item 2	Correlation
cheese	mayonnaise	0.345
bulmers	fosters	0.335
cheese	bread	0.320
lasagna	pizza	0.316
pepsi	coke	0.309
red.wine	fosters	0.308
milk	muesli	0.302
mars	twix	0.301
red.wine	bulmers	0.298
bulmers	kronenbourg	0.289
milk	tea	0.288
red.wine	kronenbourg	0.286
7up	coke	0.282
spinach	broccoli	0.282
mayonnaise	bread	0.278
peas	potatoes	0.271
peas	carrots	0.270
tea	instant.coffee	0.270
milk	instant.coffee	0.267
bread	lettuce	0.264
twix	kitkat	0.259
mars	kitkat	0.255
muesli	instant.coffee	0.251

$42(42-1)/2 = 861$ scatterplots! Fortunately, better approaches exist to help represent our data using a smaller dimension.

The PCA method was first published in 1901 (Pearson, 1901) and has been a staple procedure for dimension reduction for decades. PCA examines the covariance among features and combines multiple features into a smaller set of uncorrelated variables. These new features, which are weighted combinations of the original predictor set, are called *principal components* (PCs) and hopefully a small subset of them explain most of the variability of the full feature set. The weights used to form the PCs reveal the relative contributions of the original features to the new PCs.

17.3 Finding principal components

The *first principal component* of a set of features X_1, X_2, ..., X_p is the linear combination of the features

$$Z_1 = \phi_{11}X_1 + \phi_{21}X_2 + ... + \phi_{p1}X_p, \tag{17.1}$$

that has the largest variance. Here $\phi_1 = (\phi_{11}, \phi_{21}, \dots, \phi_{p1})$ is the *loading vector* for the first principal component. The ϕ are *normalized* so that $\sum_{j=1}^{p} \phi_{j1}^2 = 1$. After the first principal component Z_1 has been determined, we can find the second principal component Z_2. The second principal component is the linear combination of $X_1, \dots, X_p$ that has maximal variance out of all linear combinations that are ***uncorrelated*** with Z_1:

$$Z_2 = \phi_{12}X_1 + \phi_{22}X_2 + ... + \phi_{p2}X_p \tag{17.2}$$

where again we define $\phi_2 = (\phi_{12}, \phi_{22}, \dots, \phi_{p2})$ as the loading vector for the second principal component. This process proceeds until all p principal components are computed. So how do we calculate $\phi_1, \phi_2, \dots, \phi_p$ in practice?. It can be shown, using techniques from linear algebra[1], that the *eigenvector* corresponding to the largest *eigenvalue* of the feature covariance matrix is the set of loadings that explains the greatest proportion of feature variability.[2]

An illustration provides a more intuitive grasp on principal components. Assume we have two features that have moderate (0.56, say) correlation. We can explain the covariation of these variables in two dimensions (i.e., using PC 1 and PC 2). We see that the greatest covariation falls along the first PC, which is simply the line that minimizes the total squared distance from each point to its *orthogonal projection* onto the line. Consequently, we can explain the vast majority (93% to be exact) of the variability between feature 1 and feature 2 using just the first PC.

We can extend this to three variables, assessing the relationship among features 1, 2, and 3. The first two PC directions span the plane that best fits the variability in the data. It minimizes the sum of squared distances from each point to the plan. As more dimensions are added, these visuals are not as intuitive but we'll see shortly how we can still use PCA to extract and visualize important information.

---

[1]Linear algebra is fundamental to mastering many advanced concepts in statistics and machine learning, but is out of the scope of this book.

[2]In particular, the PCs can be found using an *eigen decomposition* applied to the special matrix $X'X$ (where X is an $n \times p$ matrix consisting of the centered feature values).

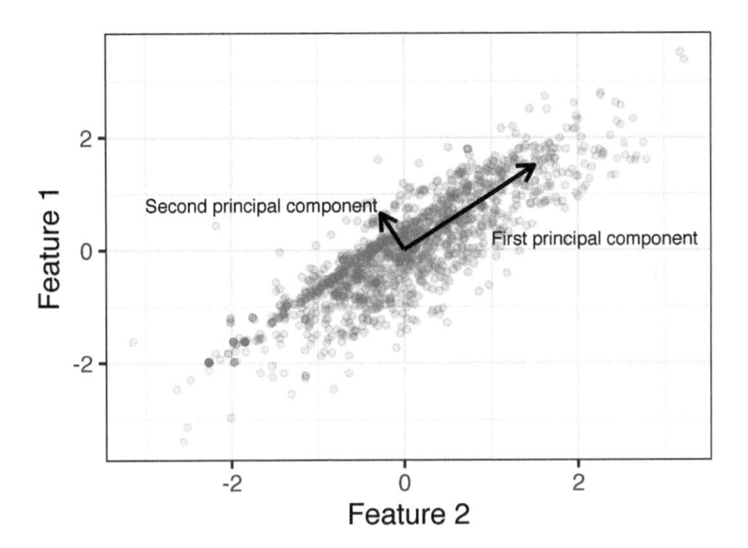

FIGURE 17.1: Principal components of two features that have 0.56 correlation.

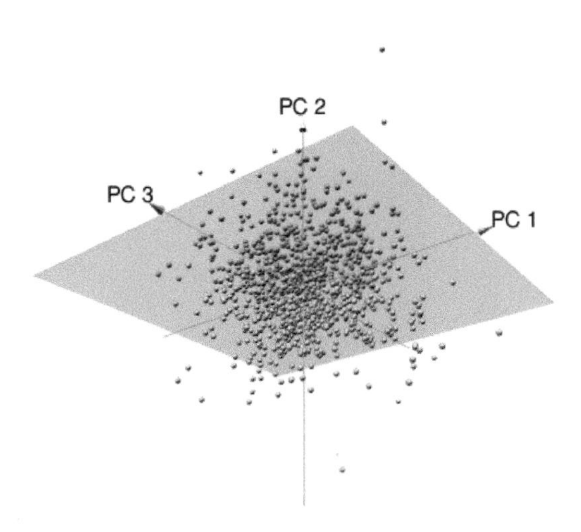

FIGURE 17.2: Principal components of three features.

17.4 Performing PCA in R

There are several built-in and external packages to perform PCA in R. We recommend to use **h2o** as it provides consistency across the dimension reduction methods we'll discuss later and it also automates much of the data preparation steps previously discussed (i.e., standardizing numeric features, imputing missing values, and encoding categorical features).

Let's go ahead and start up **h2o**:

```
h2o.no_progress()  # turn off progress bars for brevity
h2o.init(max_mem_size = "5g")  # connect to H2O instance
```

First, we convert our `my_basket` data frame to an appropriate **h2o** object and then use `h2o.prcomp()` to perform PCA. A few of the important arguments you can specify in `h2o.prcomp()` include:

- `pca_method`: Character string specifying which PC method to use. there are actually a few different approaches to calculating principal components (PCs). When your data contains mostly numeric data (such as `my_basket`), its best to use `pca_method = "GramSVD"`. When your data contain many categorical variables (or just a few categorical variables with high cardinality) we recommend you use `pca_method = "GLRM"`.
- `k`: Integer specifying how many PCs to compute. It's best to create the same number of PCs as there are features and we will see shortly how to identify the number of PCs to use, where the number of PCs is less than the number of features.
- `transform`: Character string specifying how (if at all) your data should be standardized.
- `impute_missing`: Logical specifying whether or not to impute missing values; if your data have missing values, this will impute them with the corresponding column mean.
- `max_runtime_secs`: Number specifying the max run time (in seconds); when working with large data sets this will limit the runtime for model training.

When your data contains mostly numeric data (such as `my_basket`), its best to use `pca_method = "GramSVD"`. When your data contain many categorical variables (or just a few categorical variables with high cardinality) we recommend to use `pca_method = "GLRM"`.

```
# convert data to h2o object
my_basket.h2o <- as.h2o(my_basket)

# run PCA
my_pca <- h2o.prcomp(
  training_frame = my_basket.h2o,
  pca_method = "GramSVD",
  k = ncol(my_basket.h2o),
  transform = "STANDARDIZE",
  impute_missing = TRUE,
  max_runtime_secs = 1000
)
```

Our model object (`my_pca`) contains several pieces of information that we can extract (you can view all information with `glimpse(my_pca)`). The most important information is stored in `my_pca@model$importance` (which is the same output that gets printed when looking at our object's printed output). This information includes each PC, the standard deviation of each PC, as well as the proportion and cumulative proportion of variance explained with each PC.

```
my_pca
## Model Details:
## ==============
##
## H2ODimReductionModel: pca
## Model ID:  PCA_model_R_1536152543598_1
## Importance of components:
##                             pc1      pc2      pc3      pc4 ...
## Standard deviation     1.513919 1.473768 1.459114 1.440635 ...
## Proportion of Variance 0.054570 0.051714 0.050691 0.049415 ...
## Cumulative Proportion  0.054570 0.106284 0.156975 0.206390 ...
##                            pc10     pc11     pc12     pc13 ...
## Standard deviation     1.007253 0.988724 0.985320 0.970453 ...
## Proportion of Variance 0.024156 0.023276 0.023116 0.022423 ...
## Cumulative Proportion  0.413816 0.437091 0.460207 0.482630 ...
##                            pc19     pc20     pc21     pc22 ...
## Standard deviation     0.931745 0.924207 0.917106 0.908494 ...
## Proportion of Variance 0.020670 0.020337 0.020026 0.019651 ...
## Cumulative Proportion  0.610376 0.630713 0.650739 0.670390 ...
##                            pc28     pc29     pc30     pc31 ...
## Standard deviation     0.865912 0.855036 0.845130 0.842818 ...
## Proportion of Variance 0.017852 0.017407 0.017006 0.016913 ...
```

```
## Cumulative Proportion  0.782288 0.799695 0.816701 0.833614 ...
##                            pc37     pc38     pc39     pc40 ...
## Standard deviation     0.796073 0.793781 0.780615 0.778612 ...
## Proportion of Variance 0.015089 0.015002 0.014509 0.014434 ...
## Cumulative Proportion  0.928796 0.943798 0.958307 0.972741 ...
##
##
## H2ODimReductionMetrics: pca
##
## No model metrics available for PCA
```

Naturally, the first PC (PC1) captures the most variance followed by PC2, then PC3, etc. We can identify which of our original features contribute to the PCs by assessing the loadings. The loadings for the first PC represent $\phi_{11}, \phi_{21}, \dots, \phi_{p1}$ in Equation (17.1). Thus, these loadings represent each features ***influence*** on the associated PC. If we plot the loadings for PC1 we see that the largest contributing features are mostly adult beverages (and apparently eating candy bars, smoking, and playing the lottery are also associated with drinking!).

```
my_pca@model$eigenvectors %>%
  as.data.frame() %>%
  mutate(feature = row.names(.)) %>%
  ggplot(aes(pc1, reorder(feature, pc1))) +
  geom_point()
```

We can also compare PCs against one another. For example, Figure 17.4 shows how the different features contribute to PC1 and PC2. We can see distinct groupings of features and how they contribute to both PCs. For example, adult beverages (e.g., whiskey and wine) have a positive contribution to PC1 but have a smaller and negative contribution to PC2. This means that transactions that include purchases of adult beverages tend to have larger than average values for PC1 but smaller than average for PC2.

```
my_pca@model$eigenvectors %>%
  as.data.frame() %>%
  mutate(feature = row.names(.)) %>%
  ggplot(aes(pc1, pc2, label = feature)) +
  geom_text()
```

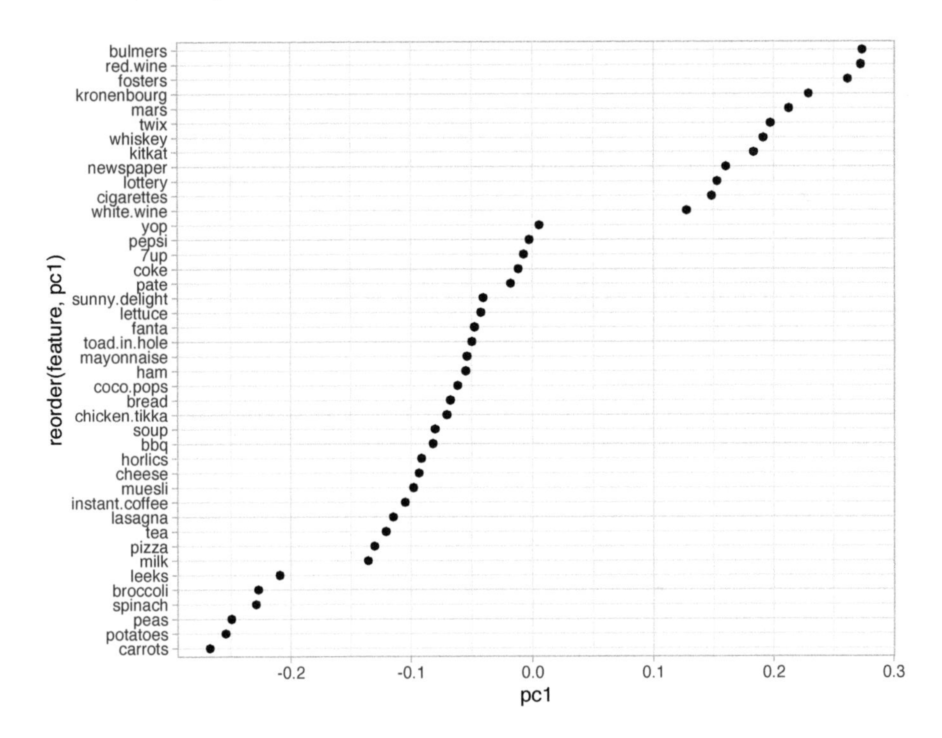

FIGURE 17.3: Feature loadings illustrating the influence that each variable has on the first principal component.

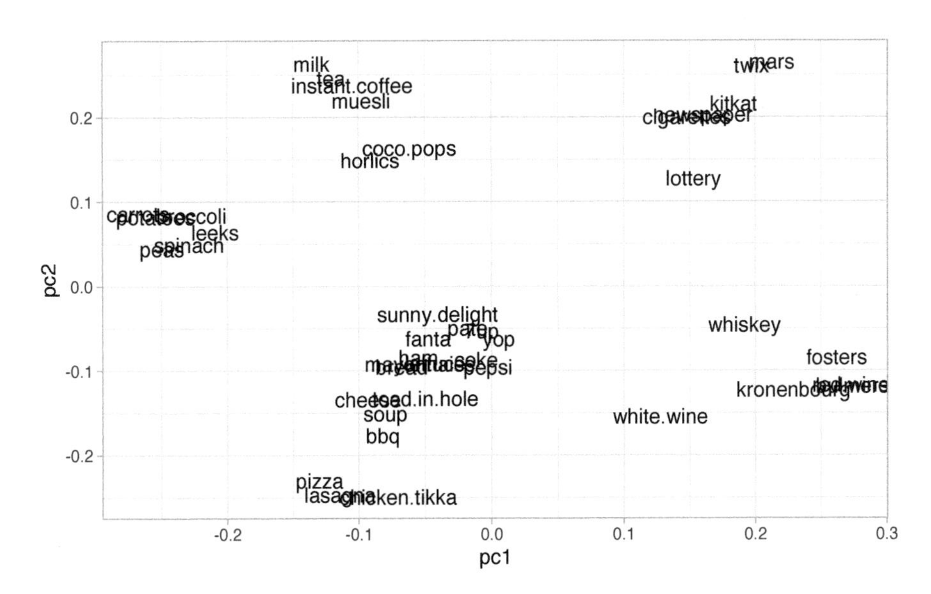

FIGURE 17.4: Feature contribution for principal components one and two.

17.5 Selecting the number of principal components

So far we have computed PCs and gained a little understanding of what the results initially tell us. However, a primary goal in PCA is dimension reduction (in this case, feature reduction). In essence, we want to come out of PCA with fewer components than original features, and with the caveat that these components explain us as much variation as possible about our data. But how do we decide how many PCs to keep? Do we keep the first 10, 20, or 40 PCs?

There are three common approaches in helping to make this decision:

1. Eigenvalue criterion
2. Proportion of variance explained criterion
3. Scree plot criterion

17.5.1 Eigenvalue criterion

The sum of the eigenvalues is equal to the number of variables entered into the PCA; however, the eigenvalues will range from greater than one to near zero. An eigenvalue of 1 means that the principal component would explain about one variable's worth of the variability. The rationale for using the eigenvalue criterion is that each component should explain at least one variable's worth of the variability, and therefore, the eigenvalue criterion states that only components with eigenvalues greater than 1 should be retained.

`h2o.prcomp()` automatically computes the standard deviations of the PCs, which is equal to the square root of the eigenvalues. Therefore, we can compute the eigenvalues easily and identify PCs where the sum of eigenvalues is greater than or equal to 1. Consequently, using this criteria would have us retain the first 10 PCs in `my_basket` (see Figure 17.5).

```
# Compute eigenvalues
eigen <- my_pca@model$importance["Standard deviation", ] %>%
  as.vector() %>%
  .^2

# Sum of all eigenvalues equals number of variables
sum(eigen)
## [1] 42

# Find PCs where the sum of eigenvalues is greater than or equal to 1
```

```
which(eigen >= 1)
##  [1]  1  2  3  4  5  6  7  8  9 10
```

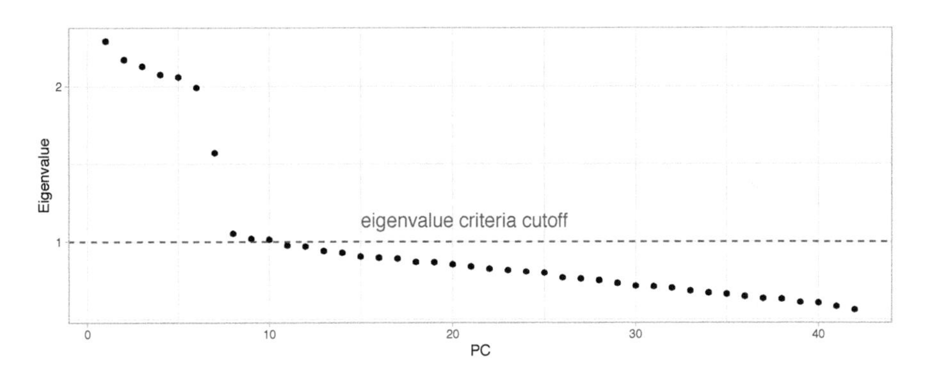

FIGURE 17.5: Eigenvalue criterion keeps all principal components where the sum of the eigenvalues are above or equal to a value of one.

17.5.2 Proportion of variance explained criterion

The *proportion of variance explained* (PVE) identifies the optimal number of PCs to keep based on the total variability that we would like to account for. Mathematically, the PVE for the m-th PC is calculated as:

$$PVE = \frac{\sum_{i=1}^{n}(\sum_{j=1}^{p}\phi_{jm}x_{ij})^2}{\sum_{j=1}^{p}\sum_{i=1}^{n}x_{ij}^2} \tag{17.3}$$

`h2o.prcomp()` provides us with the PVE and also the cumulative variance explained (CVE), so we just need to extract this information and plot it (see Figure 17.6).

```
# Extract and plot PVE and CVE
data.frame(
  PC  = my_pca@model$importance %>% seq_along(),
  PVE = my_pca@model$importance %>% .[2,] %>% unlist(),
  CVE = my_pca@model$importance %>% .[3,] %>% unlist()
) %>%
  tidyr::gather(metric, variance_explained, -PC) %>%
  ggplot(aes(PC, variance_explained)) +
  geom_point() +
  facet_wrap(~ metric, ncol = 1, scales = "free")
```

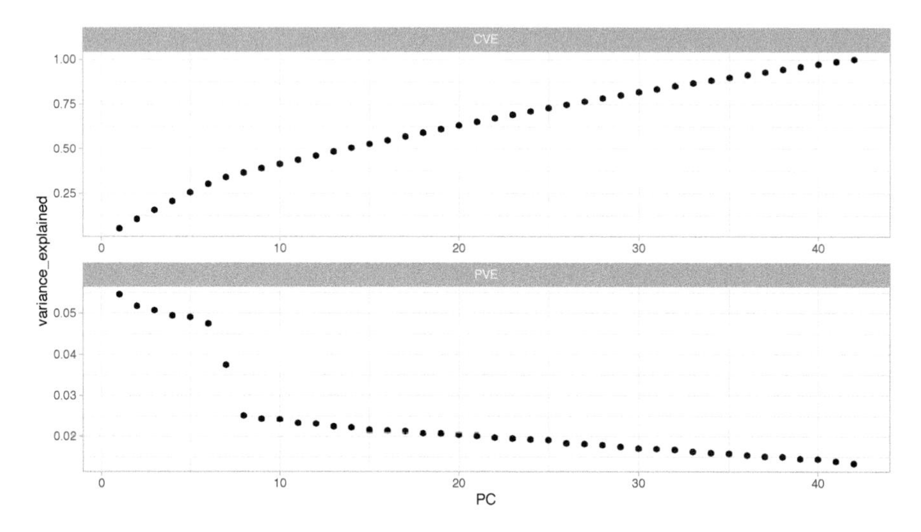

FIGURE 17.6: PVE criterion keeps all principal components that are above or equal to a pre-specified threshold of total variability explained.

The first PCt in our example explains 5.46% of the feature variability, and the second principal component explains 5.17%. Together, the first two PCs explain 10.63% of the variability. Thus, if an analyst desires to choose the number of PCs required to explain at least 75% of the variability in our original data then they would choose the first 27 components.

```
# How many PCs required to explain at least 75% of total variability
min(which(ve$CVE >= 0.75))
## [1] 27
```

What amount of variability is reasonable? This varies by application and the data being used. However, when the PCs are being used for descriptive purposes only, such as customer profiling, then the proportion of variability explained may be lower than otherwise. When the PCs are to be used as derived features for models downstream, then the PVE should be as much as can conveniently be achieved, given any constraints.

17.5.3 Scree plot criterion

A *scree plot* shows the eigenvalues or PVE for each individual PC. Most scree plots look broadly similar in shape, starting high on the left, falling rather quickly, and then flattening out at some point. This is because the first component usually explains much of the variability, the next few components explain a moderate amount, and the latter components only explain a small fraction of the overall variability. The scree plot criterion looks for the "elbow" in the curve and selects all components just before the line flattens out, which looks like eight in our example (see Figure 17.7).

```
data.frame(
  PC  = my_pca@model$importance %>% seq_along,
  PVE = my_pca@model$importance %>% .[2,] %>% unlist()
) %>%
  ggplot(aes(PC, PVE, group = 1, label = PC)) +
  geom_point() +
  geom_line() +
  geom_text(nudge_y = -.002)
```

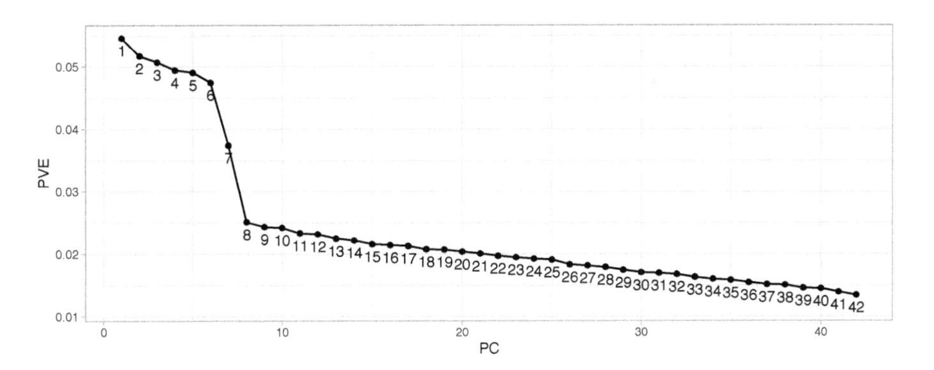

FIGURE 17.7: Scree plot criterion looks for the 'elbow' in the curve and keeps all principal components before the line flattens out.

17.6 Final thoughts

So how many PCs should we use in the `my_basket` example? The frank answer is that there is no one best method for determining how many components to use. In this case, differing criteria suggest to retain 8 (scree plot criterion), 10 (eigenvalue criterion), and 26 (based on a 75% of variance explained requirement) components. The number you go with depends on your end objective and analytic workflow. If we were merely trying to profile customers we would probably use 8 or 10, if we were performing dimension reduction to feed into a downstream predictive model we would likely retain 26 or more (the exact number being based on, for example, the CV results in the supervised modeling process). This is part of the challenge with unsupervised modeling, there is more subjectivity in modeling results and interpretation.

Traditional PCA has a few disadvantages worth keeping in mind. First, PCA can be highly affected by outliers. There have been many robust variants of PCA that act to iteratively discard data points that are poorly described by the initial components (see, for example, Luu et al. (2019) and Erichson et al. (2018)). In Chapter 18 we discuss an alternative dimension reduction procedure that takes outliers into consideration, and in Chapter 19 we illustrate a procedure to help identify outliers.

Also, note in Figures 17.1 and 17.2 that our PC directions are linear. Consequently, traditional PCA does not perform as well in very high dimensional space where complex nonlinear patterns often exist. Kernel PCA implements the kernel trick discussed in Chapter 14 and makes it possible to perform complex nonlinear projections of dimensionality reduction. See Karatzoglou et al. (2018) for an implementation of kernel PCA in R. Chapters 18 and 19 discuss two methods that allow us to reduce the feature space while also capturing nonlinearity.

18

Generalized Low Rank Models

The PCs constructed in PCA are linear in nature, which can cause deficiencies in its performance. This is much like the deficiency that linear regression has in capturing nonlinear relationships. Alternative approaches, known as matrix factorization methods have helped address this issue. More recently, however, a generalization of PCA and matrix factorization, called *generalized low rank models* (GLRMs) (Udell et al., 2016), has become a popular approach to dimension reduction.

18.1 Prerequisites

This chapter leverages the following packages:

```
# Helper packages
library(dplyr)    # for data manipulation
library(ggplot2)  # for data visualization
library(tidyr)    # for data reshaping

# Modeling packages
library(h2o)  # for fitting GLRMs
```

To illustrate GLRM concepts, we'll continue using the `my_basket` data set created in the previous chapter:

```
url <- "https://koalaverse.github.io/homlr/data/my_basket.csv"
my_basket <- readr::read_csv(url)
```

18.2 The idea

GLRMs reduce the dimension of a data set by producing a condensed vector representation for every row and column in the original data. Specifically, given a data set A with m rows and n columns, a GLRM consists of a decomposition of A into numeric matrices X and Y. The matrix X has the same number of rows as A, but only a small, *user-specified* number of columns k. The matrix Y has k rows and n columns, where n is equal to the total dimension of the embedded features in A. For example, if A has 4 numeric columns and 1 categorical column with 3 distinct levels (e.g., red, blue, and green), then Y will have 7 columns (due to one-hot encoding). When A contains only numeric features, the number of columns in A and Y are identical, as shown Eq. (18.1).

$$m\left\{\overbrace{\begin{bmatrix} \quad A \quad \end{bmatrix}}^{n}\right. \approx m\left\{\overbrace{\begin{bmatrix} X \end{bmatrix}}^{k}\right. \overbrace{\begin{bmatrix} \quad Y \quad \end{bmatrix}}^{n}\}k \tag{18.1}$$

Both X and Y have practical interpretations. Each row of Y is an archetypal feature formed from the columns of A, and each row of X corresponds to a row of A projected onto this smaller dimensional feature space. We can approximately reconstruct A from the matrix product $X \times Y$, which has rank k. The number k is chosen to be much less than both m and n (e.g., for 1 million rows and 2,000 columns of numeric data, k could equal 15). The smaller k is, the more compression we gain from our low rank representation.

To make this more concrete, lets look at an example using the `mtcars` data set (available from the built-in **datasets** package) where we have 32 rows and 11 features (see `?datasets::mtcars` for details):

```
head(mtcars)
##                    mpg cyl disp  hp drat   wt qsec vs
## Mazda RX4         21.0   6  160 110 3.90 2.62 16.5  0
## Mazda RX4 Wag     21.0   6  160 110 3.90 2.88 17.0  0
## Datsun 710        22.8   4  108  93 3.85 2.32 18.6  1
## Hornet 4 Drive    21.4   6  258 110 3.08 3.21 19.4  1
## Hornet Sportabout 18.7   8  360 175 3.15 3.44 17.0  0
## Valiant           18.1   6  225 105 2.76 3.46 20.2  1
##                   am gear carb
## Mazda RX4          1    4    4
## Mazda RX4 Wag      1    4    4
## Datsun 710         1    4    1
## Hornet 4 Drive     0    3    1
```

```
## Hornet Sportabout  0    3    2
## Valiant            0    3    1
```

`mtcars` represents our original matrix A. If we want to reduce matrix A to a rank of $k = 3$ then our objective is to produce two matrices X and Y that when we multiply them together produce a near approximation to the original values in A.

We call the condensed columns and rows in matrices X and X, respectively, "archetypes" because they are a representation of the original features and observations. The archetypes in X represent each observation projected onto the smaller dimensional space, and the archetypes in Y represent each feature projeced onto the smaller dimensional space.

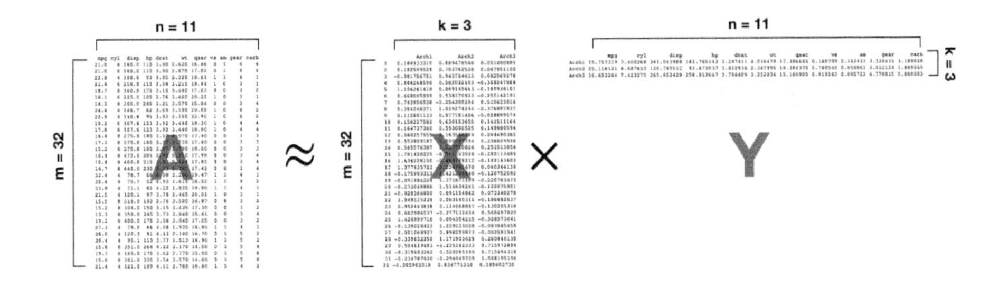

FIGURE 18.1: Example GLRM where we reduce the mtcars data set down to a rank of 3.

The resulting archetypes are similar in spirit to the PCs in PCA; as they are a reduced feature set that represents our original features. In fact, if our features truly behave in a linear and orthogonal manner than our archetypes produced by a GLRM will produce the same reduced feature set as PCA. However, if they are not linear, then GLRM will provide archetypes that are not necessarily orthogonal.

However, a few questions remain:

1. How does GLRM produce the archetype values?
2. How do you select the appropriate value for k?

We'll address these questions next.

18.3 Finding the lower ranks

18.3.1 Alternating minimization

There are a number of methods available to identify the optimal archetype values for each element in X and Y; however, the most common is based on *alternating minimization*. Alternating minimization simply alternates between minimizing some loss function for each feature in X and Y. In essence, random values are initially set for the archetype values in X and Y. The loss function is computed (more on this shortly), and then the archetype values in X are slightly adjusted via gradient descent (Section 12.2.2) and the improvement in the loss function is recorded. The archetype values in Y are then slightly adjusted and the improvement in the loss function is recorded. This process is continued until the loss function is optimized or some suitable stopping condition is reached.

18.3.2 Loss functions

As stated above, the optimal achetype values are selected based on minimizing some loss function. The loss function should reflect the intuitive notion of what it means to "fit the data well". The most common loss function is the *quadratic loss*. The quadratic loss is very similar to the SSE criterion (Section 2.6) for supervised learning models where we seek to minimize the squared difference between the actual value in our original data (matrix A) and the predicted value based on our achetypal matrices ($X \times Y$) (i.e., minimizing the squared residuals).

$$\text{quadratic loss} = minimize\left\{ \sum_{i=1}^{m} \sum_{i=1}^{n} \left(A_{i,j} - X_i Y_j\right)^2 \right\} \tag{18.2}$$

However, note that some loss functions are preferred over others in certain scenarios. For example, quadratic loss, similar to SSE, can be heavily influenced by outliers. If you do not want to emphasize outliers in your data set, or if you just want to try minimize errors for lower values in addition to higher values (e.g., trying to treat low-cost products equally as important as high-cost products) then you can use the Huber loss function. For brevity we do not show the Huber loss equation but it essentially applies quadratic loss to small errors and uses the absolute value for errors with larger values. Figure 18.2 illustrates how the quadratic and Huber loss functions differ.

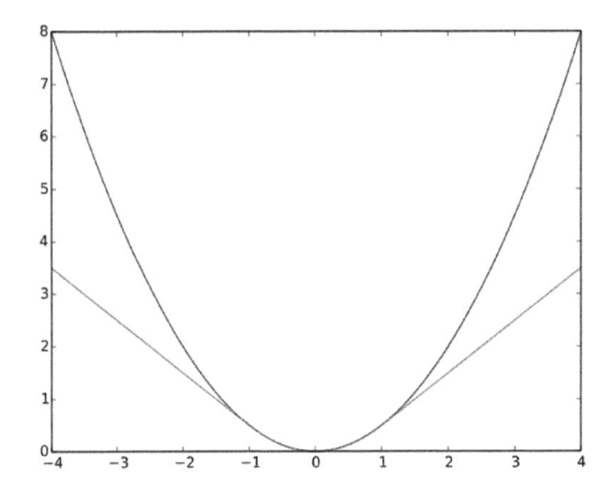

FIGURE 18.2: Huber loss (green) compared to quadratic loss (blue). The x-axis represents a particular value at $A_{i,j}$ and the y-axis represents the predicted value produced by $X_i Y_j$. Note how the Huber loss produces a linear loss while the quadratic loss produces much larger loss values as the residual value increases.

As with supervised learning, the choice of loss function should be driven by the business problem.

18.3.3 Regularization

Another important component to fitting GLRMs that you, the analyst, should consider is regularization. Much like the regularization discussed in Chapter 6, regularization applied to GLRMs can be used to constrain the size of the archetypal values in X (with $r_x(X)$ in the equation below) and/or Y (with $r_y(Y)$ in the equation below). This can help to create *sparse* X and/or Y matrices to mitigate the effect of negative features in the data (e.g., multicollinearity or excessive noise) which can help prevent overfitting.

If you're using GLRMs to merely describe your data and gain a better understanding of how observations and/or features are similar then you do not need to use regularization. If you are creating a model that will be used to assign new observations and/or features to these dimensions, or you want to use GLRMs for imputation then you should use regularization as it can make your model generalize better to unseen data.

$$\text{regularization} = minimize\left\{ \sum_{i=1}^{m} \sum_{i=1}^{n} \left(A_{i,j} - X_i Y_j\right)^2 + r_x\left(X\right) + r_y\left(Y\right) \right\} \tag{18.3}$$

As the above equation illustrates, we can regularize both matrices X and Y. However, when performing dimension reduction we are mainly concerned with finding a condensed representation of the features, or columns. Consequently, we'll be more concerned with regularizing the Y matrix ($r_y\left(Y\right)$). This regularizer encourages the Y matrix to be column-sparse so that many of the columns are all zero. Columns in Y that are zero, means that those features are likely uninformative in reproducing the original matrix A.

Even when we are focusing on dimension reduction, applying regularization to the X matrix can still improve performance. Consequently, it is good practice to compare different approaches.

There are several regularizers to choose from. You can use a ridge regularizer to retain all columns but force many of the values to be near zero. You can also use a LASSO regularizer which will help zero out many of the columns; the LASSO helps you perform automated feature selection. The non-negative regularizer can be used when your feature values should always be zero or positive (e.g., when performing market basket analysis).

The primary purpose of the regularizer is to minimize overfitting. Consequently, performing GRLMs without a regularizer will nearly always perform better than when using a regularizer if you are only focusing on a single data set. The choice of regularization should be led by statistical considerations, so that the model generalizes well to unseen data. This means you should always incorporate some form of CV to assess the performance of regularization on unseen data.

18.3.4 Selecting *k*

Lastly, how do we select the appropriate value for k? There are two main approaches, both of which will be illustrated in the section that follows. First, if you're using GLRMs to describe your data, then you can use many of the same approaches we discussed in Section 17.5 where we assess how different values of k minimize our loss function. If you are using GLRMs to produce a model that will be used to assign future observations to the reduced dimensions then you should use some form of CV.

18.4 Fitting GLRMs in R

h2o is the preferred package for fitting GLRMs in R. In fact, a few of the key researchers that developed the GLRM methodology helped develop the **h2o** implementation as well. Let's go ahead and start up **h2o**:

```
h2o.no_progress()  # turn off progress bars
h2o.init(max_mem_size = "5g")  # connect to H2O instance
```

18.4.1 Basic GLRM model

First, we convert our `my_basket` data frame to an appropriate **h2o** object before calling `h2o.glrm()`. The following performs a basic GLRM analysis with a quadratic loss function. A few arguments that `h2o.glrm()` provides includes:

- `k`: rank size desired, which declares the desired reduced dimension size of the features. This is specified by you, the analysts, but is worth tuning to see which size `k` performs best.
- `loss`: there are multiple loss functions to apply. The default is "quadratic".
- `regularization_x`: type of regularizer to apply to the X matrix.
- `regularization_y`: type of regularizer to apply to the Y matrix.
- `transform`: if your data are not already standardized this will automate this process for you. You can also normalize, demean, and descale.
- `max_iterations`: number of iterations to apply for the loss function to converge. Your goal should be to increase `max_iterations` until your loss function plot flatlines.
- `seed`: allows for reproducibility.
- `max_runtime_secs`: when working with large data sets this will limit the runtime for model training.

There are additional arguments that are worth exploring as you become more comfortable with `h2o.glrm()`. Some of the more useful ones include the magnitude of the regularizer applied (`gamma_x`, `gamma_y`). If you're working with ordinal features then `multi_loss = "Ordinal"` may be more appropriate. If you're working with very large data sets than `min_step_size` can be adjusted to speed up the learning process.

```
# convert data to h2o object
my_basket.h2o <- as.h2o(my_basket)

# run basic GLRM
basic_glrm <- h2o.glrm(
  training_frame = my_basket.h2o,
  k = 20,
  loss = "Quadratic",
  regularization_x = "None",
  regularization_y = "None",
  transform = "STANDARDIZE",
  max_iterations = 2000,
  seed = 123
)
```

We can check the results with `summary()`. Here, we see that our model converged at 901 iterations and the final quadratic loss value (SSE) is 31,004.59. We can also see how many iterations it took for our loss function to converge to its minimum:

```
# get top level summary information on our model
summary(basic_glrm)
## Model Details:
## ==============
##
## H2ODimReductionModel: glrm
## Model Key:  GLRM_model_R_1538746363268_1
## Model Summary:
##   number_of_iterations final_step_size final_objective_value
## 1                  901         0.36373           31004.59190
##
## H2ODimReductionMetrics: glrm
## ** Reported on training data. **
##
## Sum of Squared Error (Numeric):  31004.59
## Misclassification Error (Categorical):  0
## Number of Numeric Entries:  84000
## Number of Categorical Entries:  0
##
##
##
## Scoring History:
```

```
##            timestamp   duration iterations step_size   objective
## 1 2018-10-05 09:32:54  1.106 sec          0   0.66667 67533.03413
## 2 2018-10-05 09:32:54  1.149 sec          1   0.70000 49462.95972
## 3 2018-10-05 09:32:55  1.226 sec          2   0.46667 49462.95972
## 4 2018-10-05 09:32:55  1.257 sec          3   0.31111 49462.95972
## 5 2018-10-05 09:32:55  1.289 sec          4   0.32667 41215.38164
##
## ---
##              timestamp   duration iterations step_size   objective
## 896 2018-10-05 09:33:22 28.535 sec        895   0.28499 31004.59207
## 897 2018-10-05 09:33:22 28.566 sec        896   0.29924 31004.59202
## 898 2018-10-05 09:33:22 28.597 sec        897   0.31421 31004.59197
## 899 2018-10-05 09:33:22 28.626 sec        898   0.32992 31004.59193
## 900 2018-10-05 09:33:22 28.655 sec        899   0.34641 31004.59190
## 901 2018-10-05 09:33:22 28.685 sec        900   0.36373 31004.59190

# Create plot to see if results converged - if it did not converge,
# consider increasing iterations or using different algorithm
plot(basic_glrm)
```

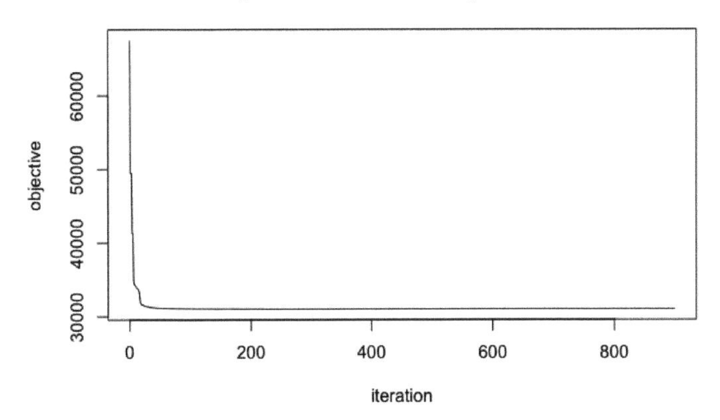

FIGURE 18.3: Loss curve for our GLRM model. The model converged at 901 iterations.

Our model object (`basic_glrm`) contains a lot of information (see everything it contains with `str(basic_glrm)`). Similar to `h2o.pca()`, we can see how much variance each archetype (aka principal component) explains by looking at the `model$importance` component:

```
# amount of variance explained by each archetype (aka "pc")
basic_glrm@model$importance
## Importance of components:
##                             pc1      pc2      pc3      pc4 ...
## Standard deviation     1.513919 1.473768 1.459114 1.440635 ...
## Proportion of Variance 0.054570 0.051714 0.050691 0.049415 ...
## Cumulative Proportion  0.054570 0.106284 0.156975 0.206390 ...
##                             pc8      pc9     pc10     pc11 ...
## Standard deviation     1.026387 1.010238 1.007253 0.988724 ...
## Proportion of Variance 0.025083 0.024300 0.024156 0.023276 ...
## Cumulative Proportion  0.365360 0.389659 0.413816 0.437091 ...
##                            pc15     pc16     pc17     pc18 ...
## Standard deviation     0.951610 0.947978 0.944826 0.932943 ...
## Proportion of Variance 0.021561 0.021397 0.021255 0.020723 ...
## Cumulative Proportion  0.526331 0.547728 0.568982 0.589706 ...
```

Consequently, we can use this information just like we did in the PCA chapter to determine how many components to keep (aka how large should our k be). For example, the following provides nearly the same results as we saw in Section 17.5.2.

When your data aligns to the linearity and orthogonal assumptions made by PCA, the default GLRM model will produce nearly the exact same results regarding variance explained. However, how features align to the archetypes will be different than how features align to the PCs in PCA.

```
data.frame(
    PC  = basic_glrm@model$importance %>% seq_along(),
    PVE = basic_glrm@model$importance %>% .[2,] %>% unlist(),
    CVE = basic_glrm@model$importance %>% .[3,] %>% unlist()
) %>%
    gather(metric, variance_explained, -PC) %>%
    ggplot(aes(PC, variance_explained)) +
    geom_point() +
    facet_wrap(~ metric, ncol = 1, scales = "free")
```

We can also extract how each feature aligns to the different archetypes by looking at the `model$archetypes` component:

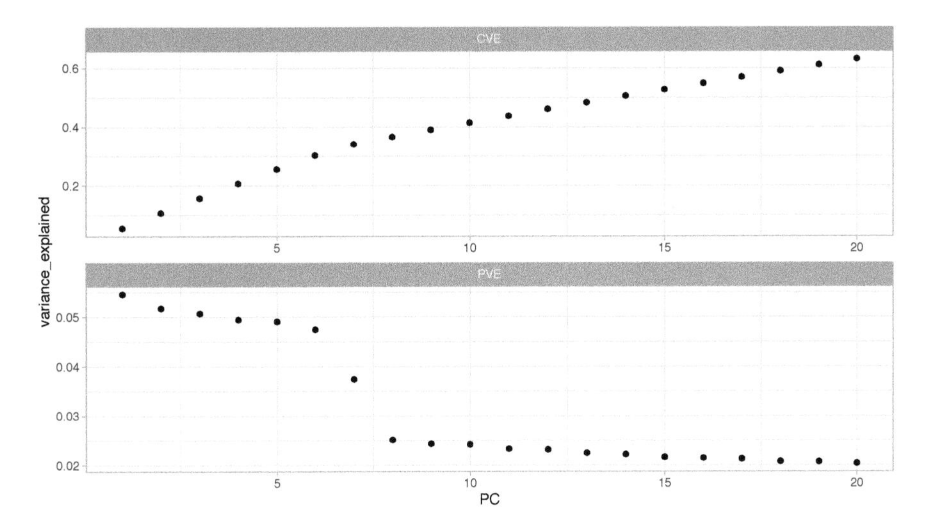

FIGURE 18.4: Variance explained by the first 20 archetypes in our GLRM model.

```
t(basic_glrm@model$archetypes)[1:5, 1:5]
##              Arch1      Arch2      Arch3      Arch4       Arch5
## 7up     -0.5783538 -1.5705325  0.9906612 -0.9306704  0.17552643
## lasagna  0.2196728  0.1213954 -0.7068851  0.8436524  3.56206178
## pepsi   -0.2504310 -0.8156136 -0.7669562 -1.2551630 -0.47632696
## yop     -0.1856632  0.4000083 -0.4855958  1.1598919 -0.26142763
## redwine -0.1372589 -0.1059148 -0.9579530  0.4641668 -0.08539977
```

We can use this information to see how the different features contribute to Archetype 1 or compare how features map to multiple Archetypes (similar to how we did this in the PCA chapter). The following shows that many liquid refreshments (e.g., instant coffee, tea, horlics, and milk) contribute positively to archetype 1. We also see that some candy bars contribute strongly to archetype 2 but minimally, or negatively, to archetype 1. The results are displayed in Figure 18.5.

```
p1 <- t(basic_glrm@model$archetypes) %>%
  as.data.frame() %>%
  mutate(feature = row.names(.)) %>%
  ggplot(aes(Arch1, reorder(feature, Arch1))) +
  geom_point()
```

```
p2 <- t(basic_glrm@model$archetypes) %>%
  as.data.frame() %>%
  mutate(feature = row.names(.)) %>%
  ggplot(aes(Arch1, Arch2, label = feature)) +
  geom_text()

gridExtra::grid.arrange(p1, p2, nrow = 1)
```

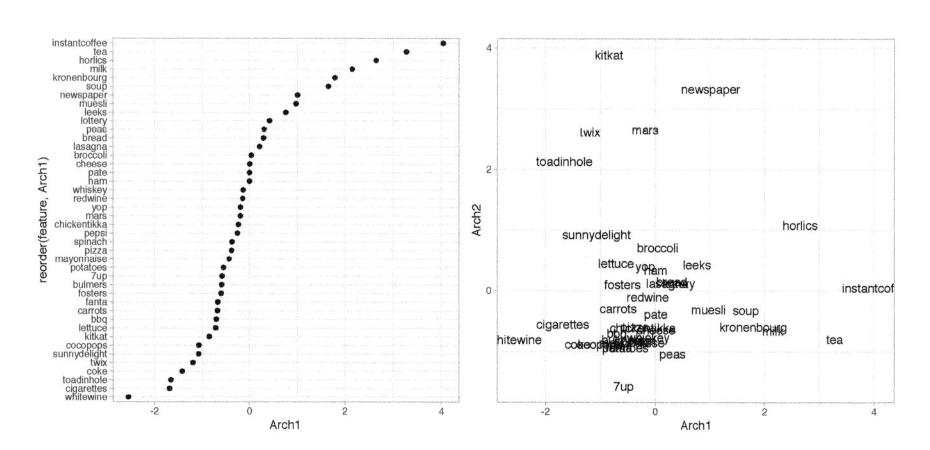

FIGURE 18.5: Feature contribution for archetype 1 and 2.

If we were to use the scree plot approach (Section 17.5.3) to determine k, we would decide on $k = 8$. Consequently, we would want to re-run our model with $k = 8$. We could then use `h2o.reconstruct()` and apply our model to a data set to see the predicted values. Below we see that our predicted values include negative numbers and non-integers. Considering our original data measures the counts of each product purchased we would need to apply some additional rounding logic to convert values to integers:

```
# Re-run model with k = 8
k8_glrm <- h2o.glrm(
  training_frame = my_basket.h2o,
  k = 8,
  loss = "Quadratic",
  regularization_x = "None",
  regularization_y = "None",
  transform = "STANDARDIZE",
  max_iterations = 2000,
  seed = 123
```

```
)

# Reconstruct to see how well the model did
my_reconstruction <- h2o.reconstruct(k8_glrm, my_basket.h2o,
                                     reverse_transform = TRUE)

# Raw predicted values
my_reconstruction[1:5, 1:4]
##   reconstr_7up reconstr_lasagna reconstr_pepsi reconstr_yop
## 1  0.025595726      -0.06657864    -0.03813350 -0.012225807
## 2 -0.041778553       0.02401056    -0.05225379 -0.052248809
## 3  0.012373600       0.04849545     0.05760424 -0.009878976
## 4  0.338875544       0.00577020     0.48763580  0.187669229
## 5  0.003869531       0.05394523     0.07655745 -0.010977765
##
## [5 rows x 4 columns]

# Round values to whole integers
my_reconstruction[1:5, 1:4] %>% round(0)
##   reconstr_7up reconstr_lasagna reconstr_pepsi reconstr_yop
## 1            0                0              0            0
## 2            0                0              0            0
## 3            0                0              0            0
## 4            0                0              0            0
## 5            0                0              0            0
##
## [5 rows x 4 columns]
```

18.4.2 Tuning to optimize for unseen data

A more sophisticated use of GLRMs is to create a model where the reduced archetypes will be used on future, unseen data. The preferred approach to deciding on a final model when you are going to use a GLRM to score future observations, is to perform a validation process to select the optimally tuned model. This will help your final model generalize better to unseen data.

As previously mentioned, when applying a GLRM model to unseen data, using a regularizer can help to reduce overfitting and help the model generalize better. Since our data represents all positive values (items purchases which can be 0 or any positive integer), we apply the non-negative regularizer. This will force all predicted values to at least be non-negative. We see this when we use `predict()` on the results.

If we compare the non-regularized GLRM model (`k8_glrm`) to our regularized model (k8_glrm_regularized), you will notice that the non-regularized model will almost always have a lower loss value. However, this is because the regularized model is being generalized more and is not overfitting to our training data, which should help improve on unseen data.

```
# Use non-negative regularization
k8_glrm_regularized <- h2o.glrm(
  training_frame = my_basket.h2o,
  k = 8,
  loss = "Quadratic",
  regularization_x = "NonNegative",
  regularization_y = "NonNegative",
  gamma_x = 0.5,
  gamma_y = 0.5,
  transform = "STANDARDIZE",
  max_iterations = 2000,
  seed = 123
)

# Show predicted values
predict(k8_glrm_regularized, my_basket.h2o)[1:5, 1:4]
##   reconstr_7up reconstr_lasagna reconstr_pepsi reconstr_yop
## 1     0.000000                0      0.0000000    0.0000000
## 2     0.000000                0      0.0000000    0.0000000
## 3     0.000000                0      0.0000000    0.0000000
## 4     0.609656                0      0.6311428    0.4565658
## 5     0.000000                0      0.0000000    0.0000000
##
## [5 rows x 4 columns]

# Compare regularized versus non-regularized loss
par(mfrow = c(1, 2))
plot(k8_glrm)
plot(k8_glrm_regularized)
```

GLRM models behave much like supervised models where there are several hyperparameters that can be tuned to optimize performance. For example, we can choose from a combination of multiple regularizers, we can adjust the magnitude of the regularization (i.e., the gamma_* parameters), and we can even tune the rank k.

Unfortunately, **h2o** does not currently provide an automated tuning grid option, such as h2o.grid() which can be applied to supervised learning models. To perform a grid search with GLRMs, we need to create our own custom process.

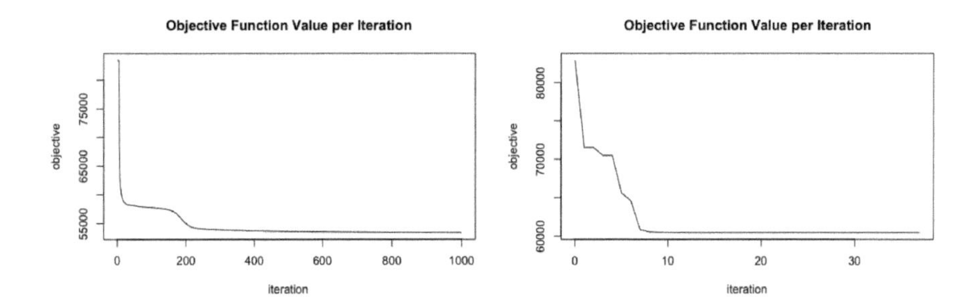

FIGURE 18.6: Loss curve for original GLRM model that does not include regularization (left) compared to a GLRM model with regularization (right).

First, we create training and validation sets so that we can use the validation data to see how well each hyperparameter setting does on unseen data. Next, we create a tuning grid that contains 225 combinations of hyperparameters. For this example, we're going to assume we want $k = 8$ and we only want to tune the type and magnitude of the regularizers. Lastly, we create a `for` loop to go through each hyperparameter combination, apply the given model, assess on the model's performance on the hold out validation set, and extract the error metric.

The squared error loss ranges from as high as 58,908 down to 13,371. This is a significant reduction in error. We see that the best models all have errors in the 13,700+ range and the majority of them have a large (signaled by `gamma_x`) L1 (LASSO) regularizer on the X matrix and also a non-negative regularizer on the Y matrix. However, the magnitude of the Y matrix regularizers (signaled by `gamma_y`) has little to no impact.

The following tuning and validation process took roughly 35 minutes to complete.

```
# Split data into train & validation
split <- h2o.splitFrame(my_basket.h2o, ratios = 0.75, seed = 123)
train <- split[[1]]
valid <- split[[2]]

# Create hyperparameter search grid
params <- expand.grid(
  regularization_x = c("None", "NonNegative", "L1"),
  regularization_y = c("None", "NonNegative", "L1"),
```

```
  gamma_x = seq(0, 1, by = .25),
  gamma_y = seq(0, 1, by = .25),
  error = 0,
  stringsAsFactors = FALSE
  )

# Perform grid search
for(i in seq_len(nrow(params))) {

  # Create model
  glrm_model <- h2o.glrm(
    training_frame = train,
    k = 8,
    loss = "Quadratic",
    regularization_x = params$regularization_x[i],
    regularization_y = params$regularization_y[i],
    gamma_x = params$gamma_x[i],
    gamma_y = params$gamma_y[i],
    transform = "STANDARDIZE",
    max_runtime_secs = 1000,
    seed = 123
  )

  # Predict on validation set and extract error
  validate <- h2o.performance(glrm_model, valid)
  params$error[i] <- validate@metrics$numerr
}

# Look at the top 10 models with the lowest error rate
params %>%
  arrange(error) %>%
  head(10)
##    regularization_x regularization_y gamma_x gamma_y    error
## 1                L1      NonNegative    1.00    0.25 13731.81
## 2                L1      NonNegative    1.00    0.50 13731.81
## 3                L1      NonNegative    1.00    0.75 13731.81
## 4                L1      NonNegative    1.00    1.00 13731.81
## 5                L1      NonNegative    0.75    0.25 13746.77
## 6                L1      NonNegative    0.75    0.50 13746.77
## 7                L1      NonNegative    0.75    0.75 13746.77
## 8                L1      NonNegative    0.75    1.00 13746.77
## 9                L1             None    0.75    0.00 13750.79
## 10               L1               L1    0.75    0.00 13750.79
```

Once we identify the optimal model, we'll want to re-run this on the entire training data set. We can then score new unseen observations with this model, which tells us based on their buying behavior and how this behavior aligns to the $k = 8$ dimensions in our model, what products are they're likely to buy and would be good opportunities to market to them.

```
# Apply final model with optimal hyperparamters
final_glrm_model <- h2o.glrm(
  training_frame = my_basket.h2o,
  k = 8,
  loss = "Quadratic",
  regularization_x = "L1",
  regularization_y = "NonNegative",
  gamma_x = 1,
  gamma_y = 0.25,
  transform = "STANDARDIZE",
  max_iterations = 2000,
  seed = 123
)

# Two new observations to score
new_observations <- as.h2o(sample_n(my_basket, 2))

# Basic scoring
score <- predict(final_glrm_model, new_observations) %>% round(0)
score[, 1:4]
##   reconstr_7up reconstr_lasagna reconstr_pepsi reconstr_yop
## 1            0                0              0            0
## 2            0                1              0            0
##
## [2 rows x 4 columns]
```

18.5 Final thoughts

GLRMs are an extension of the well-known matrix factorization methods such as PCA. While PCA is limited to numeric data, GLRMs can handle mixed numeric, categorical, ordinal, and boolean data with an arbitrary number of missing values. It allows the user to apply regularization to X and Y, imposing restrictions like non-negativity appropriate to a particular data science context. Thus, it is an extremely flexible approach for analyzing and interpreting heterogeneous data sets. Although this chapter focused on using GLRMs for dimension/feature reduction, GLRMs can also be used for clustering, missing data imputation, compute memory reduction, and speed improvements.

19

Autoencoders

An autoencoder is a neural network that is trained to learn efficient representations of the input data (i.e., the features). Although a simple concept, these representations, called *codings*, can be used for a variety of dimension reduction needs, along with additional uses such as *anomaly detection* and *generative modeling*. Moreover, since autoencoders are, fundamentally, feedforward deep learning models (Chapter 13), they come with all the benefits and flexibility that deep learning models provide. Autoencoders have been around for decades (e.g., LeCun (1987); Bourlard and Kamp (1988); Hinton and Zemel (1994)) and this chapter will discuss the most popular autoencoder architectures; however, this domain continues to expand quickly so we conclude the chapter by highlighting alternative autoencoder architectures that are worth exploring on your own.

19.1 Prerequisites

For this chapter we'll use the following packages:

```
# Helper packages
library(dplyr)    # for data manipulation
library(ggplot2)  # for data visualization

# Modeling packages
library(h2o)  # for fitting autoencoders
```

To illustrate autoencoder concepts we'll continue with the `mnist` data set from previous chapters:

```
mnist <- dslabs::read_mnist()
```

```
names(mnist)
## [1] "train" "test"
```

Since we will be using **h2o** we'll also go ahead and initialize our H2O session:

```
h2o.no_progress()  # turn off progress bars
h2o.init(max_mem_size = "5g")  # initialize H2O instance
```

19.2 Undercomplete autoencoders

An autoencoder has a structure very similar to a feedforward neural network (aka multi-layer perceptron—MLP); however, the primary difference when using in an unsupervised context is that the number of neurons in the output layer are equal to the number of inputs. Consequently, in its simplest form, an autoencoder is using hidden layers to try to re-create the inputs. We can describe this algorithm in two parts: (1) an *encoder* function ($Z = f(X)$) that converts X inputs to Z codings and (2) a *decoder* function ($X' = g(Z)$) that produces a reconstruction of the inputs (X').

For dimension reduction purposes, the goal is to create a reduced set of codings that adequately represents X. Consequently, we constrain the hidden layers so that the number of neurons is less than the number of inputs. An autoencoder whose internal representation has a smaller dimensionality than the input data is known as an *undercomplete autoencoder*, represented in Figure 19.1. This compression of the hidden layers forces the autoencoder to capture the most dominant features of the input data and the representation of these signals are captured in the codings.

To learn the neuron weights and, thus the codings, the autoencoder seeks to minimize some loss function, such as mean squared error (MSE), that penalizes X' for being dissimilar from X:

$$\text{minimize } L = f(X, X') \tag{19.1}$$

19.2.1 Comparing PCA to an autoencoder

When the autoencoder uses only linear activation functions (reference Section 13.4.2.1) and the loss function is MSE, then it can be shown that the

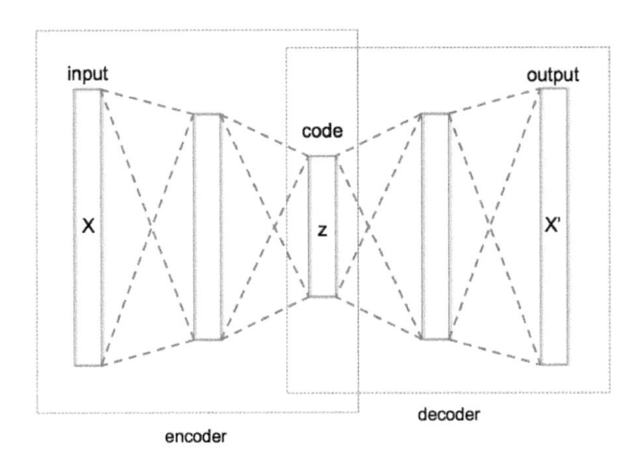

FIGURE 19.1: Schematic structure of an undercomplete autoencoder with three fully connected hidden layers (Wikipedia contributors, nda).

autoencoder reduces to PCA. When nonlinear activation functions are used, autoencoders provide nonlinear generalizations of PCA.

The following demonstrates our first implementation of a basic autoencoder. When using **h2o** you use the same `h2o.deeplearning()` function that you would use to train a neural network; however, you need to set `autoencoder = TRUE`. We use a single hidden layer with only two codings. This is reducing 784 features down to two dimensions; although not very realistic, it allows us to visualize the results and gain some intuition on the algorithm. In this example we use a hyperbolic tangent activation function which has a nonlinear sigmoidal shape. To extract the reduced dimension codings, we use `h2o.deepfeatures()` and specify the layer of codings to extract.

The MNIST data set is very sparse; in fact, over 80% of the elements in the MNIST data set are zeros. When you have sparse data such as this, using `sparse = TRUE` enables **h2o** to more efficiently handle the input data and speed up computation.

```
# Convert mnist features to an h2o input data set
features <- as.h2o(mnist$train$images)

# Train an autoencoder
```

```
ae1 <- h2o.deeplearning(
  x = seq_along(features),
  training_frame = features,
  autoencoder = TRUE,
  hidden = 2,
  activation = 'Tanh',
  sparse = TRUE
)

# Extract the deep features
ae1_codings <- h2o.deepfeatures(ae1, features, layer = 1)
ae1_codings
##      DF.L1.C1    DF.L1.C2
## 1 -0.1558956 -0.06456967
## 2  0.3778544 -0.61518649
## 3  0.2002303  0.31214266
## 4 -0.6955515  0.13225607
## 5  0.1912538  0.59865392
## 6  0.2310982  0.20322605
##
## [60000 rows x 2 columns]
```

The reduced codings we extract are sometimes referred to as deep features (DF) and they are similar in nature to the principal components for PCA and archetypes for GLRMs. In fact, we can project the MNIST response variable onto the reduced feature space and compare our autoencoder to PCA. Figure 19.2 illustrates how the nonlinearity of autoencoders can help to isolate the signals in the features better than PCA.

19.2.2 Stacked autoencoders

Autoencoders are often trained with only a single hidden layer; however, this is not a requirement. Just as we illustrated with feedforward neural networks, autoencoders can have multiple hidden layers. We refer to autoencoders with more than one layer as *stacked autoencoders* (or *deep autoencoders*). Adding additional layers to autoencoders can have advantages. Adding additional depth can allow the codings to represent more complex, nonlinear relationships at a reduced computational cost. In fact, Hinton and Salakhutdinov (2006) show that deeper autoencoders often yield better data compression than shallower, or linear autoencoders. However, this is not always the case as we'll see shortly.

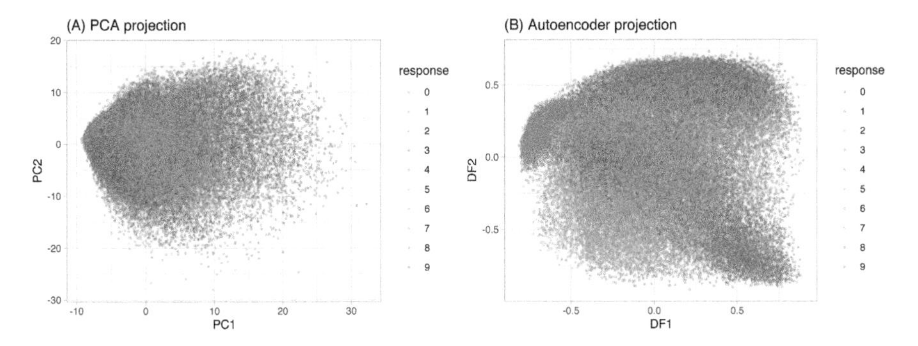

FIGURE 19.2: MNIST response variable projected onto a reduced feature space containin only two dimensions. PCA (left) forces a linear projection whereas an autoencoder with non-linear activation functions allows non-linear project.

One must be careful not to make the autoencoder too complex and powerful as you can run the risk of nearly reconstructing the inputs perfectly while not identifying the salient features that generalize well.

As you increase the depth of an autoencoder, the architecture typically follows a symmetrical pattern.[1] For example, Figure 19.3 illustrates three different undercomplete autoencoder architectures exhibiting symmetric hidden layers.

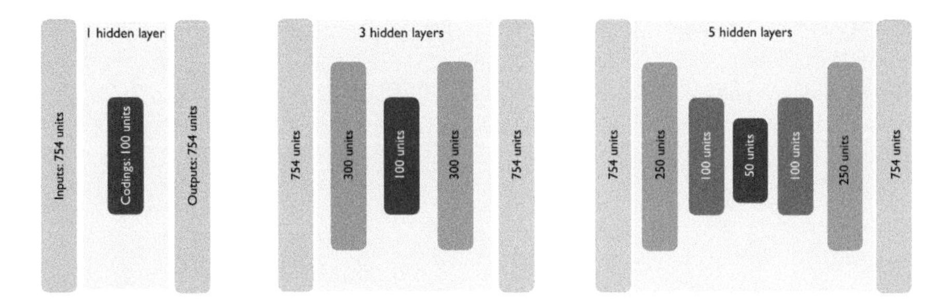

FIGURE 19.3: As you add hidden layers to autoencoders, it is common practice to have symmetric hidden layer sizes between the encoder and decoder layers.

So how does one find the right autencoder architecture? We can use the same grid search procedures we've discussed throughout the supervised learning section of the book. To illustrate, the following code examines five undercomplete

[1] This is not a hard requirement but, rather, a commonly applied practice that can be considered a good starting point.

autoencoder architectures. In this example we find that less depth provides the optimal MSE as a single hidden layer with 100 deep features has the lowest MSE of 0.007.

The following grid search took a little over 9 minutes.

```
# Hyperparameter search grid
hyper_grid <- list(hidden = list(
  c(50),
  c(100),
  c(300, 100, 300),
  c(100, 50, 100),
  c(250, 100, 50, 100, 250)
))

# Execute grid search
ae_grid <- h2o.grid(
  algorithm = 'deeplearning',
  x = seq_along(features),
  training_frame = features,
  grid_id = 'autoencoder_grid',
  autoencoder = TRUE,
  activation = 'Tanh',
  hyper_params = hyper_grid,
  sparse = TRUE,
  ignore_const_cols = FALSE,
  seed = 123
)

# Print grid details
h2o.getGrid('autoencoder_grid', sort_by = 'mse', decreasing = FALSE)
## H2O Grid Details
## ================
##
## Grid ID: autoencoder_grid
## Used hyper parameters:
##   -  hidden
## Number of models: 5
## Number of failed models: 0
##
## Hyper-Parameter Search Summary: ordered by increasing mse
```

```
##                         hidden                   model_ids        mse
## 1                        [100] autoencoder_grid3_model_2  0.0067464
## 2              [300, 100, 300] autoencoder_grid3_model_3  0.0083050
## 3               [100, 50, 100] autoencoder_grid3_model_4 0.01121531
## 4                         [50] autoencoder_grid3_model_1 0.01245011
## 5 [250, 100, 50, 100, 250] autoencoder_grid3_model_5 0.01441028
```

19.2.3 Visualizing the reconstruction

So how well does our autoencoder reconstruct the original inputs? The MSE provides us an overall error assessment but we can also directly compare the inputs and reconstructed outputs. Figure 19.4 illustrates this comparison by sampling a few test images, predicting the reconstructed pixel values based on our optimal autoencoder, and plotting the original versus reconstructed digits. The objective of the autoencoder is to capture the salient features of the images where any differences should be negligible; Figure 19.4 illustrates that our autencoder does a pretty good job of this.

```
# Get sampled test images
index <- sample(1:nrow(mnist$test$images), 4)
sampled_digits <- mnist$test$images[index, ]
colnames(sampled_digits) <- paste0("V", seq_len(ncol(sampled_digits)))

# Predict reconstructed pixel values
best_model_id <- grid_perf@model_ids[[1]]
best_model <- h2o.getModel(best_model_id)
recon_digits <- predict(best_model, as.h2o(sampled_digits))
names(recon_digits) <- paste0("V", seq_len(ncol(recon_digits)))

combine <- rbind(sampled_digits, as.matrix(recon_digits))

# Plot original versus reconstructed
par(mfrow = c(1, 3), mar=c(1, 1, 1, 1))
layout(matrix(seq_len(nrow(combine)), 4, 2, byrow = FALSE))
for(i in seq_len(nrow(combine))) {
  image(matrix(combine[i, ], 28, 28)[, 28:1], xaxt="n", yaxt="n")
}
```

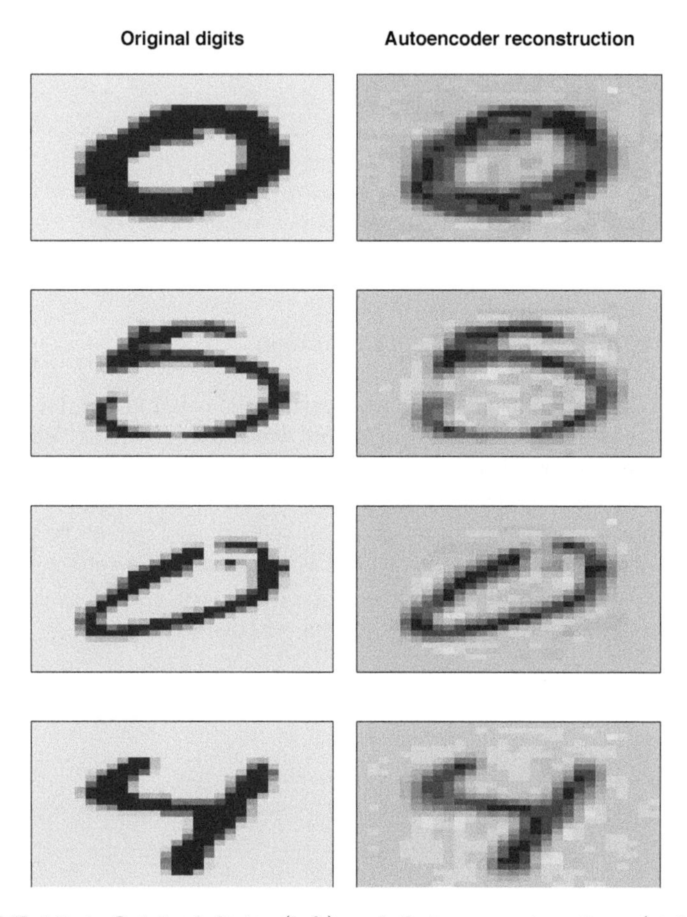

FIGURE 19.4: Original digits (left) and their reconstructions (right).

19.3 Sparse autoencoders

Sparse autoencoders are used to pull out the most influential feature representations. This is beneficial when trying to understand what are the most unique features of a data set. It's useful when using autoencoders as inputs to downstream supervised models as it helps to highlight the unique signals across the features.

Recall that neurons in a network are considered active if the threshold exceeds certain capacity. Since a Tanh activation function is S-curved from -1 to 1, we consider a neuron active if the output value is closer to 1 and inactive if its output is closer to -1.[2] Incorporating *sparsity* forces more neurons to be inactive. This requires the autoencoder to represent each input as a combination of a smaller number of activations. To incorporate sparsity, we must first understand the actual sparsity of the coding layer. This is simply the average activation of the coding layer as a function of the activation used (A) and the inputs supplied (X) as illustrated in Equation (19.2).

$$\hat{\rho} = \frac{1}{m}\sum_{i=1}^{m} A(X) \tag{19.2}$$

For our current `best_model` with 100 codings, the sparsity level is approximately zero:

```
ae100_codings <- h2o.deepfeatures(best_model, features, layer = 1)
ae100_codings %>%
    as.data.frame() %>%
    tidyr::gather() %>%
    summarize(average_activation = mean(value))
##   average_activation
## 1        -0.00677801
```

This means, on average, the coding neurons are active half the time which is illustrated in Figure 19.5.

Sparse autoencoders attempt to enforce the constraint $\hat{\rho} = \rho$ where ρ is a *sparsity parameter*. This penalizes the neurons that are too active, forcing them to activate less. To achieve this we add an extra penalty term to our objective function in Equation (19.1). The most commonly used penalty is known as the *Kullback-Leibler divergence* (KL divergence), which will measure the divergence between the target probability ρ that a neuron in the coding layer will activate, and the actual probability as illustrated in Equation (19.3).

$$\sum \rho \log \frac{\rho}{\hat{\rho}} + (1-\rho)\log\frac{1-\rho}{1-\hat{\rho}} \tag{19.3}$$

This penalty term is commonly written as Equation (19.4)

$$\sum\sum \mathrm{KL}(\rho || \hat{\rho}). \tag{19.4}$$

Similar to the ridge and LASSO penalties discussed in Section 6.2, we add this penalty to our objective function and incorporate a parameter (β) to control

[2]If using a sigmoid function then we can think of a neuron as being inactive when its output values are close to 0.

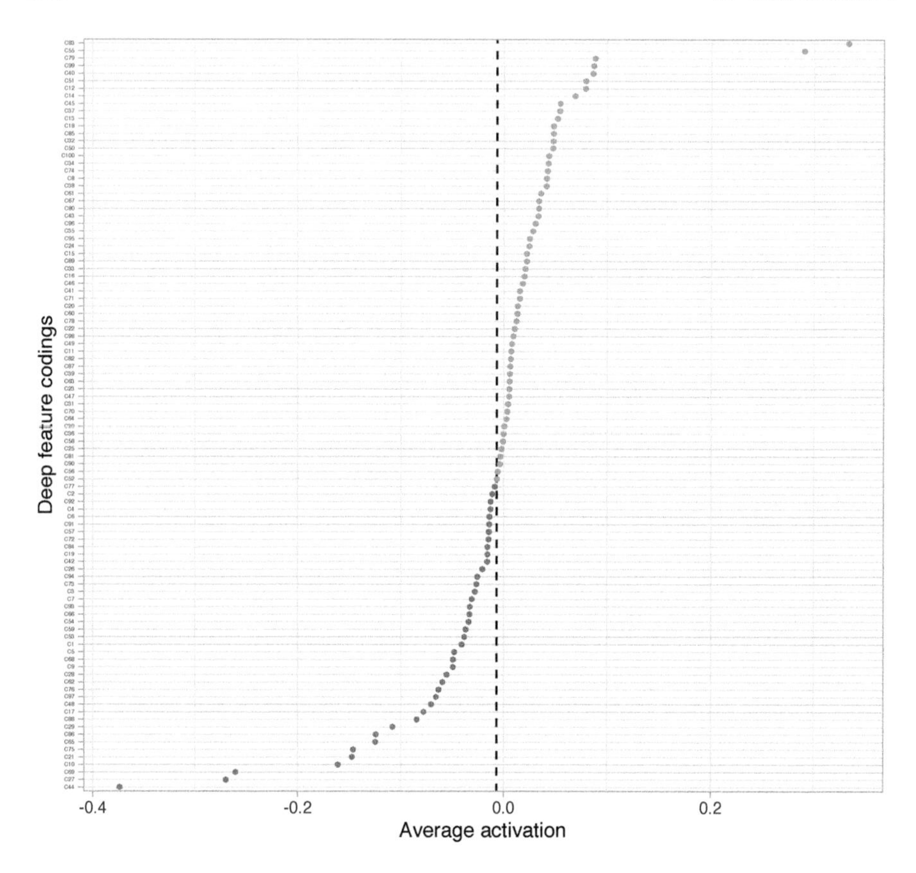

FIGURE 19.5: The average activation of the coding neurons in our default autoencoder using a Tanh activation function.

the weight of the penalty. Consequently, our revised loss function with sparsity induced is

$$\text{minimize}\left(L = f(X, X') + \beta \sum \text{KL}(\rho || \hat{\rho})\right). \tag{19.5}$$

Assume we want to induce sparsity with our current autoencoder that contains 100 codings. We need to specify two parameters: ρ and β. In this example, we'll just induce a little sparsity and specify $\rho = -0.1$ by including `average_activation = -0.1`. And since β could take on multiple values we'll do a grid search across different `sparsity_beta` values. Our results indicate that $\beta = 0.01$ performs best in reconstructing the original inputs.

The weight that controls the relative importance of the sparsity loss (β) is a hyperparameter that needs to be tuned. If this weight is too high, the model will stick closely to the target sparsity but suboptimally reconstruct the inputs. If the weight is too low, the model will mostly ignore the sparsity objective. A grid search helps to find the right balance.

```
# Hyperparameter search grid
hyper_grid <- list(sparsity_beta = c(0.01, 0.05, 0.1, 0.2))

# Execute grid search
ae_sparsity_grid <- h2o.grid(
  algorithm = 'deeplearning',
  x = seq_along(features),
  training_frame = features,
  grid_id = 'sparsity_grid',
  autoencoder = TRUE,
  hidden = 100,
  activation = 'Tanh',
  hyper_params = hyper_grid,
  sparse = TRUE,
  average_activation = -0.1,
  ignore_const_cols = FALSE,
  seed = 123
)

# Print grid details
h2o.getGrid('sparsity_grid', sort_by = 'mse', decreasing = FALSE)
## H2O Grid Details
## ================
##
## Grid ID: sparsity_grid
## Used hyper parameters:
##   -  sparsity_beta
## Number of models: 4
## Number of failed models: 0
##
## Hyper-Parameter Search Summary: ordered by increasing mse
##   sparsity_beta             model_ids                  mse
## 1          0.01 sparsity_grid_model_1 0.012982916169006953
## 2           0.2 sparsity_grid_model_4  0.01321464889160263
## 3          0.05 sparsity_grid_model_2  0.01337749148043942
## 4           0.1 sparsity_grid_model_3 0.013516631653257992
```

If we look at the average activation across our neurons now we see that it shifted to the left compared to Figure 19.5; it is now -0.108 as illustrated in Figure 19.6.

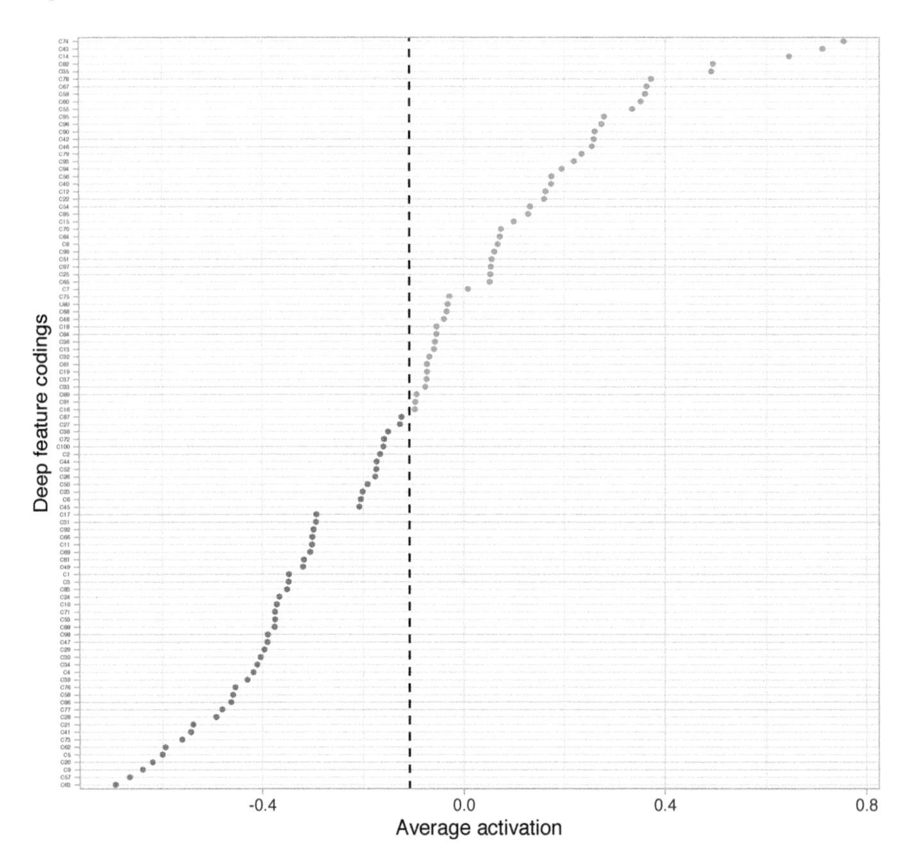

FIGURE 19.6: The average activation of the coding neurons in our sparse autoencoder is now -0.108.

The amount of sparsity you apply is dependent on multiple factors. When using autoencoders for descriptive dimension reduction, the level of sparsity is dependent on the level of insight you want to gain behind the most unique statistical features. If you're trying to understand the most essential characteristics that explain the features or images then a lower sparsity value is preferred. For example, Figure 19.7 compares the four sampled digits from the MNIST test set with a non-sparse autoencoder with a single layer of 100 codings using Tanh activation functions and a sparse autoencoder that constrains $\rho = -0.75$. Adding sparsity helps to highlight the features that are driving the uniqueness of these sampled digits. This is most pronounced with the number 5 where the sparse autoencoder reveals the primary focus is on the upper portion of the glyph.

If you are using autoencoders as a feature engineering step prior to downstream supervised modeling, then the level of sparsity can be considered a hyperparameter that can be optimized with a search grid.

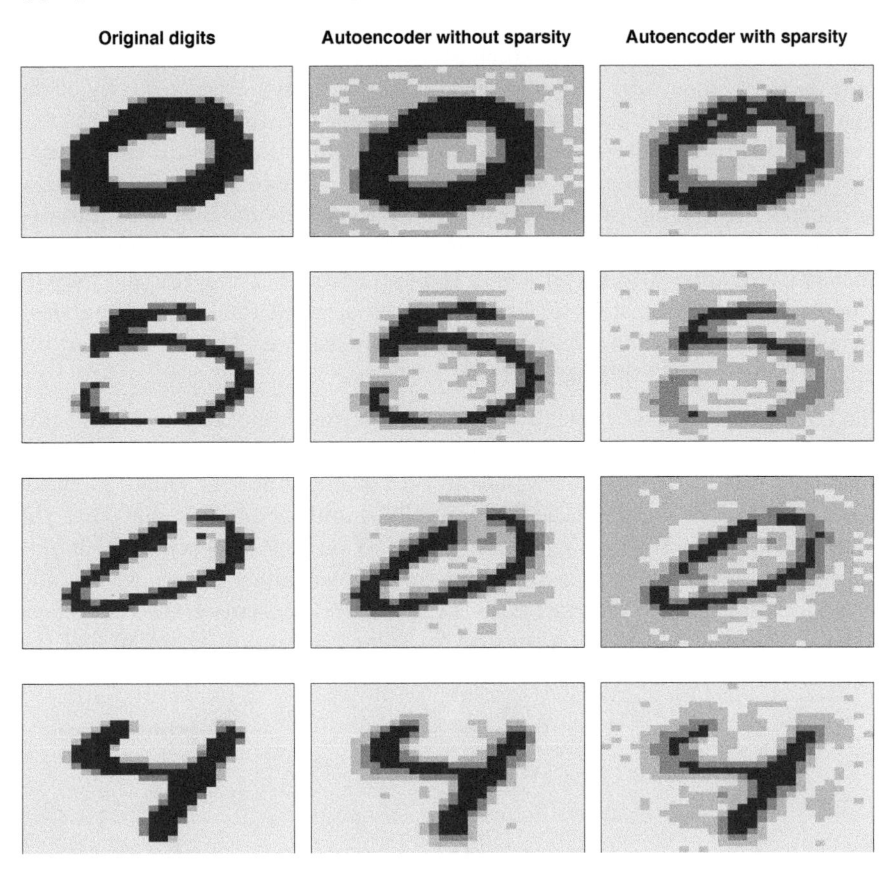

FIGURE 19.7: Original digits sampled from the MNIST test set (left), reconstruction of sampled digits with a non-sparse autoencoder (middle), and reconstruction with a sparse autoencoder (right).

In Section 19.2, we discussed how an undercomplete autoencoder is used to constrain the number of codings to be less than the number of inputs. This constraint prevents the autoencoder from learning the identify function, which would just create a perfect mapping of inputs to outputs and not learn anything about the features' salient characteristics. However, there are ways to prevent an autoencoder with more hidden units than inputs (known as an *overcomplete autoencoder*) from learning the identity function. Adding sparsity is one such approach (Poultney et al., 2007; Lee et al., 2008) and another is to add randomness in the transformation from input to reconstruction, which we discuss next.

19.4 Denoising autoencoders

The denoising autoencoder is a stochastic version of the autoencoder in which we train the autoencoder to reconstruct the input from a *corrupted* copy of the inputs. This forces the codings to learn more robust features of the inputs and prevents them from merely learning the identity function; even if the number of codings is greater than the number of inputs. We can think of a denoising autoencoder as having two objectives: (i) try to encode the inputs to preserve the essential signals, and (ii) try to undo the effects of a corruption process stochastically applied to the inputs of the autoencoder. The latter can only be done by capturing the statistical dependencies between the inputs. Combined, this denoising procedure allows us to implicitly learn useful properties of the inputs (Bengio et al., 2013).

The corruption process typically follows one of two approaches. We can randomly set some of the inputs (as many as half of them) to zero or one; most commonly it is setting random values to zero to imply missing values (Vincent et al., 2008). This can be done by manually imputing zeros or ones into the inputs or adding a dropout layer (reference Section 13.7.3) between the inputs and first hidden layer. Alternatively, for continuous-valued inputs, we can add pure Gaussian noise (Vincent, 2011). Figure 19.8 illustrates the differences between these two corruption options for a sampled input where 30% of the inputs were corrupted.

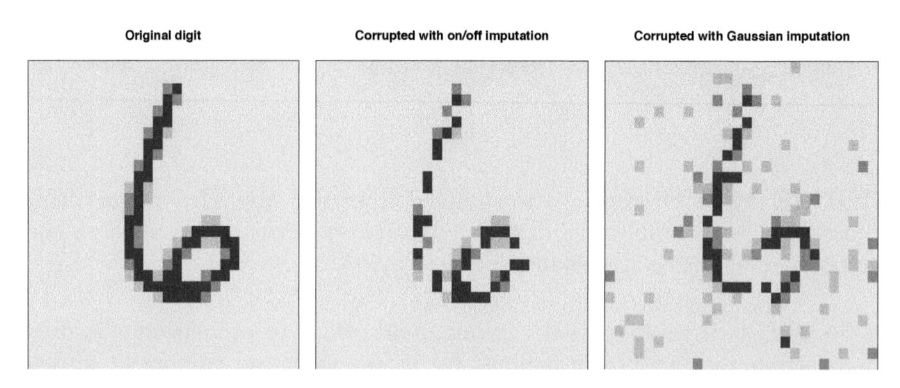

FIGURE 19.8: Original digit sampled from the MNIST test set (left), corrupted data with on/off imputation (middle), and corrupted data with Gaussian imputation (right).

Training a denoising autoencoder is nearly the same process as training a regular autoencoder. The only difference is we supply our corrupted inputs to `training_frame` and supply the non-corrupted inputs to `validation_frame`. The following code illustrates where we supply `training_frame` with inputs that have been corrupted with Gaussian noise (`inputs_currupted_gaussian`) and supply the original input data frame (`features`) to `validation_frame`. The remaining process stays, essentially, the same. We see that the validation MSE is 0.02 where in comparison our MSE of the same model without corrupted inputs was 0.006.

```
# Train a denoise autoencoder
denoise_ae <- h2o.deeplearning(
  x = seq_along(features),
  training_frame = inputs_currupted_gaussian,
  validation_frame = features,
  autoencoder = TRUE,
  hidden = 100,
  activation = 'Tanh',
  sparse = TRUE
)

# Print performance
h2o.performance(denoise_ae, valid = TRUE)
## H2OAutoEncoderMetrics: deeplearning
## ** Reported on validation data. **
##
## Validation Set Metrics:
## =====================
##
## MSE: (Extract with `h2o.mse`) 0.02048465
## RMSE: (Extract with `h2o.rmse`) 0.1431246
```

Figure 19.9 visualizes the effect of a denoising autoencoder. The left column shows a sample of the original digits, which are used as the validation data set. The middle column shows the Gaussian corrupted inputs used to train the model, and the right column shows the reconstructed digits after denoising. As expected, the denoising autoencoder does a pretty good job of mapping the corrupted data back to the original input.

19.5 Anomaly detection

We can also use autoencoders for anomaly detection (Sakurada and Yairi, 2014; Zhou and Paffenroth, 2017). Since the loss function of an autoencoder measures the reconstruction error, we can extract this information to identify

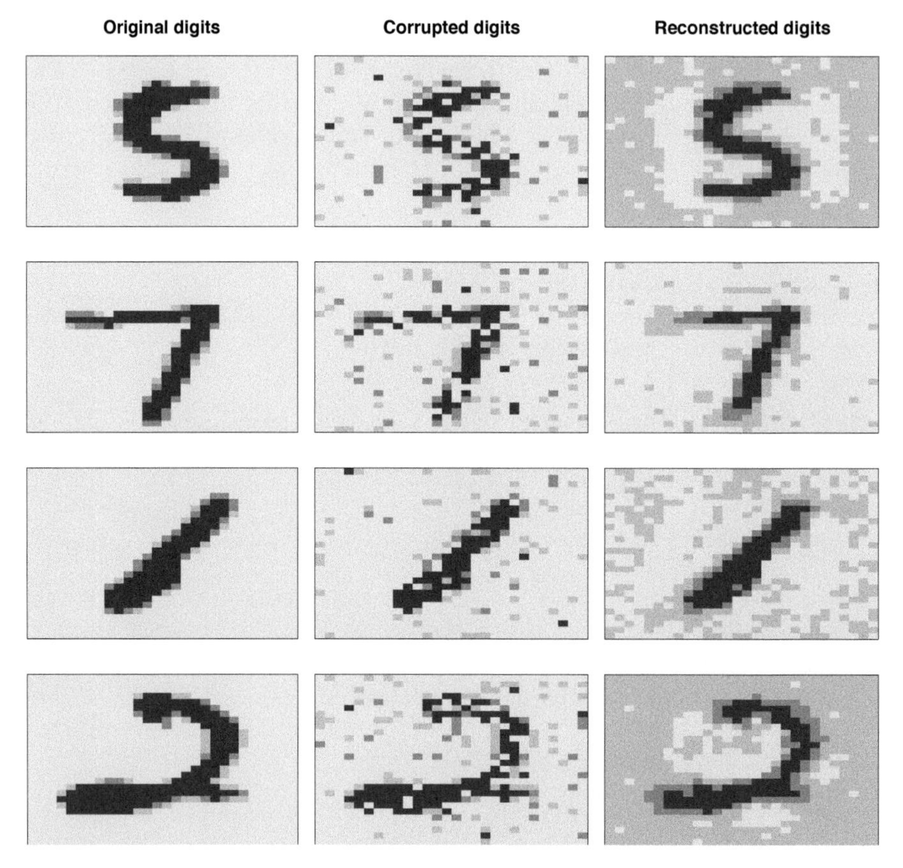

FIGURE 19.9: Original digits sampled from the MNIST test set (left), corrupted input digits (middle), and reconstructed outputs (right).

those observations that have larger error rates. These observations have feature attributes that differ significantly from the other features. We might consider such features as anomalous, or outliers.

To extract the reconstruction error with **h2o**, we use `h2o.anomaly()`. The following uses our undercomplete autoencoder with 100 codings from Section 19.2.2. We can see that the distribution of reconstruction errors range from near zero to over 0.03 with the average error being 0.006.

```
# Extract reconstruction errors
(reconstruction_errors <- h2o.anomaly(best_model, features))
##   Reconstruction.MSE
## 1        0.009879666
## 2        0.006485201
```

```
## 3          0.017470110
## 4          0.002339352
## 5          0.006077669
## 6          0.007171287
##
## [60000 rows x 1 column]

# Plot distribution
reconstruction_errors <- as.data.frame(reconstruction_errors)
ggplot(reconstruction_errors, aes(Reconstruction.MSE)) +
  geom_histogram()
```

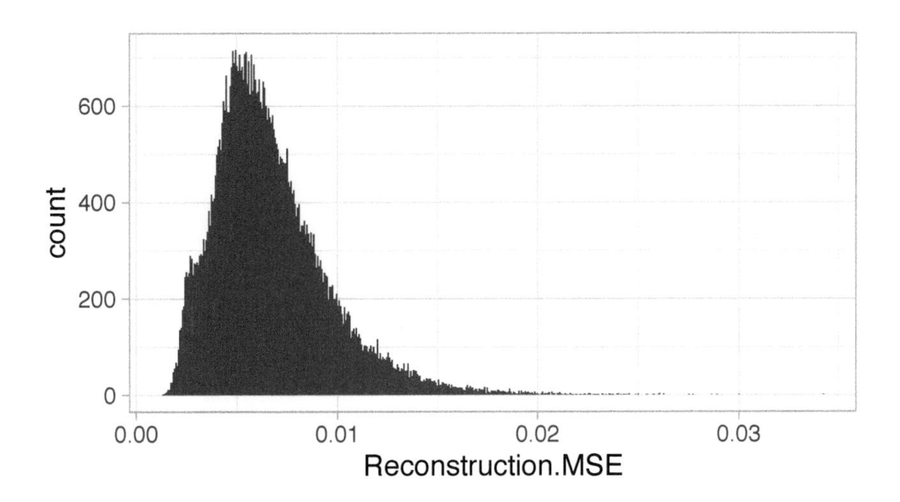

FIGURE 19.10: Distribution of reconstruction errors.

Figure 19.11 illustrates the actual and reconstructed digits for the observations with the five worst reconstruction errors. It is fairly intuitive why these observations have such large reconstruction errors as the corresponding input digits are poorly written.

In addition to identifying outliers, we can also use anomaly detection to identify unusual inputs such as fraudulent credit card transactions and manufacturing defects. Often, when performing anomaly detection, we retrain the autoencoder on a subset of the inputs that we've determined are a good representation of high quality inputs. For example, we may include all inputs that achieved a reconstruction error within the 75-th percentile and exclude the rest. We would then retrain an autoencoder, use that autoencoder on new input data, and if it exceeds a certain percentile declare the inputs as anomalous. However,

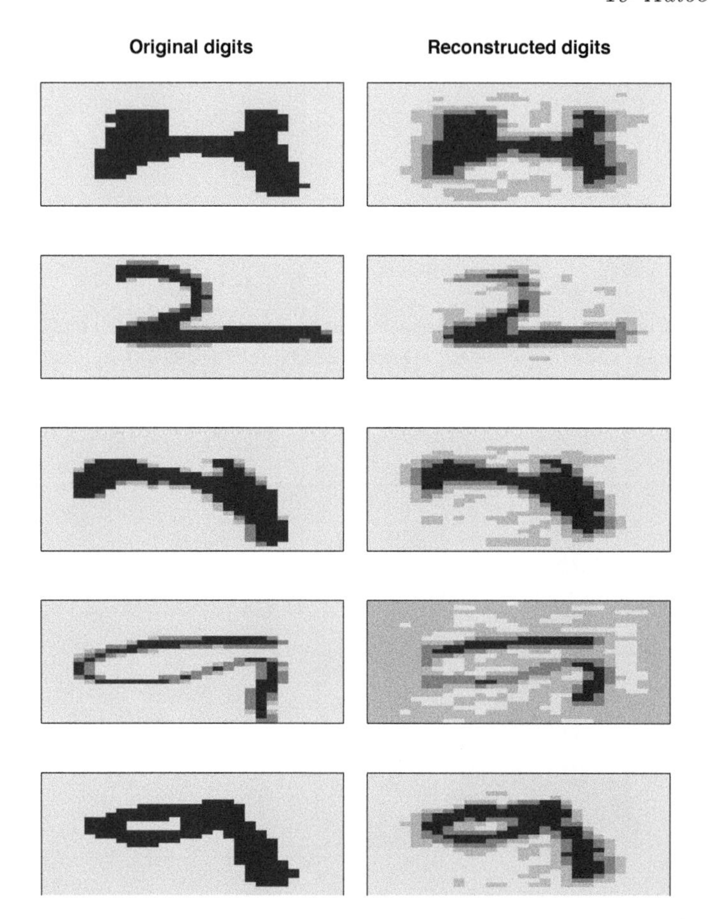

FIGURE 19.11: Original digits (left) and their reconstructions (right) for the observations with the five largest reconstruction errors.

deciding on the threshold that determines an input as anomalous is subjective and often relies on the business purpose.

19.6 Final thoughts

As we mentioned at the beginning of this chapter, autoencoders are receiving a lot of attention and many advancements have been made over the past decade. We discussed a few of the fundamental implementations of autoencoders; however, more exist. The following is an incomplete list of alternative autoencoders that are worthy of your attention.

- *Variational autoencoders* are a form of generative autoencoders, which means they can be used to create new instances that closely resemble the input data but are completely generated from the coding distributions (Doersch, 2016).
- *Contractive autoencoders* constrain the derivative of the hidden layer(s) activations to be small with respect to the inputs. This has a similar effect as denoising autoencoders in the sense that small perturbations to the input are essentially considered noise, which makes our codings more robust (Rifai et al., 2011).
- *Stacked convolutional autoencoders* are designed to reconstruct visual features processed through convolutional layers (Masci et al., 2011). They do not require manual vectorization of the image so they work well if you need to do dimension reduction or feature extraction on realistic-sized high-dimensional images.
- *Winner-take-all autoencoders* leverage only the top X% activations for each neuron, while the rest are set to zero (Makhzani and Frey, 2015). This leads to sparse codings. This approach has also been adapted to work with convolutional autoencoders (Makhzani and Frey, 2014).
- *Adversarial autoencoders* train two networks - a generator network to reconstruct the inputs similar to a regular autoencoder and then a discriminator network to compute where the inputs lie on a probabilistic distribution. Similar to variational autoencoders, adversarial autoencoders are often used to generate new data and have also been used for semi-supervised and supervised tasks (Makhzani et al., 2015).

Part IV

Clustering

20

K-means Clustering

In PART III of this book we focused on methods for reducing the dimension of our feature space (p). The remaining chapters concern methods for reducing the dimension of our observation space (n); these methods are commonly referred to as *clustering*. K-means clustering is one of the most commonly used clustering algorithms for partitioning observations into a set of k groups (i.e. k clusters), where k is pre-specified by the analyst. k-means, like other clustering algorithms, tries to classify observations into mutually exclusive groups (or clusters), such that observations within the same cluster are as similar as possible (i.e., high intra-class similarity), whereas observations from different clusters are as dissimilar as possible (i.e., low inter-class similarity). In k-means clustering, each cluster is represented by its center (i.e, centroid) which corresponds to the mean of the observation values assigned to the cluster. The procedure used to find these clusters is similar to the k-nearest neighbor (KNN) algorithm discussed in Chapter 8; albeit, without the need to predict an average response value.

20.1 Prerequisites

For this chapter we'll use the following packages (note that the primary function to perform k-means, `kmeans()`, is provided in the **stats** package that comes with your basic R installation):

```
# Helper packages
library(dplyr)       # for data manipulation
library(ggplot2)     # for data visualization
library(stringr)     # for string functionality

# Modeling packages
library(cluster)     # for general clustering algorithms
library(factoextra)  # for visualizing cluster results
```

To illustrate k-means concepts we'll use the `mnist` and `my_basket` data sets from previous chapters. We'll also discuss clustering with mixed data (e.g., data with both numeric and categorical data types) using the Ames housing data later in the chapter.

```
mnist <- dslabs::read_mnist()

url <- "https://koalaverse.github.io/homlr/data/my_basket.csv"
my_basket <- readr::read_csv(url)
```

20.2 Distance measures

The classification of observations into groups requires some method for computing the distance or the (dis)similarity between each pair of observations which form a distance or dissimilarity or matrix.

There are many approaches to calculating these distances; the choice of distance measure is a critical step in clustering (as it was with KNN). It defines how the similarity of two observations (x_a and x_b for all j features) is calculated and it will influence the shape and size of the clusters. Recall from Section 8.2.1 that the classical methods for distance measures are the Euclidean and Manhattan distances; however, alternative distance measures exist such as correlation-based distances, which are widely used for gene expression data; the *Gower distance* measure (discuss later in Section 20.7), which is commonly used for data sets containing categorical and ordinal features; and *cosine distance*, which is commonly used in the field of *text mining*. So how do you decide on a particular distance measure? Unfortunately, there is no straightforward answer and several considerations come into play.

Euclidean distance (i.e., straight line distance, or *as the crow flies*) is very sensitive to outliers; when they exist they can skew the cluster results which gives false confidence in the compactness of the cluster. If your features follow an approximate Gaussian distribution then Euclidean distance is a reasonable measure to use. However, if your features deviate significantly from normality or if you just want to be more robust to existing outliers, then Manhattan, Minkowski, or Gower distances are often better choices.

If you are analyzing unscaled data where observations may have large differences in magnitude but similar behavior then a correlation-based distance is preferred. For example, say you want to cluster customers based on common purchasing characteristics. It is possible for large volume and low volume customers to exhibit similar behaviors; however, due to their purchasing magnitude the scale of the data may skew the clusters if not using a correlation-based distance measure. Figure 20.1 illustrates this phenomenon where observation one and two purchase similar quantities of items; however, observation two and three have nearly perfect correlation in their purchasing behavior. A non-correlation distance measure would group observations one and two together whereas a correlation-based distance measure would group observations two and three together.

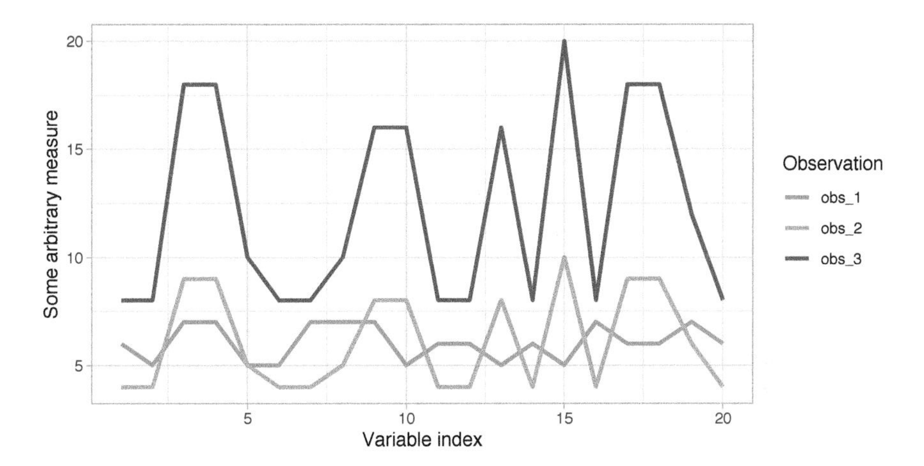

FIGURE 20.1: Correlation-based distance measures will capture the correlation between two observations better than a non-correlation-based distance measure; regardless of magnitude differences.

20.3 Defining clusters

The basic idea behind k-means clustering is constructing clusters so that the total *within-cluster variation* is minimized. There are several k-means algorithms available for doing this. The standard algorithm is the *Hartigan-Wong algorithm* (Hartigan and Wong, 1979), which defines the total within-cluster variation as the sum of the Euclidean distances between observation i's feature values and the corresponding centroid:

$$W(C_k) = \sum_{x_i \in C_k} (x_i - \mu_k)^2, \tag{20.1}$$

where:

- x_i is an observation belonging to the cluster C_k;
- μ_k is the mean value of the points assigned to the cluster C_k.

Each observation (x_i) is assigned to a given cluster such that the sum of squared (SS) distances of each observation to their assigned cluster centers (μ_k) is minimized.

We define the total within-cluster variation as follows:

$$SS_{within} = \sum_{k=1}^{k} W(C_k) = \sum_{k=1}^{k} \sum_{x_i \in C_k} (x_i - \mu_k)^2 \tag{20.2}$$

The SS_{within} measures the compactness (i.e., goodness) of the resulting clusters and we want it to be as small as possible as illustrated in Figure 20.2.

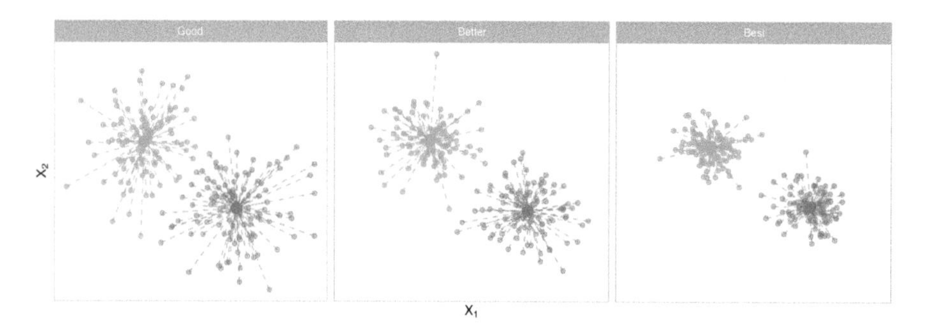

FIGURE 20.2: Total within-cluster variation captures the total distances between a cluster's centroid and the individual observations assigned to that cluster. The more compact the these distances, the more defined and isolated the clusters are.

The underlying assumptions of k-means requires points to be closer to their own cluster center than to others. This assumption can be ineffective when the clusters have complicated geometries as k-means requires convex boundaries. For example, consider the data in Figure 20.3 (A). These data are clearly grouped; however, their groupings do not have nice convex boundaries (like the convex boundaries used to illustrate the hard margin classifier in Chapter 14). Consequently, k-means clustering does not capture the appropriate groups as Figure 20.3 (B) illustrates. However, *spectral clustering methods* apply the same kernal trick discussed in Chapter 14 to allow k-means to discover non-convex boundaries (Figure 20.3 (C)). See Friedman et al. (2001) for a thorough discussion of spectral clustering and the **kernlab** package for an R implementation. We'll also discuss model-based clustering methods in Chapter 22 which provide an alternative approach to capture non-convex cluster shapes.

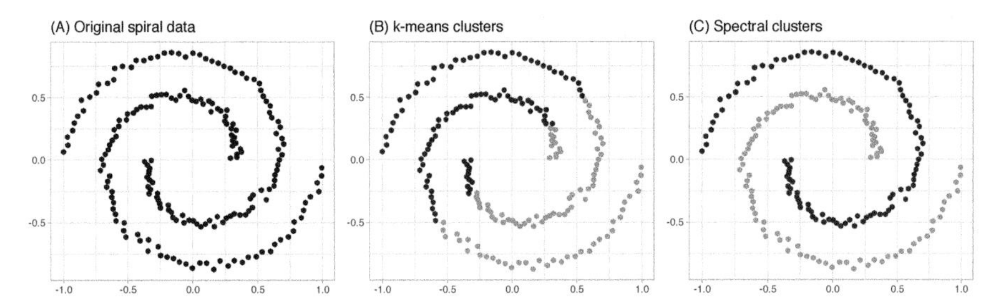

FIGURE 20.3: The assumptions of k-means lends it ineffective in capturing complex geometric groupings; however, spectral clustering allows you to cluster data that is connected but not necessarily clustered within convex boundaries.

20.4 *k*-means algorithm

The first step when using *k*-means clustering is to indicate the number of clusters (k) that will be generated in the final solution. Unfortunately, unless our data set is very small, we cannot evaluate every possible cluster combination because there are almost k^n ways to partition n observations into k clusters. Consequently, we need to estimate a *greedy local optimum* solution for our specified k (Hartigan and Wong, 1979). To do so, the algorithm starts by randomly selecting k observations from the data set to serve as the initial centers for the clusters (i.e., centroids). Next, each of the remaining observations are assigned to its closest centroid, where closest is defined using the distance between the object and the cluster mean (based on the selected distance measure). This is called the *cluster assignment step.*

Next, the algorithm computes the new center (i.e., mean value) of each cluster. The term *centroid update* is used to define this step. Now that the centers have been recalculated, every observation is checked again to see if it might be closer to a different cluster. All the objects are reassigned again using the updated cluster means. The cluster assignment and centroid update steps are iteratively repeated until the cluster assignments stop changing (i.e., when convergence is achieved). That is, the clusters formed in the current iteration are the same as those obtained in the previous iteration.

Due to randomization of the initial k observations used as the starting centroids, we can get slightly different results each time we apply the procedure. Consequently, most algorithms use several *random starts* and choose the iteration with the lowest $W(C_k)$ (Equation (20.1)). Figure 20.4 illustrates the variation in $W(C_k)$ for different random starts.

A good rule for the number of random starts to apply is 10–20.

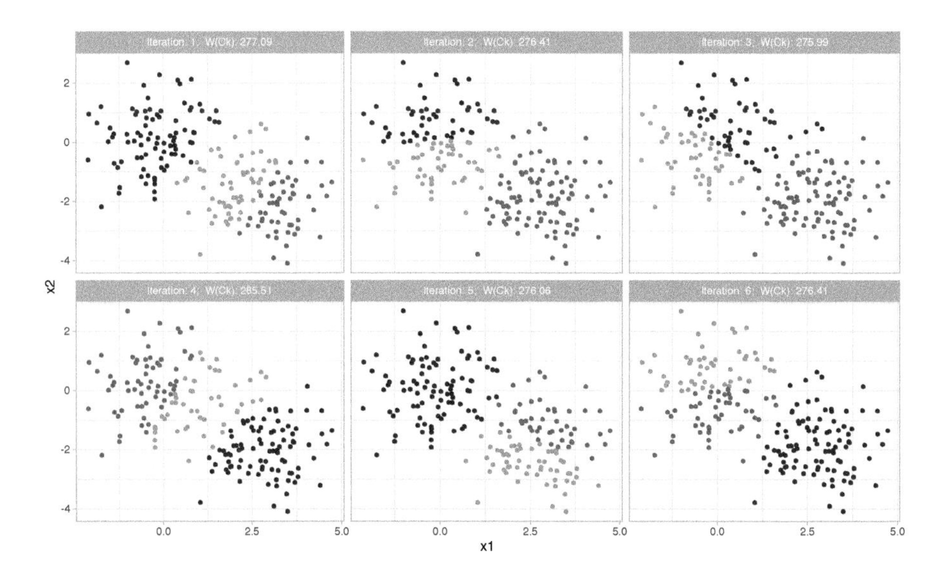

FIGURE 20.4: Each application of the k-means algorithm can achieve slight differences in the final results based on the random start.

The k-means algorithm can be summarized as follows:

1. Specify the number of clusters (k) to be created (this is done by the analyst).
2. Select k observations at random from the data set to use as the initial cluster centroids.
3. Assign each observation to their closest centroid based on the distance measure selected.
4. For each of the k clusters update the cluster centroid by calculating the new mean values of all the data points in the cluster. The centroid for the i-th cluster is a vector of length p containing the means of all p features for the observations in cluster i.
5. Iteratively minimize SS_{within}. That is, iterate steps 3–4 until the cluster assignments stop changing (beyond some threshold) or the maximum number of iterations is reached. A good rule of thumb is to perform 10–20 iterations.

20.5 Clustering digits

Let's illustrate an example by performing k-means clustering on the MNIST pixel features and see if we can identify unique clusters of digits without using the response variable. Here, we declare $k = 10$ only because we already know there are 10 unique digits represented in the data. We also use 10 random starts (`nstart = 10`). The output of our model contains many of the metrics we've already discussed such as total within-cluster variation (`withinss`), total within-cluster sum of squares (`tot.withinss`), the size of each cluster (`size`), and the iteration out of our 10 random starts used (`iter`). It also includes the `cluster` each observation is assigned to and the `centers` of each cluster.

Training k-means on the MNIST data with 10 random starts took about 4.5 minutes for us using the code below.

```
features <- mnist$train$images

# Use k-means model with 10 centers and 10 random starts
mnist_clustering <- kmeans(features, centers = 10, nstart = 10)

# Print contents of the model output
str(mnist_clustering)
## List of 9
##  $ cluster     : int [1:60000] 5 9 3 8 10 7 4 5 4 6 ...
##  $ centers     : num [1:10, 1:784] 0 0 0 0 0 0 0 0 0 0 ...
##   ..- attr(*, "dimnames")=List of 2
##   .. ..$ : chr [1:10] "1" "2" "3" "4" ...
##   .. ..$ : NULL
##  $ totss       : num 205706725984
##  $ withinss    : num [1:10] 23123576673 14119007546 ...
##  $ tot.withinss: num 153017742761
##  $ betweenss   : num 52688983223
##  $ size        : int [1:10] 7786 5384 5380 5515 7051 6706 ...
##  $ iter        : int 8
##  $ ifault      : int 0
##  - attr(*, "class")= chr "kmeans"
```

The `centers` output is a 10x784 matrix. This matrix contains the average value of each of the 784 features for the 10 clusters. We can plot this as in

Figure 20.5 which shows us what the typical digit is in each cluster. We clearly see recognizable digits even though k-means had no insight into the response variable.

```
# Extract cluster centers
mnist_centers <- mnist_clustering$centers

# Plot typical cluster digits
par(mfrow = c(2, 5), mar=c(0.5, 0.5, 0.5, 0.5))
layout(matrix(seq_len(nrow(mnist_centers)), 2, 5, byrow = FALSE))
for(i in seq_len(nrow(mnist_centers))) {
  image(matrix(mnist_centers[i, ], 28, 28)[, 28:1],
        col = gray.colors(12, rev = TRUE), xaxt="n", yaxt="n")
}
```

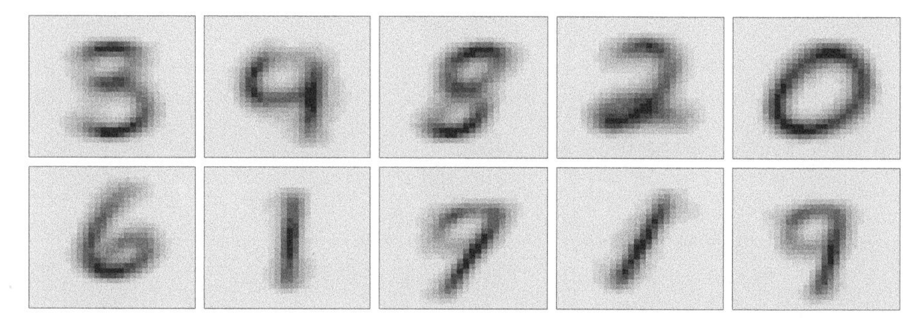

FIGURE 20.5: Cluster centers for the 10 clusters identified in the MNIST training data.

We can compare the cluster digits with the actual digit labels to see how well our clustering is performing. To do so, we compare the most common digit in each cluster (i.e., with the mode) to the actual training labels. Figure 20.6 illustrates the results. We see that k-means does a decent job of clustering some of the digits. In fact, most of the digits are clustered more often with like digits than with different digits.

However, we also see some digits are grouped often with different digits (e.g., 6s are often grouped with 0s and 9s are often grouped with 7s). We also see that 0s and 5s are never the dominant digit in a cluster. Consequently, our clustering is grouping many digits that have some resemblance (3s, 5s, and 8s are often grouped together) and since this is an unsupervised task, there is no mechanism to supervise the algorithm otherwise.

```
# Create mode function
mode_fun <- function(x){
  which.max(tabulate(x))
}

mnist_comparison <- data.frame(
  cluster = mnist_clustering$cluster,
  actual = mnist$train$labels
) %>%
  group_by(cluster) %>%
  mutate(mode = mode_fun(actual)) %>%
  ungroup() %>%
  mutate_all(factor, levels = 0:9)

# Create confusion matrix and plot results
yardstick::conf_mat(
  mnist_comparison,
  truth = actual,
  estimate = mode
) %>%
  autoplot(type = 'heatmap')
```

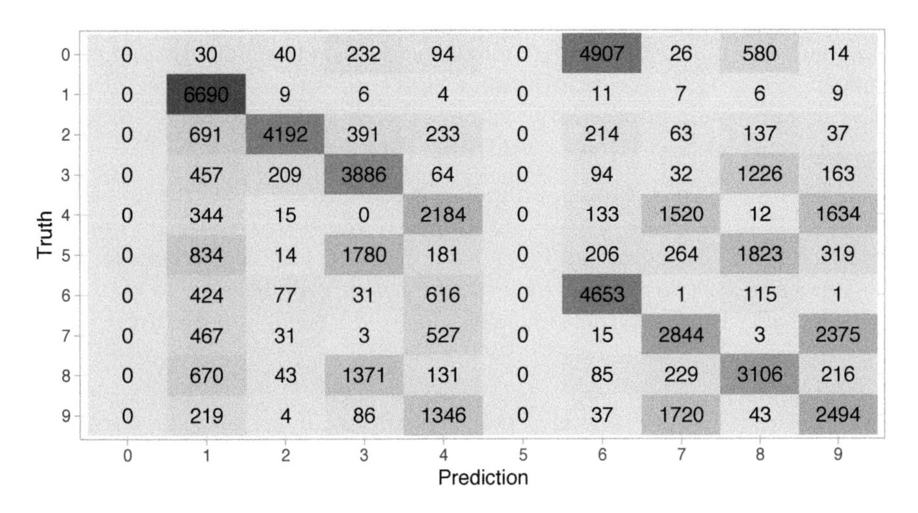

FIGURE 20.6: Confusion matrix illustrating how the k-means algorithm clustered the digits (x-axis) and the actual labels (y-axis).

20.6 How many clusters?

When clustering the MNIST data, the number of clusters we specified was based on prior knowledge of the data. However, often we do not have this kind of *a priori* information and the reason we are performing cluster analysis is to identify what clusters may exist. So how do we go about determining the right number of k?

Choosing the number of clusters requires a delicate balance. Larger values of k can improve homogeneity of the clusters; however it risks overfitting.

Best case (or maybe we should say easiest case) scenario, k is predetermined. This often occurs when we have deterministic resources to allocate. For example, a company may employ k sales people and they would like to partition their customers into one of k segments so that they can be assigned to one of the sales folks. In this case k is predetermined by external resources or knowledge. A more common case is that k is unknown; however, we can often still apply *a priori* knowledge for potential groupings. For example, maybe you need to cluster customer experience survey responses for an automobile sales company. You may start by setting k to the number of car brands the company carries. If you lack any *a priori* knowledge for setting k, then a commonly used rule of thumb is $k = \sqrt{n/2}$, where n is the number of observations to cluster. However, this rule can result in very large values of k for larger data sets (e.g., this would have us use $k = 173$ for the MNIST data set).

When the goal of the clustering procedure is to ascertain what natural distinct groups exist in the data, without any *a priori* knowledge, there are multiple statistical methods we can apply. However, many of these measures suffer from the *curse of dimensionality* as they require multiple iterations and clustering large data sets is not efficient, especially when clustering repeatedly.

See Charrad et al. (2015) for a thorough review of the vast assortment of measures of cluster performance. The **NbClust** package implements many of these methods, providing you with over 30 indices to determine the optimal k.

One of the more popular methods is the *elbow method*. Recall that the basic idea behind cluster partitioning methods, such as k-means clustering, is to define clusters such that the total within-cluster variation is minimized (Equation (20.2)). The total within-cluster sum of squares measures the compactness of the clustering and we want it to be as small as possible. Thus, we can use the following approach to define the optimal clusters:

1. Compute k-means clustering for different values of k. For instance, by varying k from 1–20 clusters.
2. For each k, calculate the total within-cluster sum of squares (WSS).
3. Plot the curve of WSS according to the number of clusters k.
4. The location of a bend (i.e., elbow) in the plot is generally considered as an indicator of the appropriate number of clusters.

When using small to moderate sized data sets this process can be performed conveniently with `factoextra::fviz_nbclust()`. However, this function requires you to specify a single max k value and it will train k-means models for $1 - k$ clusters. When dealing with large data sets, such as MNIST, this is unreasonable so you will want to manually implement the procedure (e.g., with a `for` loop and specify the values of k to assess).

The following assesses clustering the `my_basket` data into 1–25 clusters. The `method = 'wss'` argument specifies that our search criteria is using the elbow method discussed above and since we are assessing quantities across different baskets of goods we use the non-parametric Spearman correlation-based distance measure. The results show the "elbow" appears to happen when $k = 5$.

```
fviz_nbclust(
  my_basket,
  kmeans,
  k.max = 25,
  method = "wss",
  diss = get_dist(my_basket, method = "spearman")
)
```

`fviz_nbclust()` also implements other popular methods such as the Silhouette method (Rousseeuw, 1987) and Gap statistic (Tibshirani et al., 2001). Luckily, applications requiring the exact optimal set of clusters is fairly rare. In most applications, it suffices to choose a k based on convenience rather than strict performance requirements. But if necessary, the elbow method and other performance metrics can point you in the right direction.

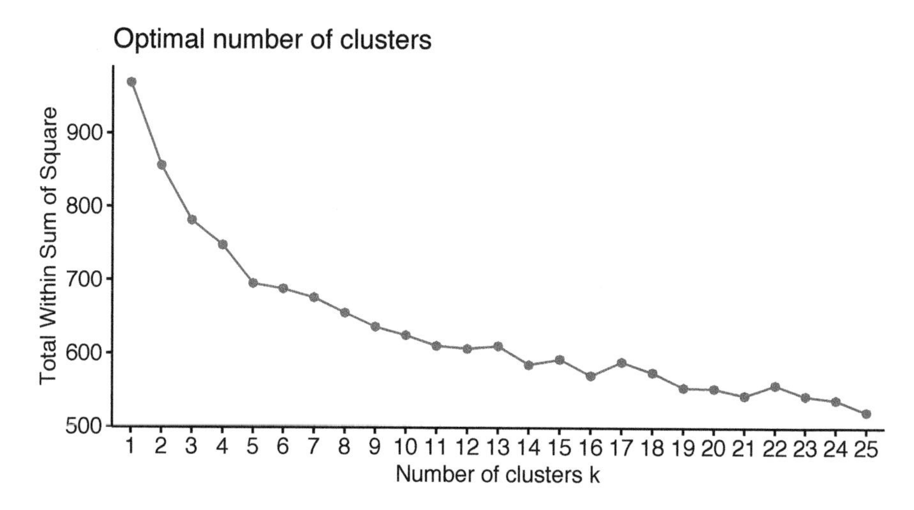

FIGURE 20.7: Using the elbow method to identify the preferred number of clusters in the my basket data set.

20.7 Clustering with mixed data

Often textbook examples of clustering include only numeric data. However, most real life data sets contain a mixture of numeric, categorical, and ordinal variables; and whether an observation is similar to another observation should depend on these data type attributes. There are a few options for performing clustering with mixed data and we'll demonstrate on the full Ames housing data set (minus the response variable `Sale_Price`). To perform k-means clustering on mixed data we can convert any ordinal categorical variables to numeric and one-hot encode the remaining nominal categorical variables.

```
# Full ames data set --> recode ordinal variables to numeric
ames_full <- AmesHousing::make_ames() %>%
  mutate_if(str_detect(names(.), 'Qual|Cond|QC|Qu'), as.numeric)

# One-hot encode --> retain only the features and not sale price
full_rank  <- caret::dummyVars(Sale_Price ~ ., data = ames_full,
                               fullRank = TRUE)
ames_1hot <- predict(full_rank, ames_full)

# Scale data
ames_1hot_scaled <- scale(ames_1hot)
```

```
# New dimensions
dim(ames_1hot_scaled)
## [1] 2930  240
```

Now that all our variables are represented numerically, we can perform k-means clustering as we did in the previous sections. Using the elbow method, there does not appear to be a definitive number of clusters to use.

```
set.seed(123)

fviz_nbclust(
  ames_1hot_scaled,
  kmeans,
  method = "wss",
  k.max = 25,
  verbose = FALSE
)
```

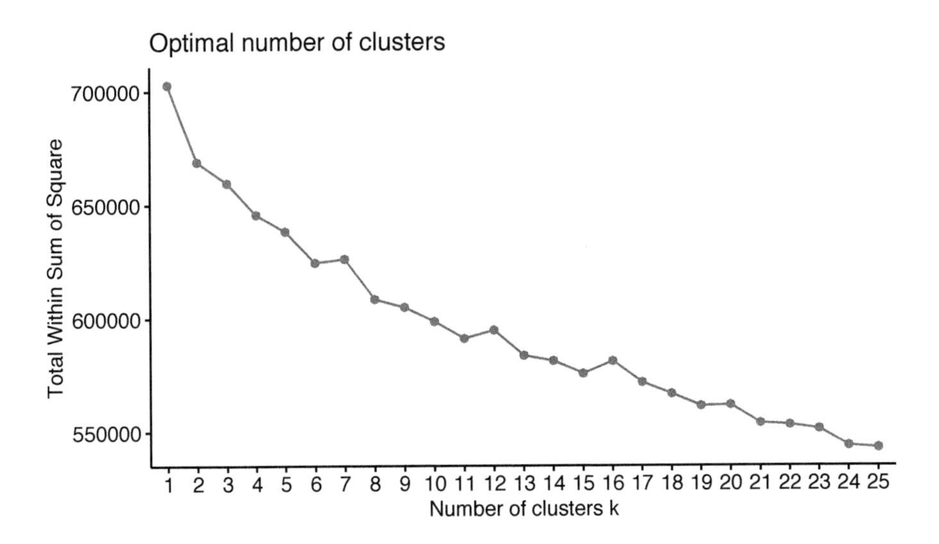

FIGURE 20.8: Suggested number of clusters for one-hot encoded Ames data using k-means clustering and the elbow criterion.

Unfortunately, this is a common issue. As the number of features expand, performance of k-means tends to break down and both k-means and hierarchical clustering (Chapter 21) approaches become slow and ineffective. This happens, typically, as your data becomes more sparse. An additional option for heavily mixed data is to use the Gower distance (Gower, 1971) measure, which applies a particular distance calculation that works well for each data type. The metrics used for each data type include:

- **quantitative (interval)**: range-normalized Manhattan distance;
- **ordinal**: variable is first ranked, then Manhattan distance is used with a special adjustment for ties;
- **nominal**: variables with k categories are first converted into k binary columns (i.e., one-hot encoded) and then the *Dice coefficient* is used. To compute the dice metric for two observations (X, Y), the algorithm looks across all one-hot encoded categorical variables and scores them as:
  - **a** — number of dummies 1 for both observations
  - **b** — number of dummies 1 for X and 0 for Y
  - **c** — number of dummies 0 for X and 1 for Y
  - **d** — number of dummies 0 for both

and then uses the following formula:

$$D = \frac{2a}{2a + b + c} \tag{20.3}$$

We can use the `cluster::daisy()` function to create a Gower distance matrix from our data; this function performs the categorical data transformations so you can supply the data in the original format.

```
# Original data minus Sale_Price
ames_full <- AmesHousing::make_ames() %>% select(-Sale_Price)

# Compute Gower distance for original data
gower_dst <- daisy(ames_full, metric = "gower")
```

We can now feed the results into any clustering algorithm that accepts a distance matrix. This primarily includes `cluster::pam()`, `cluster::diana()`, and `cluster::agnes()` (`stats::kmeans()` and `cluster::clara()` do not accept distance matrices as inputs).

`cluster::diana()` and `cluster::agnes()` are hierarchical clustering algorithms that you will learn about in Chapter 21. `cluster::pam()` and `cluster::clara()` are discussed in the next section.

```
# You can supply the Gower distance matrix to several clustering algos
pam_gower <- pam(x = gower_dst, k = 8, diss = TRUE)
diana_gower <- diana(x = gower_dst, diss = TRUE)
agnes_gower <- agnes(x = gower_dst, diss = TRUE)
```

20.8 Alternative partitioning methods

As your data grow in dimensions you are likely to introduce more outliers; since k-means uses the mean, it is not robust to outliers. An alternative to this is to use *partitioning around medians* (PAM), which has the same algorithmic steps as k-means but uses the median rather than the mean to determine the centroid; making it more robust to outliers. Unfortunately, this robustness comes with an added computational expense (Friedman et al., 2001).

To perform PAM clustering use `cluster::pam()` instead of `kmeans()`.

If you compare k-means and PAM clustering results for a given criterion and experience common results then that is a good indication that outliers are not effecting your results. Figure 20.9 illustrates the total within sum of squares for 1–25 clusters using PAM clustering on the one-hot encoded Ames data. We see very similar results as with k-means (Figure 20.8), which tells us that outliers are not negatively influencing the k-means results.

```
fviz_nbclust(
  ames_1hot_scaled,
  pam,
  method = "wss",
  k.max = 25,
  verbose = FALSE
)
```

As your data set becomes larger both hierarchical, k-means, and PAM clustering become slower. An alternative is *clustering large applications* (CLARA), which performs the same algorithmic process as PAM; however, instead of finding the *medoids* for the entire data set it considers a small sample size and applies k-means or PAM.

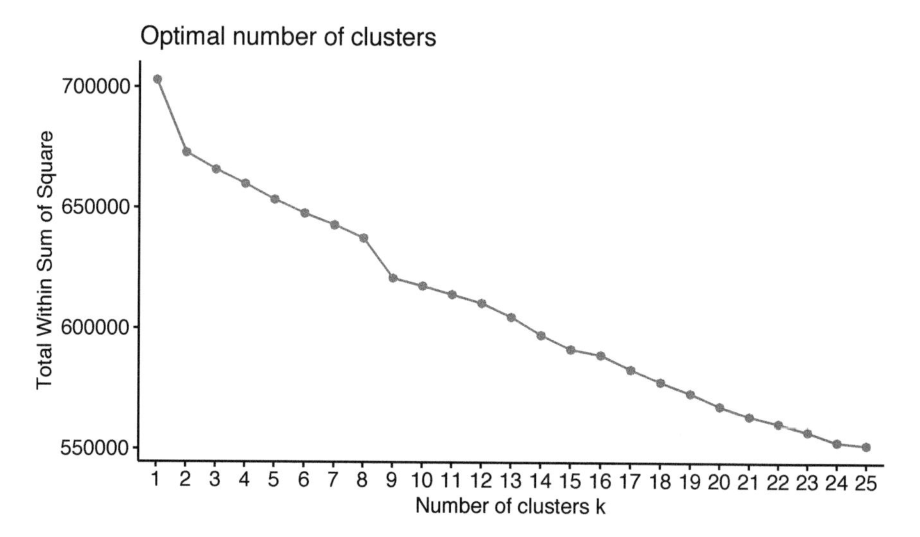

FIGURE 20.9: Total within sum of squares for 1-25 clusters using PAM clustering.

Medoids are similar in spirit to the cluster centers or means, but medoids are always restricted to be members of the data set (similar to the difference between the sample mean and median when you have an odd number of observations and no ties).

CLARA performs the following algorithmic steps:

1. Randomly split the data set into multiple subsets with fixed size.
2. Compute PAM algorithm on each subset and choose the corresponding k medoids. Assign each observation of the entire data set to the closest medoid.
3. Calculate the mean (or sum) of the dissimilarities of the observations to their closest medoid. This is used as a measure of the goodness of fit of the clustering.
4. Retain the sub-data set for which the mean (or sum) is minimal.

To perform CLARA clustering use `cluster::clara()` instead of `cluster::pam()` and `kmeans()`.

If you compute CLARA on the Ames mixed data or on the MNIST data you will find very similar results to both k-means and PAM; however, as the below code illustrates it takes less than $\frac{1}{5}$-th of the time!

```
# k-means computation time on MNIST data
system.time(kmeans(features, centers = 10))
##     user  system elapsed
## 230.875   4.659 237.404

# CLARA computation time on MNIST data
system.time(clara(features, k = 10))
##    user  system elapsed
## 37.975   0.286  38.966
```

20.9 Final thoughts

K-means clustering is probably the most popular clustering algorithm and usually the first applied when solving clustering tasks. Although there have been methods to help analysts identify the optimal number of k clusters, this task is still largely based on subjective inputs and decisions by the analyst considering the unsupervised nature of the algorithm. In the next two chapters we'll explore alternative approaches that help reduce the burden of the analyst needing to define k. These methods also address other limitations of k-means such as how the algorithm primarily performs well only if the clusters have convex, non-overlapping boundaries; attributes that rarely exists in real data sets.

21

Hierarchical Clustering

Hierarchical clustering is an alternative approach to k-means clustering for identifying groups in a data set. In contrast to k-means, hierarchical clustering will create a hierarchy of clusters and therefore does not require us to pre-specify the number of clusters. Furthermore, hierarchical clustering has an added advantage over k-means clustering in that its results can be easily visualized using an attractive tree-based representation called a *dendrogram*.

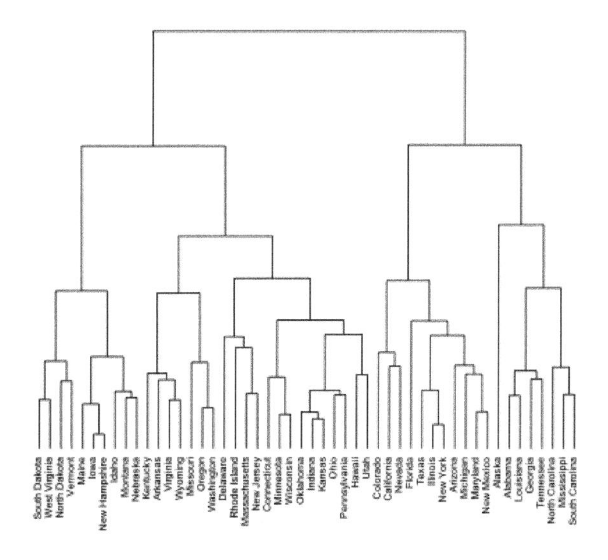

FIGURE 21.1: Illustrative dendrogram.

21.1 Prerequisites

For this chapter we'll use the following packages:

```
# Helper packages
library(dplyr)       # for data manipulation
library(ggplot2)     # for data visualization

# Modeling packages
library(cluster)     # for general clustering algorithms
library(factoextra)  # for visualizing cluster results
```

The major concepts of hierarchical clustering will be illustrated using the Ames housing data. For simplicity we'll just use the 34 numeric features but refer to our discussion in Section 20.7 if you'd like to replicate this analysis with the full set of features. Since these features are measured on significantly different magnitudes we standardize the data first:

```
ames_scale <- AmesHousing::make_ames() %>%
  select_if(is.numeric) %>%  # select numeric columns
  select(-Sale_Price) %>%    # remove target column
  mutate_all(as.double) %>%  # coerce to double type
  scale()                    # center & scale the resulting columns
```

21.2 Hierarchical clustering algorithms

Hierarchical clustering can be divided into two main types:

1. **Agglomerative clustering:** Commonly referred to as AGNES (AGglomerative NESting) works in a bottom-up manner. That is, each observation is initially considered as a single-element cluster (leaf). At each step of the algorithm, the two clusters that are the most similar are combined into a new bigger cluster (nodes). This procedure is iterated until all points are a member of just one single big cluster (root) (see Figure 21.2). The result is a tree which can be displayed using a dendrogram.
2. **Divisive hierarchical clustering:** Commonly referred to as DIANA (DIvise ANAlysis) works in a top-down manner. DIANA is like the reverse of AGNES. It begins with the root, in which all observations are included in a single cluster. At each step of the algorithm, the current cluster is split into two clusters that are considered most heterogeneous. The process is iterated until all observations are in their own cluster.

Note that agglomerative clustering is good at identifying small clusters. Divisive hierarchical clustering, on the other hand, is better at identifying large clusters.

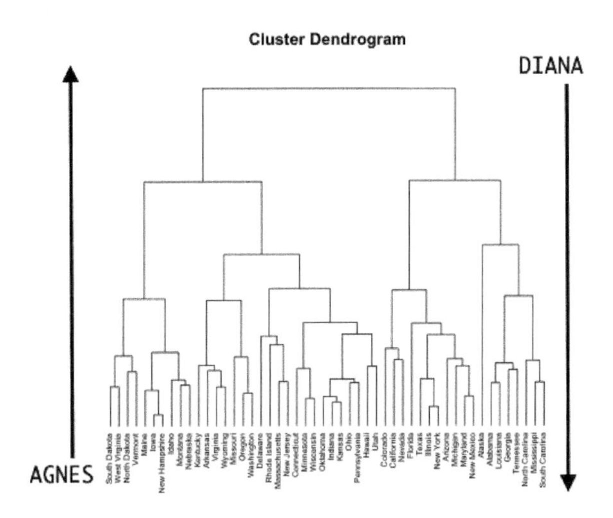

FIGURE 21.2: AGNES (bottom-up) versus DIANA (top-down) clustering.

Similar to k-means (Chapter 20), we measure the (dis)similarity of observations using distance measures (e.g., Euclidean distance, Manhattan distance, etc.); the Euclidean distance is most commonly the default. However, a fundamental question in hierarchical clustering is: *How do we measure the dissimilarity between two clusters of observations?* A number of different cluster agglomeration methods (i.e., linkage methods) have been developed to answer this question. The most common methods are:

- **Maximum or complete linkage clustering:** Computes all pairwise dissimilarities between the elements in cluster 1 and the elements in cluster 2, and considers the largest value of these dissimilarities as the distance between the two clusters. It tends to produce more compact clusters.
- **Minimum or single linkage clustering:** Computes all pairwise dissimilarities between the elements in cluster 1 and the elements in cluster 2, and considers the smallest of these dissimilarities as a linkage criterion. It tends to produce long, "loose" clusters.
- **Mean or average linkage clustering:** Computes all pairwise dissimilarities between the elements in cluster 1 and the elements in cluster 2, and considers the average of these dissimilarities as the distance between the two clusters. Can vary in the compactness of the clusters it creates.

- **Centroid linkage clustering:** Computes the dissimilarity between the centroid for cluster 1 (a mean vector of length p, one element for each variable) and the centroid for cluster 2.
- **Ward's minimum variance method:** Minimizes the total within-cluster variance. At each step the pair of clusters with the smallest between-cluster distance are merged. Tends to produce more compact clusters.

Other methods have been introduced such as measuring cluster descriptors after merging two clusters (Ma et al., 2007; Zhao and Tang, 2009; Zhang et al., 2013) but the above methods are, by far, the most popular and commonly used (Hair, 2006).

There are multiple agglomeration methods to define clusters when performing a hierarchical cluster analysis; however, complete linkage and Ward's method are often preferred for AGNES clustering. For DIANA, clusters are divided based on the maximum average dissimilarity which is very similar to the mean or average linkage clustering method outlined above. See Kaufman and Rousseeuw (2009) for details.

We can see the differences these approaches produce in the dendrograms displayed in Figure 21.3.

21.3 Hierarchical clustering in R

There are many functions available in R for hierarchical clustering. The most commonly used functions are `stats::hclust()` and `cluster::agnes()` for agglomerative hierarchical clustering (HC) and `cluster::diana()` for divisive HC.

21.3.1 Agglomerative hierarchical clustering

To perform agglomerative HC with `hclust()`, we first compute the dissimilarity values with `dist()` and then feed these values into `hclust()` and specify the agglomeration method to be used (i.e. `"complete"`, `"average"`, `"single"`, or `"ward.D"`).

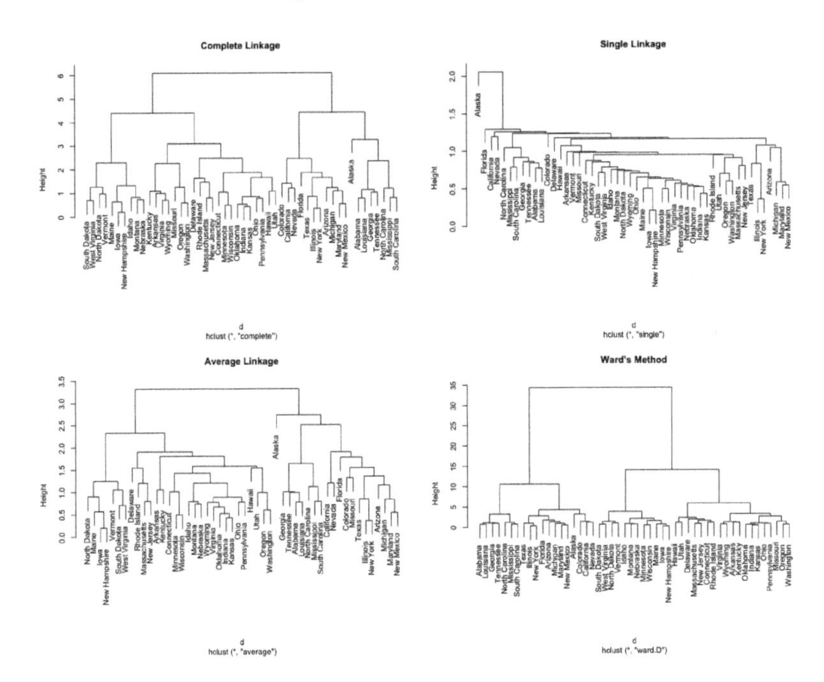

FIGURE 21.3: Differing hierarchical clustering outputs based on similarity measures.

```
# For reproducibility
set.seed(123)

# Dissimilarity matrix
d <- dist(ames_scale, method = "euclidean")

# Hierarchical clustering using Complete Linkage
hc1 <- hclust(d, method = "complete" )
```

You could plot the dendrogram with `plot(hc1, cex = 0.6, hang = -1)`; however, due to the large number of observations the output is not discernable.

Alternatively, we can use the `agnes()` function. This function behaves similar to `hclust()`; however, with the `agnes()` function you can also get the *agglomerative coefficient* (AC), which measures the amount of clustering structure found.

Generally speaking, the AC describes the strength of the clustering structure. Values closer to 1 suggest a more balanced clustering structure such as the complete linkage and Ward's method dendrograms in Figure 21.3. Values closer to 0 suggest less well-formed clusters such as the single linkage dendrogram in Figure 21.3. However, the AC tends to become larger as n increases, so it should not be used to compare across data sets of very different sizes.

```
# For reproducibility
set.seed(123)

# Compute maximum or complete linkage clustering with agnes
hc2 <- agnes(ames_scale, method = "complete")

# Agglomerative coefficient
hc2$ac
## [1] 0.927
```

This allows us to find certain hierarchical clustering methods that can identify stronger clustering structures. Here we see that Ward's method identifies the strongest clustering structure of the four methods assessed.

This grid search took a little over 3 minutes.

```
# methods to assess
m <- c( "average", "single", "complete", "ward")
names(m) <- c( "average", "single", "complete", "ward")

# function to compute coefficient
ac <- function(x) {
  agnes(ames_scale, method = x)$ac
}

# get agglomerative coefficient for each linkage method
purrr::map_dbl(m, ac)
##  average   single complete     ward
##    0.914    0.871    0.927    0.977
```

21.3.2 Divisive hierarchical clustering

The R function `diana()` in package **cluster** allows us to perform divisive hierarchical clustering. `diana()` works similar to `agnes()`; however, there is no agglomeration method to provide (see Kaufman and Rousseeuw (2009) for details). As before, a *divisive coefficient* (DC) closer to one suggests stronger group distinctions. Consequently, it appears that an agglomerative approach with Ward's linkage provides the optimal results.

```
# compute divisive hierarchical clustering
hc4 <- diana(ames_scale)

# Divise coefficient; amount of clustering structure found
hc4$dc
## [1] 0.919
```

21.4 Determining optimal clusters

Although hierarchical clustering provides a fully connected dendrogram representing the cluster relationships, you may still need to choose the preferred number of clusters to extract. Fortunately we can execute approaches similar to those discussed for k-means clustering (Section 20.6). The following compares results provided by the elbow, silhouette, and gap statistic methods. There is no definitively clear optimal number of clusters in this case; although, the silhouette method and the gap statistic suggest 8–9 clusters:

```
# Plot cluster results
p1 <- fviz_nbclust(ames_scale, FUN = hcut, method = "wss",
                   k.max = 10) +
  ggtitle("(A) Elbow method")
p2 <- fviz_nbclust(ames_scale, FUN = hcut, method = "silhouette",
                   k.max = 10) +
  ggtitle("(B) Silhouette method")
p3 <- fviz_nbclust(ames_scale, FUN = hcut, method = "gap_stat",
                   k.max = 10) +
  ggtitle("(C) Gap statistic")

# Display plots side by side
gridExtra::grid.arrange(p1, p2, p3, nrow = 1)
```

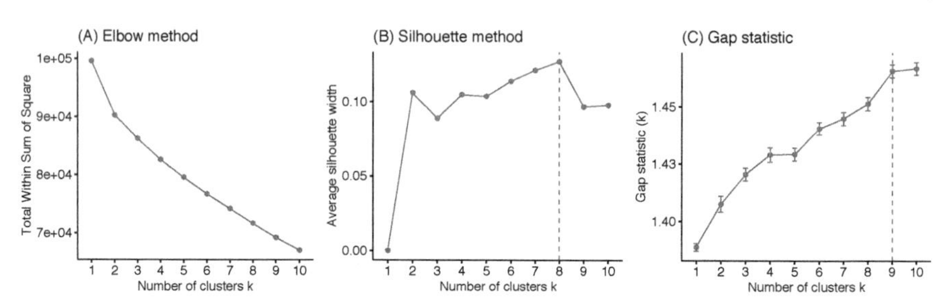

FIGURE 21.4: Comparison of three different methods to identify the optimal number of clusters.

21.5 Working with dendrograms

The nice thing about hierarchical clustering is that it provides a complete dendrogram illustrating the relationships between clusters in our data. In Figure 21.5, each leaf in the dendrogram corresponds to one observation (in our data this represents an individual house). As we move up the tree, observations that are similar to each other are combined into branches, which are themselves fused at a higher height.

```
# Construct dendorgram for the Ames housing example
hc5 <- hclust(d, method = "ward.D2" )
dend_plot <- fviz_dend(hc5)
dend_data <- attr(dend_plot, "dendrogram")
dend_cuts <- cut(dend_data, h = 8)
fviz_dend(dend_cuts$lower[[2]])
```

However, dendrograms are often misinterpreted. Conclusions about the proximity of two observations should not be implied by their relationship on the horizontal axis nor by the vertical connections. Rather, the height of the branch between an observation and the clusters of observations below them indicate the distance between the observation and that cluster it is joined to. For example, consider observation 9 & 2 in Figure 21.6. They appear close on the dendrogram (right) but, in fact, their closeness on the dendrogram imply they are approximately the same distance measure from the cluster that they are fused to (observations 5, 7, & 8). It by no means implies that observation 9 & 2 are close to one another.

In order to identify sub-groups (i.e., clusters), we can *cut* the dendrogram with `cutree()`. The height of the cut to the dendrogram controls the number of

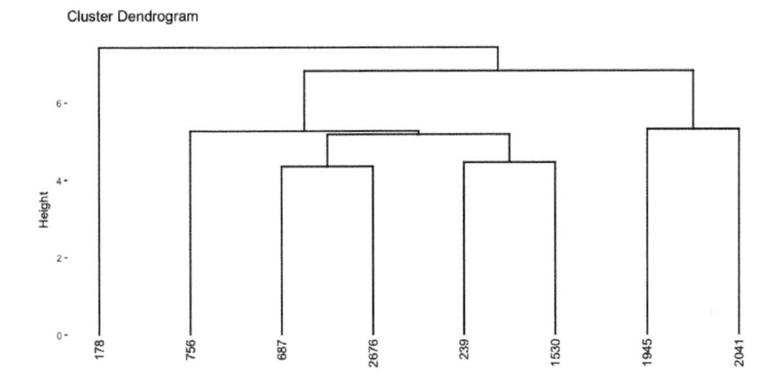

FIGURE 21.5: A subsection of the dendrogram for illustrative purposes.

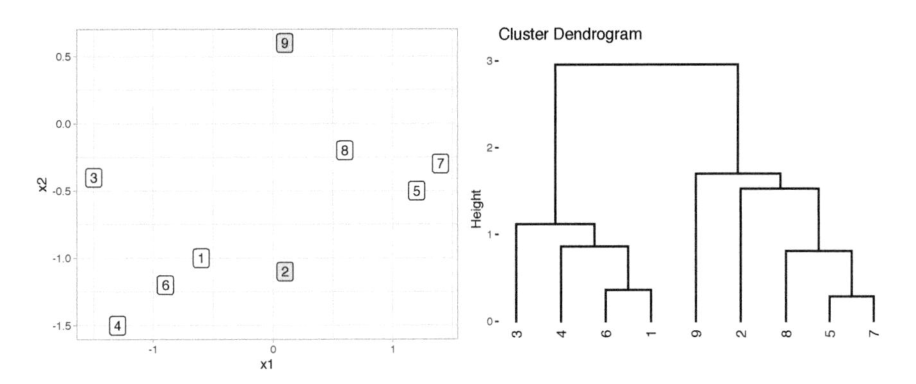

FIGURE 21.6: Comparison of nine observations measured across two features (left) and the resulting dendrogram created based on hierarchical clustering (right).

clusters obtained. It plays the same role as the k in k-means clustering. Here, we cut our agglomerative hierarchical clustering model into eight clusters. We can see that the concentration of observations are in clusters 1–3.

```
# Ward's method
hc5 <- hclust(d, method = "ward.D2" )

# Cut tree into 4 groups
sub_grp <- cutree(hc5, k = 8)

# Number of members in each cluster
table(sub_grp)
```

```
## sub_grp
##    1    2    3    4    5    6    7    8
## 1363  567  650   36  123  156   24   11
```

We can plot the entire dendrogram with `fviz_dend` and highlight the eight clusters with `k = 8`.

```
# Plot full dendogram
fviz_dend(
  hc5,
  k = 8,
  horiz = TRUE,
  rect = TRUE,
  rect_fill = TRUE,
  rect_border = "jco",
  k_colors = "jco",
  cex = 0.1
)
```

However, due to the size of the Ames housing data, the dendrogram is not very legible. Consequently, we may want to zoom into one particular region or cluster. This allows us to see which observations are most similar within a particular group.

There is no easy way to get the exact height required to capture all eight clusters. This is largely trial and error by using different heights until the output of `dend_cuts()` matches the cluster totals identified previously.

```
dend_plot <- fviz_dend(hc5)                    # create full dendogram
dend_data <- attr(dend_plot, "dendrogram") # extract plot info
dend_cuts <- cut(dend_data, h = 70.5)          # cut the dendogram at
                                               # designated height
# Create sub dendrogram plots
p1 <- fviz_dend(dend_cuts$lower[[1]])
p2 <- fviz_dend(dend_cuts$lower[[1]], type = 'circular')

# Side by side plots
gridExtra::grid.arrange(p1, p2, nrow = 1)
```

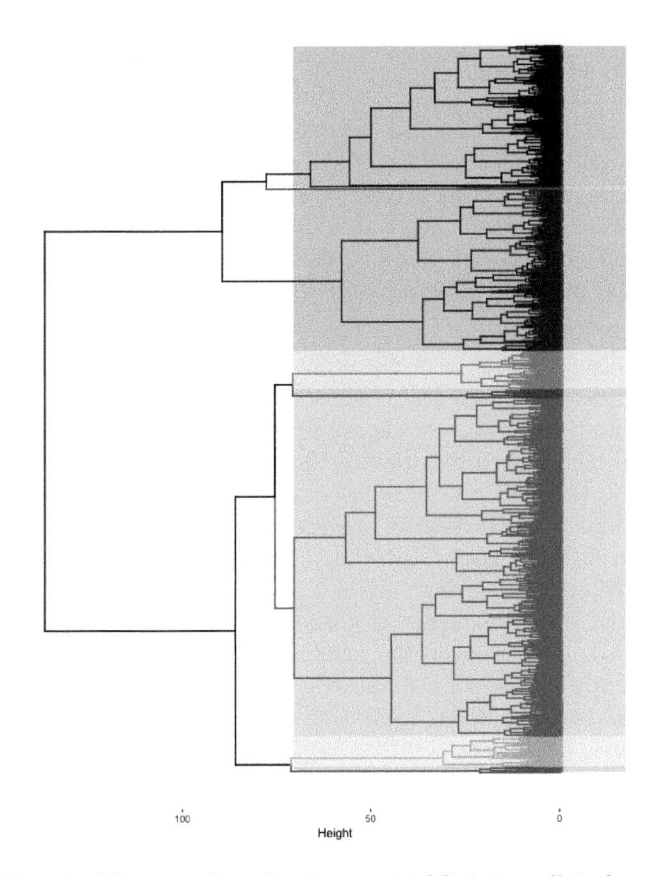

FIGURE 21.7: The complete dendogram highlighting all 8 clusters.

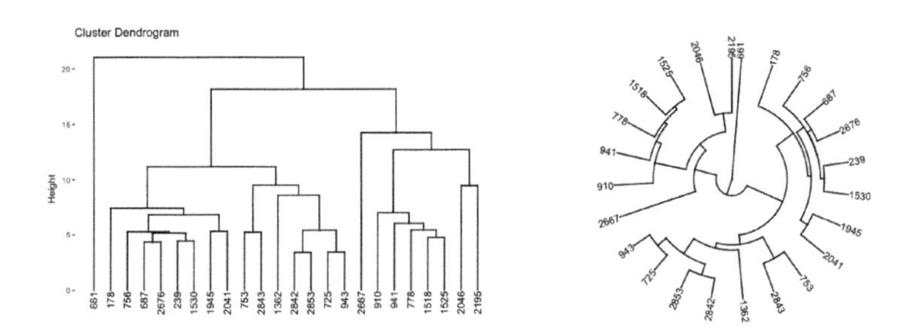

FIGURE 21.8: A subsection of the dendrogram highlighting cluster 7.

21.6 Final thoughts

Hierarchical clustering may have some benefits over k-means such as not having to pre-specify the number of clusters and the fact that it can produce a nice hierarchical illustration of the clusters (that's useful for smaller data sets). However, from a practical perspective, hierarchical clustering analysis still involves a number of decisions that can have large impacts on the interpretation of the results. First, like k-means, you still need to make a decision on the dissimilarity measure to use.

Second, you need to make a decision on the linkage method. Each linkage method has different systematic tendencies (or biases) in the way it groups observations and can result in significantly different results. For example, the centroid method has a bias toward producing irregularly shaped clusters. Ward's method tends to produce clusters with roughly the same number of observations and the solutions it provides tend to be heavily distorted by outliers. Given such tendencies, there should be a match between the algorithm selected and the underlying structure of the data (e.g., sample size, distribution of observations, and what types of variables are included-nominal, ordinal, ratio, or interval). For example, the centroid method should primarily be used when (a) data are measured with interval or ratio scales and (b) clusters are expected to be very dissimilar from each other. Likewise, Ward's method is best suited for analyses where (a) the number of observations in each cluster is expected to be approximately equal and (b) there are no outliers (Ketchen and Shook, 1996).

Third, although we do not need to pre-specify the number of clusters, we often still need to decide where to cut the dendrogram in order to obtain the final clusters to use. So the onus is still on us to decide the number of clusters, albeit in the end. In Chapter 22 we discuss a method that relieves us of this decision.

22

Model-based Clustering

Traditional clustering algorithms such as k-means (Chapter 20) and hierarchical (Chapter 21) clustering are heuristic-based algorithms that derive clusters directly based on the data rather than incorporating a measure of probability or uncertainty to the cluster assignments. Model-based clustering attempts to address this concern and provide *soft assignment* where observations have a probability of belonging to each cluster. Moreover, model-based clustering provides the added benefit of automatically identifying the optimal number of clusters. This chapter covers Gaussian mixture models, which are one of the most popular model-based clustering approaches available.

22.1 Prerequisites

For this chapter we'll use the following packages with the emphasis on **mclust** (Fraley et al., 2019):

```
# Helper packages
library(dplyr)     # for data manipulation
library(ggplot2)  # for data visualization

# Modeling packages
library(mclust)    # for fitting clustering algorithms
```

To illustrate the main concepts of model-based clustering we'll use the `geyser` data provided by the **MASS** package along with the `my_basket` data.

```
data(geyser, package = 'MASS')

url <- "https://koalaverse.github.io/homlr/data/my_basket.csv"
my_basket <- readr::read_csv(url)
```

22.2 Measuring probability and uncertainty

The key idea behind model-based clustering is that the data are considered as coming from a mixture of underlying probability distributions. The most popular approach is the *Gaussian mixture model* (GMM) (Banfield and Raftery, 1993) where each observation is assumed to be distributed as one of k multivariate-normal distributions, where k is the number of clusters (commonly referred to as *components* in model-based clustering). For a comprehensive review of model-based clustering, see Fraley and Raftery (2002).

GMMs are founded on the multivariate normal (Gaussian) distribution where p variables $(X_1, X_2, \dots, X_p)$ are assumed to have means $\mu = (\mu_1, \mu_2, \dots, \mu_p)$ and a covariance matrix Σ, which describes the joint variability (i.e., covariance) between each pair of variables:

$$\sum = \begin{bmatrix} \sigma_1^2 & \sigma_{1,2} & \cdots & \sigma_{1,p} \\ \sigma_{2,1} & \sigma_2^2 & \cdots & \sigma_{2,p} \\ \vdots & \vdots & \ddots & \vdots \\ \sigma_{p,1} & \sigma_{p,2}^2 & \cdots & \sigma_p^2 \end{bmatrix}. \tag{22.1}$$

Note that Σ contains p variances $(\sigma_1^2, \sigma_2^2, \dots, \sigma_p^2)$ and $p\,(p-1)\,/2$ unique covariances $\sigma_{i,j}$ $(i \neq j)$; note that $\sigma_{i,j} = \sigma_{j,i}$. A multivariate-normal distribution with mean μ and covariance $|sigma$ is notationally represented by Equation (22.2):

$$(X_1, X_2, \dots, X_p) \sim N_p\,(\mu, \Sigma)\,. \tag{22.2}$$

This distribution has the property that every subset of variables (say, X_1, X_5, and X_9) also has a multivariate normal distribution (albeit with a different mean and covariance structure). GMMs assume the clusters can be created using k Gaussian distributions. For example, if there are two variables (say, X and Y), then each observation (X_i, Y_i) is modeled has having been sampled from one of k distributions $(N_1\,(\mu_1, \Sigma_1)\,, N_2\,(\mu_2, \Sigma_2)\,, \dots, N_p\,(\mu_p, \Sigma_p))$. This is illustrated in Figure 22.1 which suggests variables X and Y come from three multivariate distributions. However, as the data points deviate from the center of one of the three clusters, the probability that they align to a particular cluster decreases as indicated by the fading elliptical rings around each cluster center.

We can illustrate this concretely by applying a GMM model to the `geyser` data, which is the data illustrated in Figure 22.1. To do so we apply `Mclust()` and specify three components. Plotting the output (Figure 22.2) provides a density plot (left) just like we saw in Figure 22.1 and the component assignment for each observation based on the largest probability (right). In the uncertainty plot

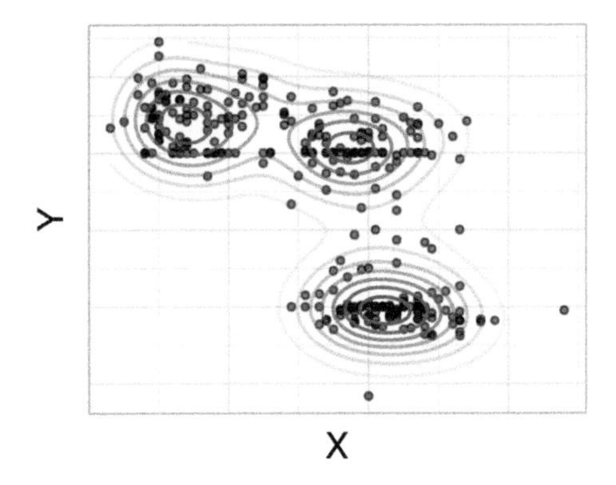

FIGURE 22.1: Data points across two features (X and Y) appear to come from three multivariate normal distributions.

(right), you'll notice the observations near the center of the densities are small, indicating small uncertainty (or high probability) of being from that respective component; however, the observations that are large represent observations with high uncertainty (or low probability) regarding the component they are aligned to.

```
# Apply GMM model with 3 components
geyser_mc <- Mclust(geyser, G = 3)

# Plot results
plot(geyser_mc, what = "density")
plot(geyser_mc, what = "uncertainty")
```

This idea of a probabilistic cluster assignment can be quite useful as it allows you to identify observations with high or low cluster uncertainty and, potentially, target them uniquely or provide alternative solutions. For example, the following six observations all have nearly 50% probability of being assigned to two different clusters. If this were an advertising data set and you were marketing to these observations you may want to provide them with a combination of marketing solutions for the two clusters they are nearest to. Or you may want to perform additional A/B testing on them to try gain additional confidence regarding which cluster they align to most.

```
# Observations with high uncertainty
sort(geyser_mc$uncertainty, decreasing = TRUE) %>% head()
```

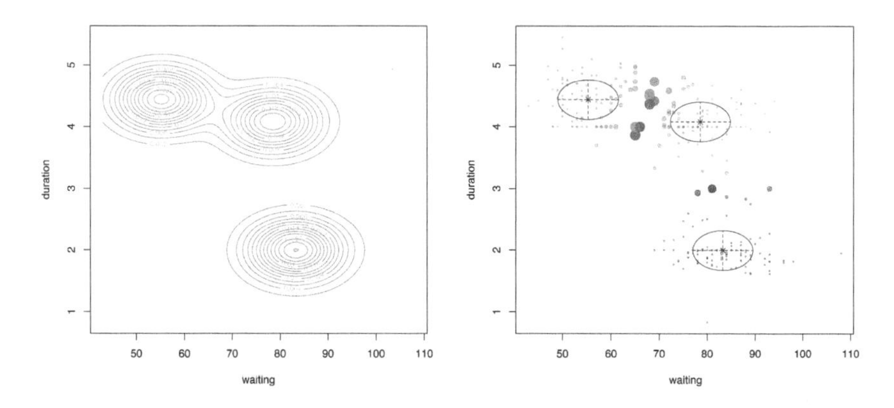

FIGURE 22.2: Multivariate density plot (left) highlighting three clusters in the 'geyser' data and an uncertainty plot (right) highlighting observations with high uncertainty of which cluster they are a member of.

```
##   187   211    85   285    28   206
## 0.469 0.454 0.436 0.436 0.431 0.417
```

22.3 Covariance types

The covariance matrix in Equation (22.1) describes the geometry of the clusters; namely, the volume, shape, and orientation of the clusters. Looking at Figure 22.2, the clusters and their densities appear approximately proportional in size and shape. However, this is not a requirement of GMMs. In fact, GMMs allow for far more flexible clustering structures.

This is done by adding constraints to the covariance matrix Σ. These constraints can be one or more of the following:

1. **volume**: each cluster has approximately the same number of observations;
2. **shape**: each cluster has approximately the same variance so that the distribution is spherical;
3. **orientation**: each cluster is forced to be axis-aligned.

The various combinations of the above constraints have been classified into three

TABLE 22.1: Parameterizations of the covariance matrix

Model	Family	Volume	Shape	Orientation	Identifier
1	Spherical	Equal	Equal	NA	EII
2	Spherical	Variable	Equal	NA	VII
3	Diagonal	Equal	Equal	Axes	EEI
4	Diagonal	Variable	Equal	Axes	VEI
5	Diagonal	Equal	Variable	Axes	EVI
6	Diagonal	Variable	Variable	Axes	VVI
7	General	Equal	Equal	Equal	EEE
8	General	Equal	Variable	Equal	EVE
9	General	Variable	Equal	Equal	VEE
10	General	Variable	Variable	Equal	VVE
11	General	Equal	Equal	Variable	EEV
12	General	Variable	Equal	Variable	VEV
13	General	Equal	Variable	Variable	EVV
14	General	Variable	Variable	Variable	VVV

main families of models: *spherical*, *diagonal*, and *general* (also referred to as *ellipsoidal*) families (Celeux and Govaert, 1995). These combinations are listed in Table 22.1. See Fraley et al. (2012) regarding the technical implementation of these covariance parameters.

These various covariance parameters allow GMMs to capture unique clustering structures in data, as illustrated in Figure 22.3.

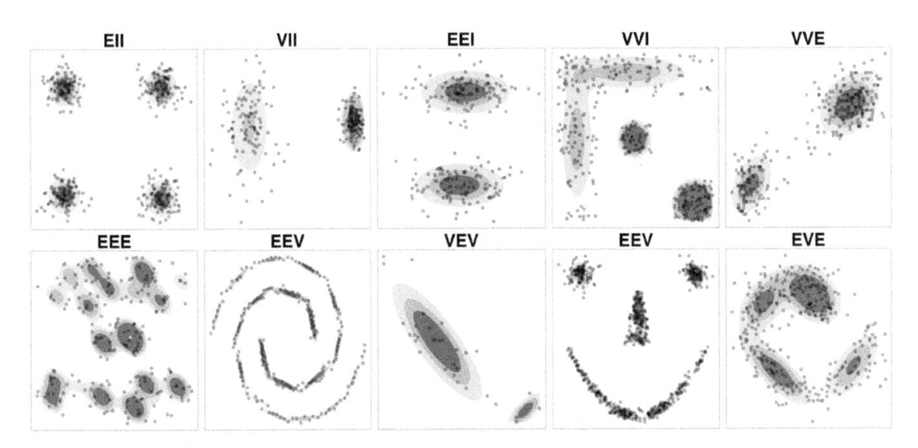

FIGURE 22.3: Graphical representation of how different covariance models allow GMMs to capture different cluster structures.

Users can optionally specify a conjugate prior if prior knowledge of the underlying probability distributions are available. By default, `Mclust()` does not apply a prior for modeling.

22.4 Model selection

If we assess the summary of our `geyser_mc` model we see that, behind the scenes, `Mclust()` applied the EEI model.

A more detailed summary, including the estimated parameters, can be obtained using `summary(geyser_mc, parameters = TRUE)`.

```
summary(geyser_mc)
## ----------------------------------------------------
## Gaussian finite mixture model fitted by EM algorithm
## ----------------------------------------------------
##
## Mclust EEI (diagonal, equal volume and shape) model
## with 3 components:
##
##  log-likelihood   n df   BIC   ICL
##           -1372 299 10 -2801 -2815
##
## Clustering table:
##   1   2   3
##  91 107 101
```

However, `Mclust()` will apply all 14 models from Table 22.3 and identify the one that best characterizes the data. To select the optimal model, any model selection procedure can be applied (e.g., Akaike information criteria (AIC), likelihood ratio, etc.); however, the Bayesian information criterion (BIC) has been shown to work well in model-based clustering (Dasgupta and Raftery, 1998; Fraley and Raftery, 1998) and is typically the default. `Mclust()` implements BIC as represented in Equation (22.3)

$$BIC = -2 \log (L) + m \log (n) , \tag{22.3}$$

where $\log(n)$ is the maximized loglikelihood for the model and data, m is the number of free parameters to be estimated in the given model, and n is the number observations in the data. This penalizes large models with many clusters.

The objective in hyperparameter tuning with `Mclust()` is to maximize BIC.

Not only can we use BIC to identify the optimal covariance parameters, but we can also use it to identify the optimal number of clusters. Rather than specify `G = 3` in `Mclust()`, leaving `G = NULL` forces `Mclust()` to evaluate 1–9 clusters and select the optimal number of components based on BIC. Alternatively, you can specify certain values to evaluate for `G`. The following performs a hyperparameter search across all 14 covariance models for 1–9 clusters on the geyser data. The left plot in Figure 22.4 shows that the EEI and VII models perform particularly poor while the rest of the models perform much better and have little differentiation. The optimal model uses the VVI covariance parameters, which identified four clusters and has a BIC of -2767.568.

```
geyser_optimal_mc <- Mclust(geyser)

summary(geyser_optimal_mc)
## ----------------------------------------------------
## Gaussian finite mixture model fitted by EM algorithm
## ----------------------------------------------------
##
## Mclust VVI (diagonal, varying volume and shape) model
## with 4 components:
##
##  log-likelihood   n df   BIC   ICL
##            -1330 299 19 -2769 -2799
##
## Clustering table:
##  1  2  3  4
## 90 17 98 94

legend_args <- list(x = "bottomright", ncol = 5)
plot(geyser_optimal_mc, what = 'BIC', legendArgs = legend_args)
plot(geyser_optimal_mc, what = 'classification')
plot(geyser_optimal_mc, what = 'uncertainty')
```

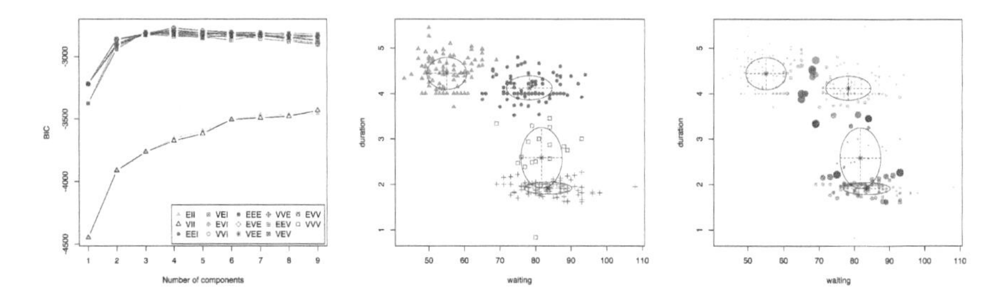

FIGURE 22.4: Identifying the optimal GMM model and number of clusters for the ‘geyser‘ data (left). The classification (center) and uncertainty (right) plots illustrate which observations are assigned to each cluster and their level of assignment uncertainty.

22.5 My basket example

Let’s turn to our `my_basket` data to demonstrate GMMs on a more modern sized data set. The following performs a search across all 14 GMM models and across 1–20 clusters. If you are following along and running the code you’ll notice that GMMs are computationally slow, especially since they are assessing 14 models for each cluster size instance. This GMM hyperparameter search took a little over a minute. Figure 22.5 illustrates the BIC scores and we see that the optimal GMM method is EEV with six clusters.

You may notice that not all models generate results for each cluster size (e.g., the VVI model only produced results for clusters 1–3). This is because the model could not converge on optimal results for those settings. This becomes more problematic as data sets get larger. Often, performing dimension reduction prior to a GMM can minimize this issue.

```r
my_basket_mc <- Mclust(my_basket, 1:20)

summary(my_basket_mc)
## ----------------------------------------------------
## Gaussian finite mixture model fitted by EM algorithm
## ----------------------------------------------------
##
```

```
## Mclust EEV (ellipsoidal, equal volume and shape) model
## with 6 components:
##
##  log-likelihood    n   df    BIC    ICL
##            8309 2000 5465 -24921 -25038
##
## Clustering table:
##   1   2   3   4   5   6
## 391 403  75 315 365 451
```

```
plot(my_basket_mc, what = 'BIC',
     legendArgs = list(x = "bottomright", ncol = 5))
```

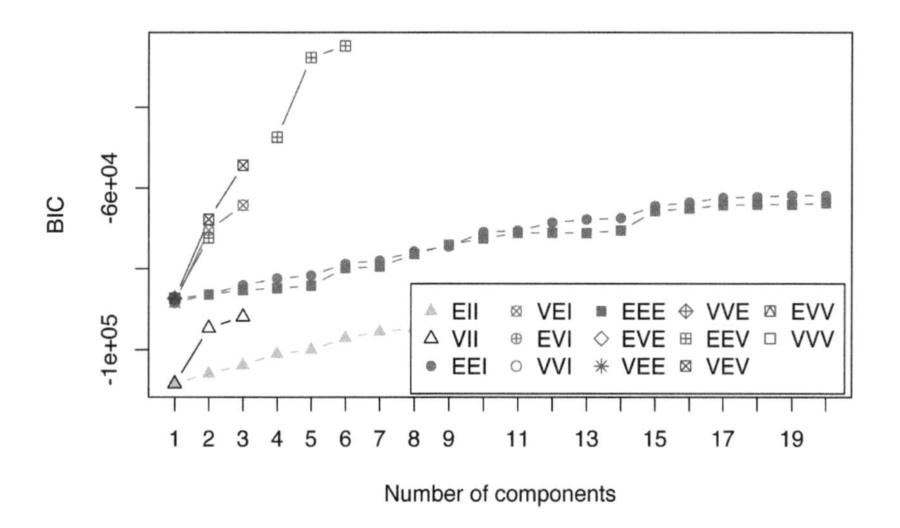

FIGURE 22.5: BIC scores for clusters (components) ranging from 1-20

We can look across our six clusters and assess the probabilities of cluster membership. Figure 22.6 illustrates very bimodal distributions of probabilities. Observations with greater than 0.50 probability will be aligned to a given cluster so this bimodality is preferred as it illustrates that observations have either a very high probability of the cluster they are aligned to or a very low probability, which means they would not be aligned to that cluster.

Looking at cluster `C3`, we see that there are very few, if any, observations in the middle of the probability range. `C3` also has far fewer observations with high probability. This means that C3 is the smallest of the clusters (confirmed using summary(my_basket_mc)) above, and that C3 is a more compact cluster. As clusters have more observations with middling levels of probability (i.e., 0.25–0.75), their clusters are usually less compact. Therefore, cluster C2 is less compact than cluster C3.

```
probabilities <- my_basket_mc$z
colnames(probabilities) <- paste0('C', 1:6)

probabilities <- probabilities %>%
  as.data.frame() %>%
  mutate(id = row_number()) %>%
  tidyr::gather(cluster, probability, -id)

ggplot(probabilities, aes(probability)) +
  geom_histogram() +
  facet_wrap(~ cluster, nrow = 2)
```

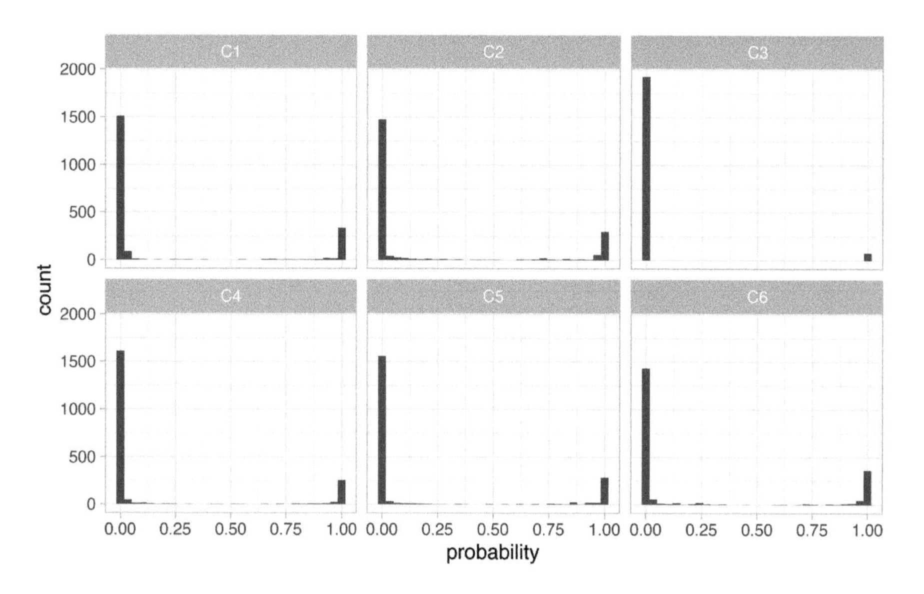

FIGURE 22.6: Distribution of probabilities for all observations aligning to each of the six clusters.

We can extract the cluster membership for each observation with my_basket_mc$classification. In Figure 22.7 we find the observations that are aligned to each cluster but the uncertainty of their membership to that particular cluster is 0.25 or greater. You may notice that cluster three is not represented. Recall from Figure 22.6 that cluster three's observations all had very strong membership probabilities so they have no observations with uncertainty greater than 0.25.

```
uncertainty <- data.frame(
  id = 1:nrow(my_basket),
  cluster = my_basket_mc$classification,
  uncertainty = my_basket_mc$uncertainty
)

uncertainty %>%
  group_by(cluster) %>%
  filter(uncertainty > 0.25) %>%
  ggplot(aes(uncertainty, reorder(id, uncertainty))) +
  geom_point() +
  facet_wrap(~ cluster, scales = 'free_y', nrow = 1)
```

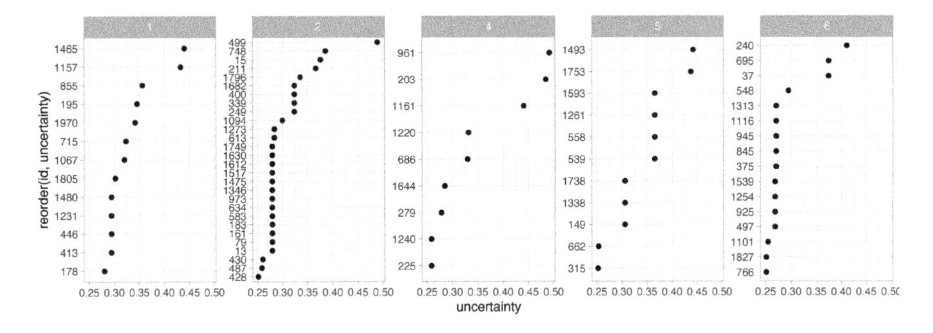

FIGURE 22.7: Observations that are aligned to each cluster but their uncertainty of membership is greater than 0.25.

When doing cluster analysis, our goal is to find those observations that are most similar to others. What defines this similarity becomes difficult as our data sets become larger. Let's take a look at cluster two. The following standardizes the count of each product across all baskets and then looks at consumption for cluster two. Figure 22.8 illustrates the results and shows that cluster two baskets have above average consumption for candy bars, lottery tickets, cigarettes, and alcohol.

Needless to say, this group may include our more unhealthy baskets—or maybe they're the recreational baskets made on the weekends when people just want to sit on the deck and relax with a drink in one hand and a candy bar in the other! Regardless, this group is likely to receive marketing ads for candy bars, alcohol, and the like rather than the items we see at the bottom of Figure

22.8, which represent the items this group consumes less than the average observations.

```
cluster2 <- my_basket %>%
  scale() %>%
  as.data.frame() %>%
  mutate(cluster = my_basket_mc$classification) %>%
  filter(cluster == 2) %>%
  select(-cluster)

cluster2 %>%
  tidyr::gather(product, std_count) %>%
  group_by(product) %>%
  summarize(avg = mean(std_count)) %>%
  ggplot(aes(avg, reorder(product, avg))) +
  geom_point() +
  labs(x = "Average standardized consumption", y = NULL)
```

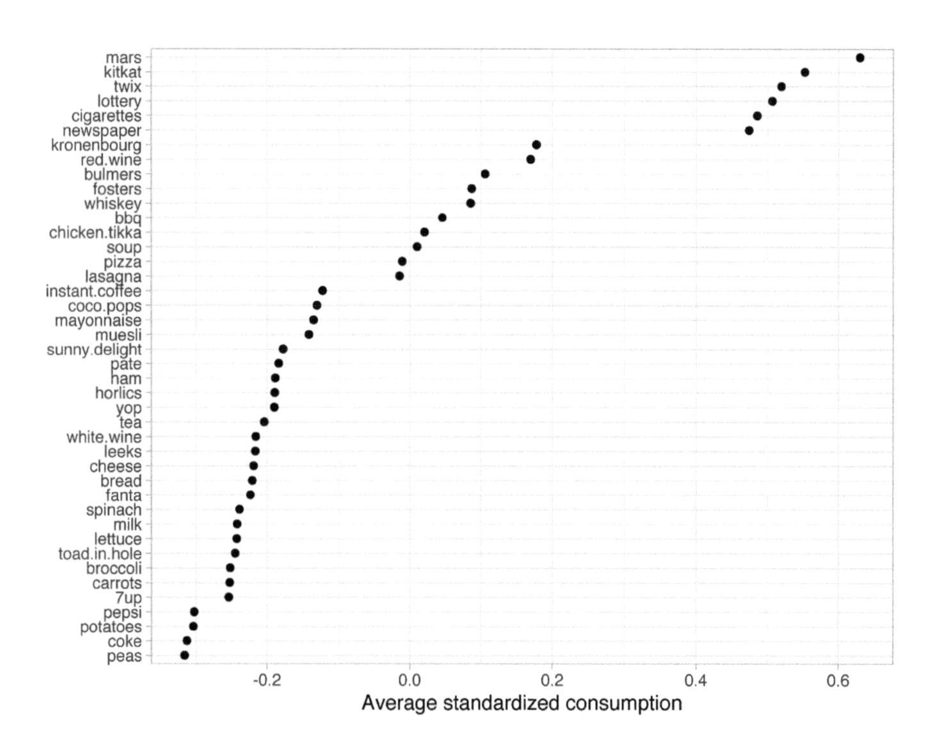

FIGURE 22.8: Average standardized consumption for cluster 2 observations compared to all observations.

22.6 Final thoughts

Model-based clustering techniques do have their limitations. The methods require an underlying model for the data (e.g., GMMs assume multivariate normality), and the cluster results are heavily dependent on this assumption. Although there have been many advancements to limit this constraint (Lee and McLachlan, 2013), software implementations are still lacking.

A more significant limitation is the computational demands. Classical model-based clustering show disappointing computational performance in high-dimensional spaces (Bouveyron and Brunet-Saumard, 2014). This is mainly due to the fact that model-based clustering methods are dramatically over-parameterized. The primary approach for dealing with this is to perform dimension reduction prior to clustering. Although this often improves computational performance, reducing the dimension without taking into consideration the clustering goal may be dangerous. Indeed, dimension reduction may yield a loss of information which could have been useful for discriminating the groups. There have been alternative solutions proposed, such as high-dimensional GMMs (Bouveyron et al., 2007), which has been implemented in the **HDclassif** package (Berge et al., 2018).

Bibliography

Agresti, A. (2003). *Categorical Data Analysis*. Wiley Series in Probability and Statistics. Wiley.

Allaire, J. (2018). *tfruns: Training Run Tools for 'TensorFlow'*. R package version 1.4.

Allaire, J. and Chollet, F. (2019). *keras: R Interface to 'Keras'*. R package version 2.2.4.1.9001.

Banfield, J. D. and Raftery, A. E. (1993). Model-based gaussian and non-gaussian clustering. *Biometrics*, pages 803–821.

Bengio, Y., Yao, L., Alain, G., and Vincent, P. (2013). Generalized denoising auto-encoders as generative models. In *Advances in Neural Information Processing Systems*, pages 899–907.

Berge, L., Bouveyron, C., and Girard, S. (2018). *HDclassif: High Dimensional Supervised Classification and Clustering*. R package version 2.1.0.

Bergstra, J. and Bengio, Y. (2012). Random search for hyper-parameter optimization. *Journal of Machine Learning Research*, 13(Feb):281–305.

Beygelzimer, A., Kakade, S., and Langford, J. (2006). Cover trees for nearest neighbor. In *Proceedings of the 23rd International Conference on Machine Learning*, pages 97–104. ACM.

Beygelzimer, A., Kakadet, S., Langford, J., Arya, S., Mount, D., and Li, S. (2019). *FNN: Fast Nearest Neighbor Search Algorithms and Applications*. R package version 1.1.3.

Biecek, P. (2019). *DALEX: Descriptive mAchine Learning EXplanations*. R package version 0.4.

Bourlard, H. and Kamp, Y. (1988). Auto-association by multilayer perceptrons and singular value decomposition. *Biological Cybernetics*, 59(4-5):291–294.

Bouveyron, C. and Brunet-Saumard, C. (2014). Model-based clustering of high-dimensional data: A review. *Computational Statistics & Data Analysis*, 71:52–78.

Bouveyron, C., Girard, S., and Schmid, C. (2007). High-dimensional data clustering. *Computational Statistics & Data Analysis*, 52(1):502–519.

Box, G. E. and Cox, D. R. (1964). An analysis of transformations. *Journal of the Royal Statistical Society. Series B (Methodological)*, pages 211–252.

Breiman, L. (1984). *Classification and Regression Trees.* Routledge.

Breiman, L. (1996a). Bagging predictors. *Machine Learning*, 24(2):123–140.

Breiman, L. (1996b). Stacked regressions. *Machine Learning*, 24(1):49–64.

Breiman, L. (2001). Random forests. *Machine Learning*, 45(1):5–32.

Breiman, L. et al. (2001). Statistical modeling: The two cultures (with comments and a rejoinder by the author). *Statistical Science*, 16(3):199–231.

Breiman, L. and Ihaka, R. (1984). *Nonlinear discriminant analysis via scaling and ACE.* Department of Statistics, University of California.

Bruce, P. and Bruce, A. (2017). *Practical Statistics for Data Scientists: 50 Essential Concepts.* O'Reilly Media, Inc.

Carroll, R. J. and Ruppert, D. (1981). On prediction and the power transformation family. *Biometrika*, 68(3):609–615.

Celeux, G. and Govaert, G. (1995). Gaussian parsimonious clustering models. *Pattern Recognition*, 28(5):781–793.

Charrad, M., Ghazzali, N., Boiteau, V., and Niknafs, A. (2015). *NbClust: Determining the Best Number of Clusters in a Data Set.* R package version 3.0.

Chawla, N. V., Bowyer, K. W., Hall, L. O., and Kegelmeyer, W. P. (2002). Smote: synthetic minority over-sampling technique. *Journal of Artificial Intelligence Research*, 16:321–357.

Chen, T. and Guestrin, C. (2016). XGBoost: A scalable tree boosting system. In *Proceedings of the 22nd ACM SIGKDD International Conference on Knowledge Discovery and Data Mining*, KDD '16, pages 785–794, New York, NY, USA. ACM.

Chen, T., He, T., Benesty, M., Khotilovich, V., Tang, Y., Cho, H., Chen, K., Mitchell, R., Cano, I., Zhou, T., Li, M., Xie, J., Lin, M., Geng, Y., and Li, Y. (2018). *xgboost: Extreme Gradient Boosting.* R package version 0.71.2.

Chollet, F. and Allaire, J. J. (2018). *Deep Learning with R.* Manning Publications Company.

Cireşan, D., Meier, U., and Schmidhuber, J. (2012). Multi-column deep neural networks for image classification. *arXiv preprint arXiv:1202.2745.*

Cunningham, P. and Delany, S. J. (2007). k-nearest neighbour classifiers. *Multiple Classifier Systems*, 34(8):1–17.

Dasgupta, A. and Raftery, A. E. (1998). Detecting features in spatial point processes with clutter via model-based clustering. *Journal of the American Statistical Association*, 93(441):294–302.

Davison, A. C., Hinkley, D. V., et al. (1997). *Bootstrap Methods and their Application*, volume 1. Cambridge University Press.

De Cock, D. (2011). Ames, Iowa: Alternative to the Boston housing data as an end of semester regression project. *Journal of Statistics Education*, 19(3).

De Maesschalck, R., Jouan-Rimbaud, D., and Massart, D. L. (2000). The mahalanobis distance. *Chemometrics and Intelligent Laboratory Systems*, 50(1):1–18.

Deane-Mayer, Z. A. and Knowles, J. E. (2016). *caretEnsemble: Ensembles of Caret Models*. R package version 2.0.0.

Díaz-Uriarte, R. and De Andres, S. A. (2006). Gene selection and classification of microarray data using random forest. *BMC Bioinformatics*, 7(1):3.

Dietterich, T. G. (2000a). Ensemble methods in machine learning. In *International Workshop on Multiple Classifier Systems*, pages 1–15. Springer.

Dietterich, T. G. (2000b). An experimental comparison of three methods for constructing ensembles of decision trees: Bagging, boosting, and randomization. *Machine Learning*, 40(2):139–157.

Doersch, C. (2016). Tutorial on variational autoencoders. *arXiv preprint arXiv:1606.05908*.

Dorogush, A. V., Ershov, V., and Gulin, A. (2018). Catboost: gradient boosting with categorical features support. *arXiv preprint arXiv:1810.11363*.

Doshi-Velez, F. and Kim, B. (2017). Towards a rigorous science of interpretable machine learning. *arXiv preprint arXiv:1702.08608*.

Efron, B. (1983). Estimating the error rate of a prediction rule: improvement on cross-validation. *Journal of the American Statistical Association*, 78(382):316–331.

Efron, B. and Hastie, T. (2016). *Computer Age Statistical Inference*, volume 5. Cambridge University Press.

Efron, B. and Tibshirani, R. (1986). Bootstrap methods for standard errors, confidence intervals, and other measures of statistical accuracy. *Statistical Science*, pages 54–75.

Efron, B. and Tibshirani, R. (1997). Improvements on cross-validation: the 632+ bootstrap method. *Journal of the American Statistical Association*, 92(438):548–560.

Erichson, N. B., Zheng, P., and Aravkin, S. (2018). *sparsepca: Sparse Principal Component Analysis (SPCA)*. R package version 0.1.2.

Faraway, J. J. (2016a). *Extending the Linear Model with R: Generalized Linear, Mixed Effects and Nonparametric Regression Models*, volume 124. CRC press.

Faraway, J. J. (2016b). *Linear Models with R*. Chapman and Hall/CRC.

Fisher, A., Rudin, C., and Dominici, F. (2018). Model class reliance: Variable importance measures for any machine learning model class, from the" rashomon" perspective. *arXiv preprint arXiv:1801.01489*.

Fisher, R. A. (1936). The use of multiple measurements in taxonomic problems. *Annals of Eugenics*, 7(2):179–188.

Fisher, W. D. (1958). On grouping for maximum homogeneity. *Journal of the American Statistical Association*, 53(284):789–798.

Fraley, C. and Raftery, A. E. (1998). How many clusters? which clustering method? answers via model-based cluster analysis. *The Computer Journal*, 41(8):578–588.

Fraley, C. and Raftery, A. E. (2002). Model-based clustering, discriminant analysis, and density estimation. *Journal of the American Statistical Association*, 97(458):611–631.

Fraley, C., Raftery, A. E., Murphy, T. B., and Scrucca, L. (2012). mclust Version 4 for R: Normal Mixture Modeling for Model-based Clustering, Classification, and Density Estimation. Technical report, University of Washington.

Fraley, C., Raftery, A. E., and Scrucca, L. (2019). *mclust: Gaussian Mixture Modelling for Model-Based Clustering, Classification, and Density Estimation*. R package version 5.4.3.

Freund, Y. and Schapire, R. E. (1999). Adaptive game playing using multiplicative weights. *Games and Economic Behavior*, 29(1-2):79–103.

Friedman, J., Hastie, T., and Tibshirani, R. (2001). *The Elements of Statistical Learning*, volume 1. Springer Series in Statistics New York, NY, USA:.

Friedman, J., Hastie, T., Tibshirani, R., Simon, N., Narasimhan, B., and Qian, J. (2018). *glmnet: Lasso and Elastic-Net Regularized Generalized Linear Models*. R package version 2.0-16.

Friedman, J. H. (1991). Multivariate adaptive regression splines. *The Annals of Statistics*, pages 1–67.

Friedman, J. H. (2001). Greedy function approximation: a gradient boosting machine. *Annals of Statistics*, pages 1189–1232.

Friedman, J. H. (2002). Stochastic gradient boosting. *Computational Statistics & Data Analysis*, 38(4):367–378.

Friedman, J. H., Popescu, B. E., et al. (2008). Predictive learning via rule ensembles. *The Annals of Applied Statistics*, 2(3):916–954.

from mda:mars by Trevor Hastie, S. M. D. and utilities with Thomas Lumley's leaps wrapper., R. T. U. A. M. F. (2019). *earth: Multivariate Adaptive Regression Splines*. R package version 5.1.1.

Geladi, P. and Kowalski, B. R. (1986). Partial least-squares regression: a tutorial. *Analytica chimica acta*, 185:1–17.

Géron, A. (2017). *Hands-on Machine Learning with Scikit-Learn and TensorFlow: Concepts, Tools, and Techniques to Build Intelligent Systems*. O'Reilly Media, Inc.

Geurts, P., Ernst, D., and Wehenkel, L. (2006). Extremely randomized trees. *Machine Learning*, 63(1):3–42.

Goldstein, A., Kapelner, A., Bleich, J., and Pitkin, E. (2015). Peeking inside the black box: Visualizing statistical learning with plots of individual conditional expectation. *Journal of Computational and Graphical Statistics*, 24(1):44–65.

Goldstein, B. A., Polley, E. C., and Briggs, F. B. (2011). Random forests for genetic association studies. *Statistical Applications in Genetics and Molecular Biology*, 10(1).

Golub, G. H., Heath, M., and Wahba, G. (1979). Generalized cross-validation as a method for choosing a good ridge parameter. *Technometrics*, 21(2):215–223.

Goodfellow, I., Bengio, Y., and Courville, A. (2016). *Deep Learning*, volume 1. MIT Press Cambridge.

Gower, J. C. (1971). A general coefficient of similarity and some of its properties. *Biometrics*, pages 857–871.

Granitto, P. M., Furlanello, C., Biasioli, F., and Gasperi, F. (2006). Recursive feature elimination with random forest for ptr-ms analysis of agroindustrial products. *Chemometrics and Intelligent Laboratory Systems*, 83(2):83–90.

Greenwell, B. (2018). *pdp: Partial Dependence Plots*. R package version 0.7.0.

Greenwell, B., Boehmke, B., Cunningham, J., and Developers, G. (2018a). gbm: Generalized boosted regression models. *R Package Version 2.1*, 4.

Greenwell, B. M., Boehmke, B. C., and McCarthy, A. J. (2018b). A simple and effective model-based variable importance measure. *arXiv preprint arXiv:1805.04755.*

Greenwell, B. M., McCarthy, A. J., Boehmke, B. C., and Lui, D. (2018c). Residuals and diagnostics for binary and ordinal regression models: An introduction to the sure package. *The R Journal*, 10(1):1–14.

Greenwell, Brandon M. and Boehmke, Bradley C. (2019). Quantifying the strength of potential interaction effects. https://koalaverse.github.io/vip/articles/vip-interaction.html.

Guo, C. and Berkhahn, F. (2016). Entity embeddings of categorical variables. *arXiv preprint arXiv:1604.06737.*

Hair, J. F. (2006). *Multivariate Data Analysis.* Pearson Education India.

Hall, Patrick (2018). Awesome machine learning interpretability: A curated, but probably biased and incomplete, list of awesome machine learning interpretability resources. https://github.com/jphall663/awesome-machine-learning-interpretability.

Han, J., Pei, J., and Kamber, M. (2011). *Data Mining: Concepts and Techniques.* Elsevier.

Harrell, F. E. (2015). *Regression Modeling Strategies: With Applications to Linear Models, Logistic and Ordinal Regression, and Survival Analysis.* Springer Series in Statistics. Springer International Publishing.

Harrison Jr, D. and Rubinfeld, D. L. (1978). Hedonic housing prices and the demand for clean air. *Journal of Environmental Economics and Management*, 5(1):81–102.

Hartigan, J. A. and Wong, M. A. (1979). Algorithm as 136: A k-means clustering algorithm. *Journal of the Royal Statistical Society. Series C (Applied Statistics)*, 28(1):100–108.

Hastie, T. (2016). *svmpath: The SVM Path Algorithm.* R package version 0.955.

Hastie, T., Tibshirani, R., and Wainwright, M. (2015). *Statistical Learning with Sparsity: The Lasso and Generalizations.* Chapman & Hall/CRC Monographs on Statistics & Applied Probability. Taylor & Francis.

Hawkins, D. M., Basak, S. C., and Mills, D. (2003). Assessing model fit by cross-validation. *Journal of Chemical Information and Computer Sciences*, 43(2):579–586.

Hinton, G. E. and Salakhutdinov, R. R. (2006). Reducing the dimensionality of data with neural networks. *Science*, 313(5786):504–507.

Hinton, G. E., Srivastava, N., Krizhevsky, A., Sutskever, I., and Salakhutdinov, R. R. (2012). Improving neural networks by preventing co-adaptation of feature detectors. *arXiv preprint arXiv:1207.0580.*

Hinton, G. E. and Zemel, R. S. (1994). Autoencoders, minimum description length and helmholtz free energy. In *Advances in Neural Information Processing Systems*, pages 3–10.

Hoerl, A. E. and Kennard, R. W. (1970). Ridge regression: Biased estimation for nonorthogonal problems. *Technometrics*, 12(1):55–67.

Hothorn, T., Hornik, K., and Zeileis, A. (2006). Unbiased recursive partitioning: A conditional inference framework. *Journal of Computational and Graphical Statistics*, 15(3):651–674.

Hothorn, T. and Zeileis, A. (2015). partykit: A modular toolkit for recursive partytioning in r. *The Journal of Machine Learning Research*, 16(1):3905–3909.

Hunt, T. (2018). *ModelMetrics: Rapid Calculation of Model Metrics.* R package version 1.2.2.

Hyndman, R. J. and Athanasopoulos, G. (2018). *Forecasting: Principles and Practice.* OTexts.

Ioffe, S. and Szegedy, C. (2015). Batch normalization: Accelerating deep network training by reducing internal covariate shift. *arXiv preprint arXiv:1502.03167.*

Irizarry, R. A. (2018). *dslabs: Data Science Labs.* R package version 0.5.2.

Janitza, S., Binder, H., and Boulesteix, A.-L. (2016). Pitfalls of hypothesis tests and model selection on bootstrap samples: causes and consequences in biometrical applications. *Biometrical Journal*, 58(3):447–473.

Jiang, S., Pang, G., Wu, M., and Kuang, L. (2012). An improved k-nearest-neighbor algorithm for text categorization. *Expert Systems with Applications*, 39(1):1503–1509.

Karatzoglou, A., Smola, A., and Hornik, K. (2018). *kernlab: Kernel-Based Machine Learning Lab.* R package version 0.9-27.

Karatzoglou, A., Smola, A., Hornik, K., and Zeileis, A. (2004). kernlab – an S4 package for kernel methods in R. *Journal of Statistical Software*, 11(9):1–20.

Kass, G. V. (1980). An exploratory technique for investigating large quantities of categorical data. *Applied Statistics*, pages 119–127.

Kaufman, L. and Rousseeuw, P. J. (2009). *Finding Groups in Data: an Introduction to Cluster Analysis*, volume 344. John Wiley & Sons.

Ke, G., Meng, Q., Finley, T., Wang, T., Chen, W., Ma, W., Ye, Q., and Liu, T.-Y. (2017). Lightgbm: A highly efficient gradient boosting decision tree. In *Advances in Neural Information Processing Systems*, pages 3146–3154.

Ketchen, D. J. and Shook, C. L. (1996). The application of cluster analysis in strategic management research: an analysis and critique. *Strategic Management Journal*, 17(6):441–458.

Kim, J.-H. (2009). Estimating classification error rate: Repeated cross-validation, repeated hold-out and bootstrap. *Computational Statistics & Data Analysis*, 53(11):3735–3745.

Kingma, D. P. and Ba, J. (2014). Adam: A method for stochastic optimization. *arXiv preprint arXiv:1412.6980.*

Kuhn, M. (2014). Futility analysis in the cross-validation of machine learning models. *arXiv preprint arXiv:1405.6974.*

Kuhn, M. (2017a). *AmesHousing: The Ames Iowa Housing Data.* R package version 0.0.3.

Kuhn, M. (2017b). The R formula method: the bad parts. *R Views.*

Kuhn, M. (2018). Applied machine learning workshop. https://github.com/tidymodels/aml-training.

Kuhn, M. (2019). Applied machine learning. RStudio Conference.

Kuhn, M. and Johnson, K. (2013). *Applied Predictive Modeling*, volume 26. Springer.

Kuhn, M. and Johnson, K. (2018). *AppliedPredictiveModeling: Functions and Data Sets for 'Applied Predictive Modeling'.* R package version 1.1-7.

Kuhn, M. and Johnson, K. (2019). *Feature Engineering and Selection: A Practical Approach for Predictive Models.* Chapman & Hall/CRC.

Kuhn, M. and Wickham, H. (2019). *rsample: General Resampling Infrastructure.* R package version 0.0.4.

Kursa, M. B., Rudnicki, W. R., et al. (2010). Feature selection with the boruta package. *J Stat Softw*, 36(11):1–13.

Kutner, M. H., Nachtsheim, C. J., Neter, J., and Li, W. (2005). *Applied Linear Statistical Models.* McGraw Hill, 5th edition.

LeCun, Y. (1987). Modeles connexionnistes de l'apprentissage (connectionist learning models). Technical report, Ph.D. thesis, Universite P. et M. Curie (Paris 6).

LeCun, Y., Boser, B. E., Denker, J. S., Henderson, D., Howard, R. E., Hubbard, W. E., and Jackel, L. D. (1990). Handwritten digit recognition with a back-propagation network. In *Advances in Neural Information Processing Systems*, pages 396–404.

LeCun, Y., Bottou, L., Bengio, Y., and Haffner, P. (1998). Gradient-based learning applied to document recognition. *Proceedings of the IEEE*, 86(11):2278–2324.

LeDell, E., Sapp, S., van der Laan, M., and LeDell, M. E. (2014). *Package 'subsemble'*.

Lee, H., Ekanadham, C., and Ng, A. Y. (2008). Sparse deep belief net model for visual area v2. In *Advances in Neural Information Processing Systems*, pages 873–880.

Lee, S. X. and McLachlan, G. J. (2013). Model-based clustering and classification with non-normal mixture distributions. *Statistical Methods & Applications*, 22(4):427–454.

Liaw, A. and Wiener, M. (2002). Classification and regression by randomforest. *R News*, 2(3):18–22.

Little, R. J. and Rubin, D. B. (2014). *Statistical Analysis with Missing Data*, volume 333. John Wiley & Sons.

Liu, D. and Zhang, H. (2018). Residuals and diagnostics for ordinal regression models: A surrogate approach. *Journal of the American Statistical Association*, 113(522):845–854.

Loh, W.-Y. and Vanichsetakul, N. (1988). Tree-structured classification via generalized discriminant analysis. *Journal of the American Statistical Association*, 83(403):715–725.

Lundberg, S. and Lee, S.-I. (2016). An unexpected unity among methods for interpreting model predictions. *arXiv preprint arXiv:1611.07478*.

Lundberg, S. M. and Lee, S.-I. (2017). A unified approach to interpreting model predictions. In *Advances in Neural Information Processing Systems*, pages 4765–4774.

Luu, K., Blum, M., and Privé, F. (2019). *pcadapt: Fast Principal Component Analysis for Outlier Detection.* R package version 4.1.0.

Ma, Y., Derksen, H., Hong, W., and Wright, J. (2007). Segmentation of multivariate mixed data via lossy data coding and compression. *IEEE Transactions on Pattern Analysis and Machine Intelligence*, 29(9):1546–1562.

Makhzani, A. and Frey, B. (2014). A winner-take-all method for training sparse convolutional autoencoders. In *NIPS Deep Learning Workshop*. Citeseer.

Makhzani, A. and Frey, B. J. (2015). Winner-take-all autoencoders. In *Advances in Neural Information Processing Systems*, pages 2791–2799.

Makhzani, A., Shlens, J., Jaitly, N., Goodfellow, I., and Frey, B. (2015). Adversarial autoencoders. *arXiv preprint arXiv:1511.05644*.

Maldonado, S. and Weber, R. (2009). A wrapper method for feature selection using support vector machines. *Information Sciences*, 179(13):2208–2217.

Masci, J., Meier, U., Cireşan, D., and Schmidhuber, J. (2011). Stacked convolutional auto-encoders for hierarchical feature extraction. In *International Conference on Artificial Neural Networks*, pages 52–59. Springer.

Massy, W. F. (1965). Principal components regression in exploratory statistical research. *Journal of the American Statistical Association*, 60(309):234–256.

Mccord, M. and Chuah, M. (2011). Spam detection on twitter using traditional classifiers. In *International Conference on Autonomic and Trusted Computing*, pages 175–186. Springer.

Meyer, D., Dimitriadou, E., Hornik, K., Weingessel, A., and Leisch, F. (2019). *e1071: Misc Functions of the Department of Statistics, Probability Theory Group (Formerly: E1071), TU Wien*. R package version 1.7-1.

Micci-Barreca, D. (2001). A preprocessing scheme for high-cardinality categorical attributes in classification and prediction problems. *ACM SIGKDD Explorations Newsletter*, 3(1):27–32.

Molinaro, A. M., Simon, R., and Pfeiffer, R. M. (2005). Prediction error estimation: a comparison of resampling methods. *Bioinformatics*, 21(15):3301–3307.

Molnar, C. (2019). *iml: Interpretable Machine Learning*. R package version 0.9.0.

Molnar, C. et al. (2018). Interpretable machine learning: A guide for making black box models explainable. *E-book at< https://christophm.github.io/interpretable-ml-book/>, version dated*, 10.

Pearson, K. (1901). Liii. on lines and planes of closest fit to systems of points in space. *The London, Edinburgh, and Dublin Philosophical Magazine and Journal of Science*, 2(11):559–572.

Pedersen, T. L. and Benesty, M. (2018). *lime: Local Interpretable Model-Agnostic Explanations*. R package version 0.4.1.

Platt, J. C. (1999). Probabilistic outputs for support vector machines and comparisons to regularized likelihood methods. In *Advances in Large Margin Classifiers*, pages 61–74. MIT Press.

Polley, E., LeDell, E., Kennedy, C., and van der Laan, M. (2019). *SuperLearner: Super Learner Prediction.* R package version 2.0-25.

Poultney, C., Chopra, S., Cun, Y. L., et al. (2007). Efficient learning of sparse representations with an energy-based model. In *Advances in Neural Information Processing Systems*, pages 1137–1144.

Probst, P., Bischl, B., and Boulesteix, A.-L. (2018). Tunability: Importance of hyperparameters of machine learning algorithms. *arXiv preprint arXiv:1802.09596.*

Probst, P., Wright, M. N., and Boulesteix, A.-L. (2019). Hyperparameters and tuning strategies for random forest. *Wiley Interdisciplinary Reviews: Data Mining and Knowledge Discovery*, page e1301.

Quinlan, J. R. (1986). Induction of decision trees. *Machine Learning*, 1(1):81–106.

Quinlan, J. R. et al. (1996). Bagging, boosting, and c4. 5. In *AAAI/IAAI, Vol. 1*, pages 725–730.

Rashmi, K. V. and Gilad-Bachrach, R. (2015). Dart: Dropouts meet multiple additive regression trees. In *AISTATS*, pages 489–497.

Ribeiro, M. T., Singh, S., and Guestrin, C. (2016). Why should i trust you?: Explaining the predictions of any classifier. In *Proceedings of the 22nd ACM SIGKDD International Conference on Knowledge Discovery and Data Mining*, pages 1135–1144. ACM.

Rifai, S., Vincent, P., Muller, X., Glorot, X., and Bengio, Y. (2011). Contractive auto-encoders: Explicit invariance during feature extraction. In *Proceedings of the 28th International Conference on International Conference on Machine Learning*, pages 833–840. Omnipress.

Ripley, B. D. (2007). *Pattern Recognition and Neural Networks.* Cambridge University Press.

Robinson, J. T. (1981). The kdb-tree: a search structure for large multidimensional dynamic indexes. In *Proceedings of the 1981 ACM SIGMOD International Conference on Management of Data*, pages 10–18. ACM.

Rousseeuw, P. J. (1987). Silhouettes: a graphical aid to the interpretation and validation of cluster analysis. *Journal of Computational and Applied Mathematics*, 20:53–65.

Ruder, S. (2016). An overview of gradient descent optimization algorithms. *arXiv preprint arXiv:1609.04747.*

Saeys, Y., Inza, I., and Larrañaga, P. (2007). A review of feature selection techniques in bioinformatics. *Bioinformatics*, 23(19):2507–2517.

Sakurada, M. and Yairi, T. (2014). Anomaly detection using autoencoders with nonlinear dimensionality reduction. In *Proceedings of the MLSDA 2014 2nd Workshop on Machine Learning for Sensory Data Analysis*, page 4. ACM.

Sapp, S., van der Laan, M. J., and Canny, J. (2014). Subsemble: an ensemble method for combining subset-specific algorithm fits. *Journal of Applied Statistics*, 41(6):1247–1259.

Sarle, Warren S. (n.d.). comp.ai.neural-nets faq. [Online; accessed 116-April-2019].

Segal, M. R. (2004). Machine learning benchmarks and random forest regression. *UCSF: Center for Bioinformatics and Molecular Biostatistics*.

Shah, A. D., Bartlett, J. W., Carpenter, J., Nicholas, O., and Hemingway, H. (2014). Comparison of random forest and parametric imputation models for imputing missing data using mice: a caliber study. *American Journal of Epidemiology*, 179(6):764–774.

Srivastava, N., Hinton, G., Krizhevsky, A., Sutskever, I., and Salakhutdinov, R. (2014a). Dropout: A simple way to prevent neural networks from overfitting. *Journal of Machine Learning Research*, 15:1929–1958.

Srivastava, N., Hinton, G., Krizhevsky, A., Sutskever, I., and Salakhutdinov, R. (2014b). Dropout: a simple way to prevent neural networks from overfitting. *The Journal of Machine Learning Research*, 15(1):1929–1958.

Staniak, M. and Biecek, P. (2018). Explanations of model predictions with live and breakdown packages. *arXiv preprint arXiv:1804.01955*.

Stekhoven, D. J. (2015). missforest: Nonparametric missing value imputation using random forest. *Astrophysics Source Code Library*.

Stone, C. J., Hansen, M. H., Kooperberg, C., Truong, Y. K., et al. (1997). Polynomial splines and their tensor products in extended linear modeling: 1994 wald memorial lecture. *The Annals of Statistics*, 25(4):1371–1470.

Strobl, C., Boulesteix, A.-L., Zeileis, A., and Hothorn, T. (2007). Bias in random forest variable importance measures: Illustrations, sources and a solution. *BMC Bioinformatics*, 8(1):25.

Štrumbelj, E. and Kononenko, I. (2014). Explaining prediction models and individual predictions with feature contributions. *Knowledge and Information Systems*, 41(3):647–665.

Surowiecki, J. (2005). *The wisdom of crowds*. Anchor.

Therneau, T. M., Atkinson, E. J., et al. (1997). An introduction to recursive partitioning using the RPART routines. Technical report, Mayo Foundation. http://www.mayo.edu/hsr/techrpt/61.pdf.

Tibshirani, R. (1996). Regression shrinkage and selection via the lasso. *Journal of the Royal Statistical Society. Series B (Methodological)*, pages 267–288.

Tibshirani, R., Walther, G., and Hastie, T. (2001). Estimating the number of clusters in a data set via the gap statistic. *Journal of the Royal Statistical Society: Series B (Statistical Methodology)*, 63(2):411–423.

Tierney, N. (2019). *visdat: Preliminary Visualisation of Data*. R package version 0.5.3.

Udell, M., Horn, C., Zadeh, R., Boyd, S., et al. (2016). Generalized low rank models. *Foundations and Trends® in Machine Learning*, 9(1):1–118.

van der Laan, M. J., Polley, E. C., and Hubbard, A. E. (2003). Super learner. *Statistical Applications in Genetics and Molecular Biology*, 6(1).

Van der Laan, M. J., Polley, E. C., and Hubbard, A. E. (2007). Super learner. *Statistical Applications in Genetics and Molecular Biology*, 6(1).

Vincent, P. (2011). A connection between score matching and denoising autoencoders. *Neural Computation*, 23(7):1661–1674.

Vincent, P., Larochelle, H., Bengio, Y., and Manzagol, P.-A. (2008). Extracting and composing robust features with denoising autoencoders. In *Proceedings of the 25th Iternational Conference on Machine Learning*, pages 1096–1103. ACM.

West, B. T., Welch, K. B., and Galecki, A. T. (2014). *Linear Mixed Models: A Practical Guide Using Statistical Software*. Chapman and Hall/CRC.

Wickham, H. (2014). *Advanced R*. Chapman and Hall/CRC.

Wickham, H. et al. (2014). Tidy data. *Journal of Statistical Software*, 59(10):1–23.

Wickham, H. and Grolemund, G. (2016). *R for Data Science: Import, Tidy, Transform, Visualize, and Model Data*. O'Reilly Media, Inc.

Wikipedia contributors (n.d.a). Autoencoder. [Online; accessed 25-May-2019].

Wikipedia contributors (n.d.b). Mnist database. [Online; accessed 15-April-2019].

Wolpert, D. H. (1992). Stacked generalization. *Neural Networks*, 5:241–259.

Wolpert, D. H. (1996). The lack of a priori distinctions between learning algorithms. *Neural Computation*, 8(7):1341–1390.

Wright, M. and Ziegler, A. (2017). ranger: A fast implementation of random forests for high dimensional data in c++ and r. *Journal of Statistical Software, Articles*, 77(1):1–17.

Zhang, W., Zhao, D., and Wang, X. (2013). Agglomerative clustering via maximum incremental path integral. *Pattern Recognition*, 46(11):3056–3065.

Zhao, D. and Tang, X. (2009). Cyclizing clusters via zeta function of a graph. In *Advances in Neural Information Processing Systems*, pages 1953–1960.

Zheng, A. and Casari, A. (2018). *Feature Engineering for Machine Learning: Principles and Techniques for Data Scientists.* O'Reilly Media, Inc.

Zhou, C. and Paffenroth, R. C. (2017). Anomaly detection with robust deep autoencoders. In *Proceedings of the 23rd ACM SIGKDD International Conference on Knowledge Discovery and Data Mining*, pages 665–674. ACM.

Zou, H. and Hastie, T. (2005). Regularization and variable selection via the elastic net. *Journal of the Royal Statistical Society: Series B (Statistical Methodology)*, 67(2):301–320.

Zumel, N. and Mount, J. (2016). vtreat: a data. frame processor for predictive modeling. *arXiv preprint arXiv:1611.09477.*

Index